C. Francis Jenkins,

Pioneer of Film and Television

DONALD G. GODFREY

UNIVERSITY OF ILLINOIS PRESS
Urbana, Chicago, and Springfield

© 2014 by the Board of Trustees
of the University of Illinois
All rights reserved
Manufactured in the United States of America
C 5 4 3 2 1

♾ This book is printed on acid-free paper.

Library of Congress Cataloging-in-Publication Data
Godfrey, Donald G.
C. Francis Jenkins, pioneer of film and television / Donald G.
Godfrey.
pages   cm — (The history of communication)
Includes bibliographical references and index.
ISBN 978-0-252-03828-0 (hardcover : alk. paper) —
ISBN 978-0-252-09615-0 (e-book)
1. Jenkins, C. Francis (Charles Francis), 1867–1934. 2. Inventors—
United States—Biography. 3. Cinematographers—United States—
Biography. 4. Television—Biography. 5. Television—History.
6. Cinematography—History. I. Title. II. Title: Charles Francis
Jenkins, pioneer of film and television.
T40.J4G63   2014
621.3880092—dc23   2013038271
[B]

C. Francis Jenkins, Pioneer of Film and Television

THE HISTORY OF COMMUNICATION

Robert W. McChesney
and John C. Nerone, editors

A list of books in the series appears at the end of this book.

Dedicated to
Christina Maria Godfrey,
My children, grandchildren,
and
Virginia Stickle Roach,
Dave and Robin Vierbuchen Sproul,
with the Jenkins Family

Contents

Preface ix

Acknowledgments xiii

Prologue 1

1. Jenkins' Heritage and Youth 3
2. Early Film Experiments 15
3. A Lifetime of Struggle 23
4. Jenkins' Motion Pictures 51
5. Founding the Society of Motion Picture and Television Engineers 69
6. Visionary Entrepreneur 77
7. RadioVision: The Genesis and Promotion 95
8. Radio Pictures: Going Operational 107
9. Television: Seeing by Electricity 121
10. The Eyes of Radio 135
11. The Jenkins Television Corporation 149
12. American Visionary 169

Epilogue   181

Appendix A   U.S. Patents Issued to C. Francis Jenkins   185
Appendix B   Selected Jenkins Patents Referenced
             in Modern Patent Applications   193
Notes   195
Bibliography   265
Index   279

Preface

The name Charles Francis Jenkins (1867–1934) has been too long forgotten. He was an American original for whom inventing was a natural talent, a visionary working on the leading edge of technical discovery in film and television. He was the only inventor present at the inception of both large-screen motion-picture projection and television. A generation later, he was the only American actively working on television. He was primarily a film and television pioneer, but he also held multiple patents for a variety of creations. This is a biography of a man whose passion for life was inseparable from the process of invention.

The challenges in writing Jenkins' history were those similar to any historical biography. The papers were scattered. The primary repository of Jenkins history is at the Wayne County Historical Museum in Richmond, Indiana. The Smithsonian Institution, the American Museum of History's Clark "Radioana" Collection, the National Archives, and the Franklin Institute are the primary sources. His family records too are scattered. Even the diaries of his wife, Grace Love-Jenkins, exist in parts with different family members. This is not unusual; it is just a first challenge.

The Internet materials on Jenkins are extensive. Most online information is undocumented and full of unsubstantiated claims. It is a useful starting point, but one requiring significant supportive documentation, which is footnoted. Fortunately, the available original materials from institutional and family collections have supported a sustained historical integrity in this work.

Attention to Jenkins' life history has been limited, and this presents both a challenge and an opportunity. The only existing biography was written and published by Jenkins himself: *The Boyhood of an Inventor* (1931). There is one doctoral dissertation on Jenkins, by David Arthur Hollenback in 1983, which details his contributions to early television. There are biographies on America's two other primary television inventors: Albert Abramson's *Zworykin; Pioneer of Television* (1995) and my *Philo T. Farnsworth: The Father of Television* (2001). Jenkins' work predates both these American inventors and actually forms their foundations, yet he is barely mentioned. He is in every broadcast-history text, but all are minor references linking him simply with mechanical television.

In this biography, I have used archival, business, and family records to document the life of Jenkins. This work does not argue technological contributions or patent priority. Yet, while writing about Jenkins, one cannot avoid these topics. The objectives are to reflect his life, his work, and his contributions as of one of America's greatest individual inventors. His life was inseparable from his works. When you watch a movie or television, fly in an airplane, or even purchase a carton of milk, you have been touched by this man's work.

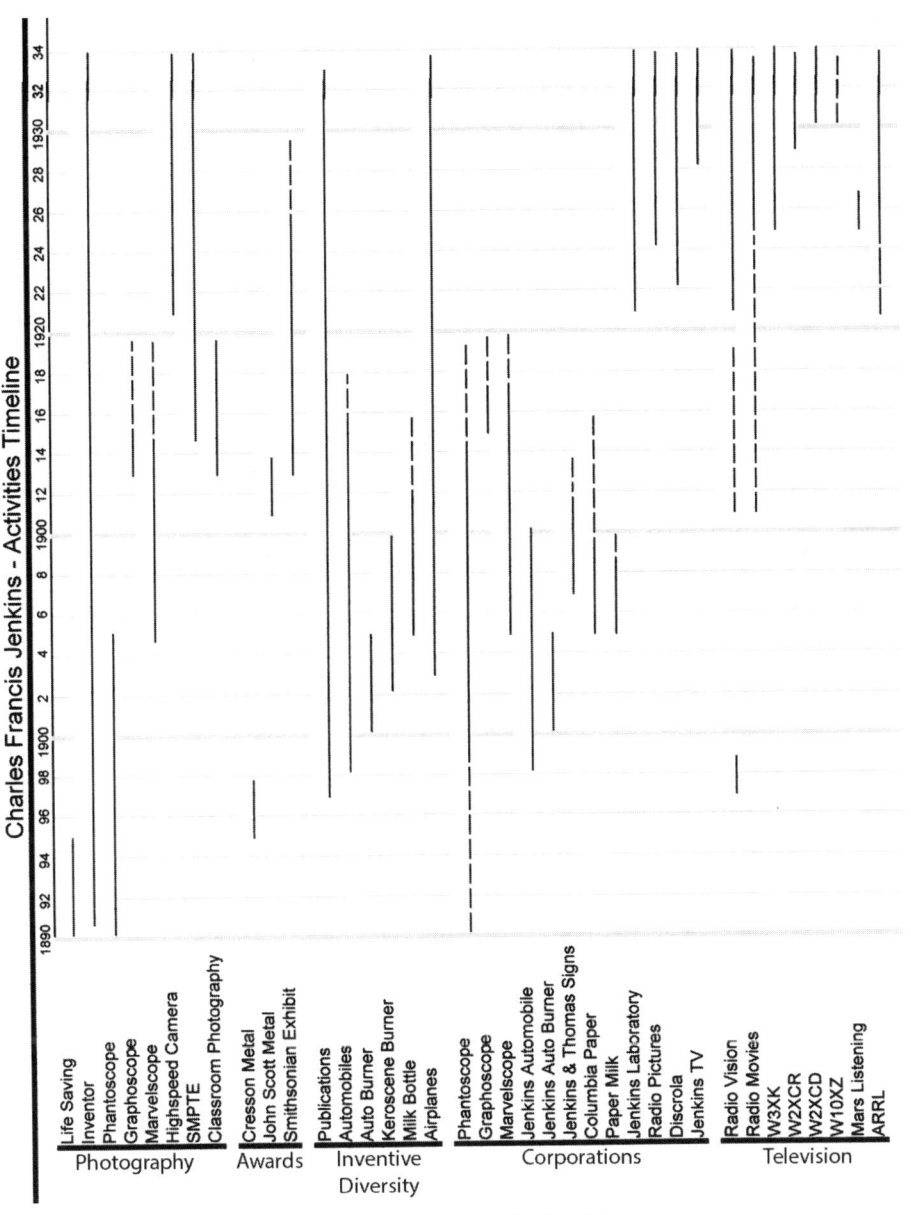

C. Francis Jenkins: a lifetime of activity. Courtesy Knight Graphics.

Acknowledgments

A book is never the work of a single author in isolation. My appreciation goes to family, friends, colleagues, and archivists who have supported this work. I extend a special thanks to my wife, Christina Maria, who patiently supports all my writing. My daughter Emma Maria Knight worked as my editorial assistant, preparing the patents, graphics, dollar-equivalents, and appendix materials; and my brother Robert M. Godfrey, a ham operator, guided me in the understanding of amateur radio. I am blessed with a supportive family.

A special thanks to Christopher H. Sterling, an associate dean at George Washington University. He, more than any single individual, has been a constant catalyst and sounding board, providing counsel, readings, and direction in this project. He is a dear friend. I thank my friends Louise Benjamin at Kansas State University and Mike Adams at San Jose State University, who read the manuscript and provided helpful suggestions in the Jenkins-DeForest connections. The late Albert Abramson, friend, author, and retired CBS engineer; and Kuman Blake, architecture engineer at the Analog and Interface Products Division of Microchip Technology Inc., helped me translate Jenkins' televison-technological vision into today's perspectives. Noah Arceneaux at San Diego State University provided information regarding the history of still-picture transmission; Paul S. Martineau helped me to understand the railroads of the late 1800s, and thus Jenkins' travels. I also thank Alison Oswold of the National Museum of American History; John V. Alviti, curator at the Franklin Institute; and James D. Harlan, executive director of the Wayne County Museum. Thanks are due to John S. Eustis, retired

Washington, D.C., civil-service employee, who read preliminary drafts of the Jenkins-Armat controversy and helped me understand that conflict; and David A. Hollenback for his work on Jenkins' early television operations. A special note of appreciation goes to Professor Fran Matera at Arizona State University, who is the best style editor I know. I appreciate the constant push upward, encouragement, questions, and most particularly, the friendships we share.

Finally, I thank the family of C. Francis Jenkins. Virginia Roach (1941–2003), Jenkins' great-grandniece first invited me into this project, sharing her own research and interest. I thank Robin and Dave Sproul, niece and nephew of Virginia Roach, who continued with me after her death. Phil and Barbara Jenkins and Ann McKee Coffin kept a continual supply of family photos and documents coming my way. Lewis Janney, a nephew who worked at the Jenkins' television station, provided photos and early insight into station operations.

Prologue

It all started with an inquisitive young mind on the American frontier of the late 1880s. Charles Francis Jenkins was a teenager when he left home to work in the lumber and mining industries of the West. His midwestern Indiana family had a hard time imagining the beauty of what their young traveler described—the Rocky Mountains, the Pacific Ocean, the Southwest Sonoran desert—and Jenkins wanted to share. This challenge set his mind to working. Photography at the time was awkward, still-camera technology.

Jenkins' inventions would capture motion and project it onto a large screen. He premiered his film device on June 6, 1894, for a small group of family and friends at the Jenkins and Company Jewelry Store at 726 Main Street in Richmond, Indiana. They were attending the world's first large-screen motion-picture exhibition. These were the days of the popular nickelodeon film-parlor theaters, where one person at a time peered into a cabinet to see the marvels of lifelike moving pictures using animated card drawings. These days were about to change. The store had been closed for Jenkins' demonstration. He shut the curtains over the windows, darkened the room, and hung a canvas screen on the wall. The audience was distracted by the noisy equipment as he readied it for the viewing.[1] Then the film began rolling, and lifelike images appeared, depicting "Annabelle the Dancing Girl," a beautiful young lady dressed in a butterfly costume.[2] She danced across the screen waving her arms as the whole group watched in amazement. This showing produced what might be considered the first film protest. As the ballerina lifted her skirt, she revealed her ankle, and the ladies in Jenkins' audience, all Quakers, stormed out of the store in protest over such a display of nudity.

They all "left, returned to [the] Fountain City [chapel], and prayed for [Jenkins'] soul. . . . [The] women left, but the men stayed on to see the show."³ This demonstration changed the world of motion-picture film and eventually would take Jenkins into television.

The first motion pictures ever shown were on a makeshift screen inside the Jenkins and Company Jewelry Store. Sketch Courtesy *Richmond Palladium Item.*

1

Jenkins' Heritage and Youth

An Age of Things Mechanical

Today's world is predicated on the inventive ingenuity of those who preceded us. Imagine life without the movies, television, telephone, airplanes, or automobiles. The twentieth century added those things and more into our lives. The inventors of past centuries, including Charles Francis Jenkins, made these things possible. They worked as independents and often clashed with established industry giants, as Jenkins did. It was challenging for lone inventors to make a living, fund their work, and promote acceptance of a new device, and Jenkins had to meet each of these challenges. He was brilliant, gifted, mechanically inclined, and intuitive. His life spanned six decades of American history, seeing the birth of photography, radio, television, the automobile, and the airplane. He was an amateur photographer who loved to travel, a personality in demand, and a pilot. His mechanical television devices are gone, but his concepts in film projection, using intermittent motion, and his theories related to optical signals in television remain a viable force.

Jenkins lived in an age of things mechanical. He was a product of the Industrial Revolution motivated by the famous Rev. Russell Conwell's sermon "Acres of Diamonds," which reflected the American Dream: "rags to riches . . . onward and upward."[1] Conwell set this national opportunistic tone, preaching this sermon more than six thousand times across the nation.[2] His ancestors had left Europe, settled in rural America, and later moved into cities in search of better lives. Collectively, these immigrants were the labor force of the Industrial Era. In the case of Jenkins individually, Conwell was a resourceful reserve of ideas.

Jenkins lived from the Industrial and Gilded Ages through World War I, the Roaring Twenties, and the Great Depression. The 1918 Spanish Flu epidemic killed an estimated fifty million people—more than the war itself. It was the deadliest human disaster in history. Schools, businesses, and homes were closed and quarantined.[3] Jenkins' work contributed to stalling the spread of the disease by providing a new sanitary milk bottle, which, according to family lore, brought him more profits than any other invention. To Jenkins, however, all of this was a means of supporting his primary endeavors in film and later television. By the 1920s, the United States had become a global economic, industrial, and military power. Financial investments once reserved for the war effort were now back in the marketplace.[4] As a result, inventors prospered. No one anticipated the stock market crash on October 29, 1929. The market had always corrected itself, and in the 1920s investment returns had been exceptional. The crash ended all of that. It devastated the American population. Corporations crumbled, including the Jenkins Television Corporation and the Jenkins Laboratories.

So, what did Jenkins contribute during these contrasting ages? Why should he be remembered today? He should be remembered as an inventor and a man of eminent forward-thinking ideas. In his own time, he ranked among the "Remakers of Civilization."[5] He was described as "one of a trio of the nation's foremost inventors"[6] and "one of the ten greatest figures in Motion Pictures."[7] In radio, he was placed among the top one hundred men of science.[8] *Scientific American* described him as having "the mind of the practical, working inventor . . . dedicated to his profession like any banker, lawyer, medical doctor or journalist [who] would become the man who [kept] America in the front ranks in the development of radio vision or 'seeing' by radio."[9] He was known as "one of radio's most colorful personages," a pioneer of "seeing via the ether."[10] The Massachusetts Institute of Technology professor Arthur C. Hardy credited our ability to see movies in our homes as "due largely to the efforts of C. Francis Jenkins."[11] His primary inventions were in motion picture and television, but he held hundreds of patents for an amazingly wide variety of inventions. He spearheaded motion-picture and television work with mechanical systems, and then, when electronic television replaced mechanical systems, he moved into optical-electronic systems. Inventing was his natural talent (see appendix A for a list of Jenkins' patents).[12] The story of C. Francis Jenkins would have been different without the Great Depression, or if he had lived longer. Yet, even in a world permeated by inventive contrasts, worldwide epidemics, the Great War, and economic roller coasters, L. C. Porters depicted Jenkins as "a man of great vision, with courage of his convictions, a man of indomitable will and boundless energy." He was a "happy warrior ... launching a daring attempt to unite television and motion pictures."[13] This is our Charles Francis Jenkins.

Growing Up on the Farm

Charles Francis Jenkins was a Quaker farm boy, born just north of Dayton, Ohio, on August 22, 1867.[14] Two years after his birth, his parents moved to Richmond, Indiana, where he grew up through his teenage years. The 141-acre farm was ten miles north of Richmond, near the small community of Fountain City.[15] The house where he spent his youth still stands: a twelve-room, two-story brick structure that remains picturesque today. Tall pine, maple, fir, and linden trees line a long lane from the main road down to the home. In one corner of the yard there was once a circle of cedar trees, which the Jenkins children used as an imaginative playhouse during the warm summer months. Green lawns, shrubbery, and fields of beans surrounded the home. The children's nursery was on the first floor in the northeast corner of the house.[16] They were easily entertained by life on the farm—there were animals to play with and trees to climb. The honking of the Canadian geese flying overhead, in their annual migrations, fascinated the youngsters. Cows, sheep, pigs, and work animals were raised on the farm. Making soap was Jenkins' first lesson in practical chemistry. The primary crops were beans and wheat. Francis learned the meaning of hard work, driving the four horse teams at an early age.[17]

The daily chores were physically demanding, farm work was difficult, and life depended upon a successful harvest. His earliest summer chores included keeping the fire going in the smokehouse, where meat was hung and cured for the winter. The meat came from the family's small herd of grazing cattle or the surrounding forest, which was full of game and other wildlife. As a boy, he cracked the large rocks and cleared them from the fields in preparation for planting. He helped in the harvest, tended the animals, and, more reflective of his natural talents, he kept the farm in good mechanical working order. Each farmer was self-reliant and traded "in kind" with neighbors and storekeepers for their needs. The stores in town accepted farm products as trade currency.

The New Garden Friends church, where the Jenkins family worshiped, still stands. It is just south of Fountain City and was established by the Quakers in 1811. It was a centerpiece of life and one of the Underground Railroad depots where southern slaves were helped North to escape captivity.[18] The family of Amasa and Mary Jenkins were devout. The horse-and-buggy stalls, originally just north of the church, are long gone, but in the Jenkins family's day these shelters protected their horses from the cold winters during church services. Amasa drove "a carriage with a matched pair of horses, a clear sign of affluence."[19]

Farm life and his Quaker origins were the foundations of life for young Jenkins—who in his early years dropped "Charles" and became known as "Francis." Formally, he would be known as C. Francis Jenkins. His understanding of mechanics helped solve regular farming challenges. In those days, if something was broken, you fixed it. You could not readily replace it or purchase a part. Add to this independence a natural work ethic, the Quaker belief in a working Christianity and God as a personal being who encourages works over financial gain, and one can begin to understand Jenkins' lifelong devotion to his work.

The Jenkins Family

Francis came from a rural family with a rich inventive streak; extended family members say that invention was in the Jenkins blood. A family cousin, later a technical advisor to the president of the Monsanto Research Corporation, had forty-five of his own patents and enjoyed "gadgeteering." Robert Jenkins, who first owned the local jewelry store, was said to have invented the rolling cookie cutter.[20]

Francis's father Amasa was born in Marion County, Ohio. He was Welsh, and his mother English. Amasa worked on his family's farm near Dayton, Ohio, and attended public school. He later attended the Spiceland Academy and Earlham College for his advanced education. He was drafted into the military but was honorably discharged due to his religious beliefs as a devout Quaker and thus a conscientious objector. He was six feet tall, a dignified man who carried himself in a stately manner. He had a "kind face, inspiring confidence . . . gracious[ness] . . . and common sense." He was a Christian man who commanded respect.[21]

Francis's mother, Mary Ann Thomas, came from a large family of seven brothers and sisters. She was born and raised in the New Garden neighborhood and attended church and school. She too was a student at Earlham College and later taught school. Mary's brother was a friend of Amasa Jenkins, and Amasa was a frequent visitor at the Thomas farm. Mary saw life as an "opportunity for service . . . either in her own home . . . or in helping someone else." Her friends compared her to Dorcas, the New Testament woman who spent her life making clothes for the poor.[22] Mary befriended the lonely, "in trial or need. . . . [T]o visit and cheer the sick and shut-ins was an especial pleasure to her."[23] Amasa and Mary would have seven children—five boys and two girls. Francis was the firstborn, the only one born in Ohio, as his brother and sisters were all born in Indiana.

* * *

THE JENKINS FAMILY TREE[24]

Father: Amasa Jenkins (1844–1938)—Parents Robert Jenkins and Ann Pearson-Jenkins
Mother: Mary Ann Thomas (1843–1916)—Parents Luke Thomas and Mildred Fulghum-Thomas
Married: June 16, 1866
Children Born:[25]

  C. Francis, August 22, 1867 (1934)
  Atwood L., December 14, 1869 (1946)
  Olive L., February 14, 1873 (1929)
  Alice A., July 21, 1878 (1931)
  Alvin, February 12, 1881 (1882)
  Alfred "Willie" William, November 23, 1883 (1958)

* * *

Francis's grandfather, Luke Thomas, lived in a log cabin on acreage not far from the Jenkins farm.[26] Francis would later recall the open fireplace hearth and working together with his grandfather. At his grandfather's knee he heard stories of the pioneering days, Daniel Boone, Indian attacks, and lessons from the Bible. Reading and reciting Bible passages was a daily ritual, just prior to breakfast. These lessons in love and appreciation inspired Jenkins. They excited him with the possibilities of learning, travel, and his own future. He wanted to see the West that his grandfather described in the stories.

The Youthful Francis Jenkins

Young Francis was a freckled redhead, "red as a brick," as he said, but with age his hair would become steel grey. His character was humble and unassuming, with a self-assured, driving passion.[27] His eyes were blue, and at five-and-a-half feet tall, he was a short, slender man. "He had a medium square build, and weighed around 175 pounds, with the face of many who spent much of his time out of doors . . . [and the] forehead of a thinker." He was full of energy, nervous, and constantly on the move. He was a high-strung adventurer, always modest about his achievements, yet with an ingenious aptitude. Things came easily to him.[28]

One summer day, when Francis was a youngster, his mother dressed him to play in the yard. He must not have been thrilled with the outfit—a short dress and sunbonnet, to keep the fair-skinned boy from getting sunburned. The yard was surrounded by a large board fence, so his mother did not worry about her children—they could not go far, and she could see them from the windows while she worked around the house. On this particular day, when she checked the grounds, little Francis had disappeared. Panic-stricken, Mary began to search for her son—he'd gotten through the fence and out of the yard. He had borrowed a wood saw from the tool shed, cut a hole in the fence, gotten out, and headed for the barn. The evidence of his escape and ingenuity lay beside the fence—he had taken off the sunbonnet, left the saw, and was gone. He was found shortly thereafter, playing with the new suckling pigs.[29]

As Francis grew, so did his contributions to farm work. His earliest inventions facilitated farm operations. He once caught his fingers in the gears of a vibrating bean-sifter screen, and this incident left him with lifelong scars. So he invented a bean husker, which removed the seed of the bean from the outer shell. His father was not too appreciative of the boy's idea, which Francis had built in the attic of the house, because inventing diverted his son's attention from the chores of greasing the wagon axles. To make that greasing job easier, Francis designed a jack that raised the wagons so that the lubricant could be more easily applied. This invention caught the attention of his neighbors. Francis and a younger brother, likely Atwood, made several jacks and sold them in town. In selling the jacks, he learned his first lesson in marketing. The painted jacks sold first, while those not yet painted, but just as workable, were slower to sell. More of his inventiveness complemented farm work. He set up a telephone system to communicate between the barn and the house. It became his responsibility to keep the farm machinery working. "That is thy gift, and to thee it is no great credit," his father commented after Francis had fixed the field mower.[30] Amasa sought not to chastise his son but to teach him that his "gift" was from God.

Francis was always inquisitive, wanting to know how everything worked. Such things as opening a pocket watch with a hatchet and disappearing at the train station, where he was found examining the engine, were youthful explorations that did not always sit well with his parents. His uncle once brought him a small model-train engine he had made, and Francis immediately set out to make one of his own. The farm was a haven for a growing young mind—it set the pattern of "need and inventiveness" in his life, but as Francis grew, the farm could not hold him. He longed to travel West.

Traveling the Western Frontier

Francis graduated high school and attended a year at Earlham College, where he developed an interest in electricity.[31] Then, sometime between 1883 and 1887, he left home for the West Coast. He was most likely about twenty years of age when he boarded the Santa Fe railroad in Chicago for the trek to the Pacific.[32] It was a pivotal point in his life. Had he remained on the farm, he might have made different contributions to society. The record of why Francis left is unclear, but the death of his younger brother Alvin likely influenced his decision. The baby Alvin had died after having been left in Francis' care. It was a crushing blow to Francis, as he seemed to be blamed for his brother's death.[33] Thus, grieving and motivated, he left home in search of his own life.

Francis's travels for the next several years took him through the lumber mills of Washington and Oregon. He rode the logs in the mill pond where the timber was cut and loaded aboard the trains. If he fell off a log, he worked all day in cold, wet clothes. At the saw mill his mechanical aptitude proved valuable when two trains collided, knocking both off the tracks. His ideas untangled the cars and had them running again within a few hours. In Seattle, he took a course in stenography. He studied telegraphy in Newburg, Oregon, and traveled through Nevada, Arizona, and New Mexico working in various mines. He was reportedly an accountant for the Sierra Nevada silver mines of Arizona and New Mexico and helped keep their machinery in good repair.[34] He made friends with the workers on the Santa Fe railroad and did a little independent engineering of his own one evening when he took his girlfriend on an unofficial train ride. He later admitted that he had taken "a steam locomotive from a railroad terminal and drove it ten miles into the country. Why? The girl of course. Her pop's ranch fronted the railroad. I stopped the locomotive near the ranch house and pulled the whistle."[35] The kindly boss was waiting by the tracks when Francis returned the engine. He undoubtedly made an impression on the girl, but the boss was likely none too happy. Another time, while out in the desert near Cook's Peak, with a search party looking for an old miner, Jenkins suffered a flesh wound in an attack by a local Apache hunting party. He was "nicked in the shinbone before he found cover and his companion was killed a few days later by the same Indians."³⁶ Francis did not stay long in any single place. He roamed the West, as many did in his time, using the railroads and on horseback, with his rifle at his side. He worked different jobs, growing in maturity with each opportunity.³⁷

The love for his family brought him home each year for a brief visit. Their difficulty in imagining his wilderness lifestyle—the high mountains, deserts, and the Pacific Ocean—was a clear catalyst for Jenkins' interest in photography. He wanted to share his experiences. On one of his visits home, his mother encouraged him to take a civil service exam; it would be a turning point in his life. He was in New Mexico when he received a telegram notifying him that he had been appointed as a clerk for Sumner I. Kimball of the U.S. Life Saving Service—now the U.S. Coast Guard, then a part of the U.S. Treasury Department.

Moving to Washington, D.C.

In 1890, Francis gave his guns and holsters to his friend. He turned eastward to report in Washington, D.C., for his new job. His first home was a boarding house at 317 East Capitol Street, where he lived until 1898.³⁸ In the Life Saving Service, Francis experienced his adventures vicariously, as his job was to write stories of people guarding the coast. He wrote them by hand and then typed them into annual reports. Often he rode in the rescue boats with seaman seeking firsthand accounts as well as adventure. His work with the Life Saving Service honed his writing skills, and his association with Kimball further instilled a humanitarian spirit complementing his Quaker commitment to service. He enjoyed his work, good fortune, and the prospects of a promotion.³⁹ He loved Washington, D.C. He was fascinated by the mechanics and the technological progress that flourished around him. He felt that he was in the center of growing opportunity. The city was a bustling center of "charming femininity and handsome manhood."⁴⁰ It was a stark contrast to the farm and western chores, but it created a turning point in his life with service connections that he would use throughout his career.

During his first years with the Life Saving Service, Jenkins' experiments in photography began in earnest. They initially took up the young bachelor's evenings and weekends, but it would not be long before he was working full-time with film and creating "devices for recording and reproducing motion."⁴¹ One of his friends, a young lady with whom he shared his enthusiasm, would find out "what it means to be the wife of a great inventor."⁴²

Romancing Grace Love

A beautiful young woman entered Francis Jenkins' life shortly after he arrived in Washington. Her name was Grace Hannah Love. He and Grace met in the winter of 1891.[43] She worked in the Treasury Department cafeteria and later the Log Cabin Café, both owned by her family. Francis ate at the cafeteria regularly while he worked at the Life Saving Service.[44] Their relationship evolved slowly, from their early restaurant exchanges to pleasant conversation, dating, and marriage. Indeed, they courted eleven years before they were married.

Grace was born on January 19, 1868, and raised near Conowingo in Cecil County, Maryland, on a farm not far from the Susquehanna River. Nature, the woods, and the river were her youthful playgrounds. Her ancestors were Scottish immigrants who came to the American colonies before the Revolutionary War. Her father, Samuel Love, was a Methodist, and her mother, Sophia Jane Taylor-Love, was a Quaker. She had four sisters and three brothers. Grace was the youngest daughter. The family relationship was close, and the household regularly attended Methodist services. The sisters were devoted to each other. The "Love girls" were noted for their musical talent, a reputation gained from singing in church and at congregational socials.[45] Grace was described as placid by nature—calm, even-tempered, and likable. She was sympathetic and understanding, winning the hearts and confidences of all who associated with her.[46] She was a little taller than Francis, with "brown eyes and a steady straightforward look like the captain of a ship. . . . She dressed simply and inconspicuously."[47] Francis "treated her, both in public and in private, as a youthful lover."[48] He was young at heart, and "Mrs. J was a motherly, understanding type who mothered him and kidded him along."[49] She was raised a strict Methodist, and while she was certainly interested in the young inventor, it took some time before she yielded to his proposals.

Grace had originally moved to Washington, D.C., to start a job, be with family, and get away from a jilted love affair.[50] Two of her older sisters, Annie and Sadie, were already living in the city and married to brothers Henry and Lewis Breuninger.[51] Grace was engaged to the third brother, George, but George had an affair with the "innkeeper's daughter."[52] Heartbroken, Grace moved to the District of Columbia and lived with her sister Annie. Eventually, Grace's whole family would follow. In the city, she worked at her grandfather's eatery, the Log Cabin Café, at 1337 F. Street, NW.[53] It was fall when she met Jenkins. His evenings and spare time would now be divided between full-time work with the Life Saving Service, part-time work on his photographic gadgets, and dating.[54]

Grace and Francis enjoyed being together. A picnic, boating, cycling, or just walking along the banks of the creek were their most common dates. The two "love[d] to pack a little basket of lunch and fare forth to the banks of Rocky Creek . . . and just sit there among growing things . . . talking over their plans and experiences till the great yellow moon [came] up."[55] These were years before the social revolution in morals and manners of the 1920s, so Grace and Francis were always chaperoned by a member of the family or a friend. They talked about his inventions and just savored the natural state of the riverside park—it was pleasurable recreation that would be repeated many times in their lives together. In her diaries, Grace provided a flavor of life's simple joys, her character, and the cold Washington winter as she wrote simply, "beautiful day, I went to church, Rev. Wilson preached. I met CFJ [Charles Francis Jenkins] downtown and we took a walk in the sunshine."[56] She was receptive, supportive, and interested—a comfortable audience, and always willing to walk in the warmth of the sun. Francis attributed his personal success to her "business wisdom" and "personal genius."[57] By 1899, her brother-in-law Lewis Breuninger had opened a second restaurant in the Treasury Department, and Grace became its manager.[58]

One of the reasons Francis and Grace's romance progressed so slowly was their families. They were initially uncertain whether Francis was good husband material. He had resigned his job as a clerk in the Life Saving Service to become a full-time inventor. Since 1895, Francis had listed his profession in the Washington, D.C., directory as "inventor." The family criticized him. What kind of a man lists "inventor" as his occupation in the city directory? Even Francis's own mother did not like the title. From his youth she had encouraged him to "stop messing" and get to work. Years later, Jenkins described inventing as "a profession like a tramp, having no visible means of support." However, after his later years of success, he asked his mother for a title that would describe him, and she replied, "a finder-outer."[59] The family's questions and comments about Francis' avocation were really more a reflection of worry and misunderstanding than a condemnation of the man's character. At the center of concern was the vexation of how he would support himself, a wife, and a family. However, Francis and Grace's affection for one another was unquestionable, as they spent increasing amounts of time together. During the winter of 1899, due to the inclement weather, Jenkins even stayed overnight in the home of Annie and Henry, where Grace lived. The next morning, he walked her to work in one of the coldest winters on record.[60] Grace would never admit to how much of her influence had persuaded him "to give up a regularly monthly [government] check for certain poverty [of an inventor] over a long period on the chance that he would make good eventually, but she was always supportive."[61]

The wedding took place on January 30, 1902.[62] The family gathered at the home of Grace's sister Sadie and Lewis Breuninger. The union was performed by Grace's Methodist minister, Rev. J. T. Heisse. Francis's mother commented to the new bride, "I fear thee hast married my black sheep"[63] No one was quite sure what she meant. Was it Francis's declaration to become an inventor, his participation in the Methodist faith, a reference to family differences, finances, or just her own uncertainties?

As a wedding gift, Grace gave Francis her life savings of three thousand dollars ($78,353 in 2012), and he promised her half of all he made from his inventions.[64] They invested the cash in a laboratory and in a modest little home at 1103 Harvard Street, NW—the cost was $25.50 ($666) a month. Grace's mother lived there with them until her death in 1906.

Francis credited his success to his wife: "[I]t [was] to her kindly help and business wisdom, rather than to any personal 'genius,' that this inventor attributes such success as has attended his efforts."[65] Grace was totally unselfish, even in hard times. Their relationship was "the most beautiful relationship in human life."[66]

Francis and Grace had no children of their own, so her sister's children had multiple mothers: Grace, Sadie, Annie, and their grandmother. The sisters lived close to each other, so they were constantly together with the extended families. Francis' sense of humor endeared him to his extended family and children. They all took part in his photography. He loved his nieces and nephews as his own children. He was continually doing things for them as they grew older, helping them get a financial start in life.[67] Family was the foundation of Jenkins' life and his work. Throughout his life, he retained strong relationships with his Indiana family and his wife's family. He was an endlessly hard worker and a man of strong religious values. His inquisitive mind and exploring nature motivated a lifetime of inquiry across a broad spectrum of ideas. Inventing was what he liked to do, and his family was a part of it.

2

Early Film Experiments

The large-screen projector and the concept of intermittent motion remain Jenkins' most lasting contributions to the industries of film and motion pictures in television. His projector was the first to freeze a frame of film for a fraction of a second, as it flowed past the lamp and lens of the projector. It is a recognition that he would begrudgingly share with Thomas Armat, but its roots begin with Charles Francis Jenkins.

The late 1870s into the early 1900s was a time of growing industrial opportunity, contrasted with what critics and historians have described as a time of "conspicuous waste, a great barbeque, an age of bad taste."[1] The motion-picture business was at its beginnings during this time, and it was hardly a business at all. But it was attracting attention.[2] Film titles and those appearing in films had no impact. The subject and the science reflected the content and experiment, not star power. The nickelodeons grew in interest toward the end of the century. Even then, not everyone was getting rich, but the prospects were encouraging. These were the years in which Jenkins grew to maturity. He was a transient youthful worker out West at the end of the 1880s and took his first full-time job at the U.S. Live Saving Service in 1890. Photography had become a passionate interest, and he would be among its full-time inventors by 1895.

Within this colorful context and rapid changes in society, Jenkins was not alone. Still photography was becoming a popular hobby, while inventors were working on ways to add motion. In 1892, Etienne Marey and his associate Georges Demeny in France invented a camera system to capture the flight of birds.³ In 1895, the Lumiere brothers, Louis and Auguste, took out a French patent on the first film-projection machine.⁴ By 1888, another Frenchman, Emile Reynaud, provided movement to pictures with the Reynaud praxinoscope—praxis meaning "action." Reynaud demonstrated the apparatus throughout the theaters of Paris.⁵ These experiments did not go unnoticed by their American counterparts Thomas Edison, George Eastman, Thomas Armat, and C. Francis Jenkins.

In America, Thomas A. Edison (1847–1931) eventually created a strong portfolio of 1,093 inventions, including common items we now take for granted, such as the light bulb, sound recording, and motion pictures.⁶ Edison, twenty years Jenkins' senior, was the dominant force in the infancy of the motion picture. He was the leader technologically and in business. He directed industry.⁷ Edison and his assistant, W. K. L. Dickson, were among the primary motion-picture pioneers. But their earliest pictures were "absolutely microscopic," and there was much room for improvement.⁸ Dickson was working on a kinetograph camera and a viewing apparatus called the kinetoscope. These allowed the viewer to crank an endless film loop through the peep-show device. The Edison Kinetoscope did not allow for screen projection. It was premiered at the 1893 Chicago World's Fair, and by 1895 Edison was selling them to the retail stores, which installed them as penny-arcade parlors to attract customers. They typically featured short, one-minute filmed subjects such as vaudeville acts, dancers, and acrobats. The films were created at the Edison motion-picture "studio" in West Orange, New Jersey, called the Black Maria.⁹ The Edison Kinetoscope was the apparatus creating the excitement, but it suffered a major shortcoming in that it could only be viewed by one person at a time.¹⁰

George Eastman's (1854–1932) Kodak still-picture camera was released to the public in 1888. The cost of the camera and film was twenty-five dollars ($629). More important to the future of motion in film pictures was his patent on the film roll, which Jenkins would utilize in his experimentation.¹¹ Eastman's rolls replaced heavy glass-plate negatives and became the standard in still and motion-picture photography for the next century.¹² The Eastman Kodak Company was founded in Rochester, New York, in 1892.

Surprisingly few photographic pioneers had yet considered large-screen projection. By 1895, some inventors saw an opportunity: the Lumiere brothers (France), Robert W. Paul (Great Britain), Woodville Latham (New York), and Thomas J. Armat and C. Francis Jenkins (Washington, D.C.).¹³ Large-screen projection technology would eventually come through Jenkins and Armat.

Together, they would change motion-picture history with intermittent-motion large-screen motion-picture projection—the Jenkins Phantoscope.

Jenkins' Early Demonstrations

The years 1890 to 1902 mark the beginnings of Jenkins' many photographic inventions and private demonstrations; his friends, family, and children provided the talent for each test and technological improvement.

In 1890 and 1891, he drew the technical sketches, diagraming a camera and recorder he hoped to build, so his family could see what he was experiencing as he traveled in the West. These drawings were later shown to a boardinghouse roommate at 317 Capital Street in Washington, D.C.[14] He described the construction of these ideas as systems for "screen machines"—the projecting of the pictures.[15] In 1892, he would astonish friends and members of Grace's family by projecting moving pictures across the room.[16] He reported that the ideas emanated from clearing the lumbermen's right-of-way through the Oregon forests, as "the flashes of sunlight reflected [through] the trees and by the axes ... [giving him the idea] to produce those flashes by another [photographic] means" whereby pictures could be preserved on film and projected.

By 1892, he produced a camera with James P. Freeman, a financial underwriter.[17] Little is known about Freeman, except that he worked for Jenkins and helped financially in return for a four-fifths interest in the invention.[18] The Jenkins-Freeman partnership produced a crude rotary-lens camera and a peep-show device.[19] They used unperforated rolls of Kodak film, with the lenses moving in an arc.[20] This was the forerunner of what Jenkins called the Cabinet Phantoscopes, and it introduced him into the business, but his pursuit was toward life-sized screens.[21] This challenge led him to enroll in night classes at the Bliss School of Electricity in Washington, D.C. He was looking for new approaches to his ideas. It was this same year that he first approached Alexander Graham Bell and offered to demonstrate for him.

In the summer of 1893, while still working full-time in federal government service, Jenkins was experimenting with an oil lantern as a light source for the projector. The idea was actually that of his friend, Dwight G. Washburn. However, the lantern could not produce sufficient brightness, and the following year, electric lamps proved far more satisfactory.[22] Jenkins' large-screen pictures included a small girl dancing and children jumping, swimming, running, and doing somersaults.[23] Some films were shot by Jenkins himself, and others were rented from nickelodeons.

This early Phantoscope was created under the Freeman-Jenkins contract in 1893. Courtesy Wayne County Historical Museum.

Richmond Demonstrations

In 1894 and 1895, Jenkins conducted three reported demonstrations in his home town of Richmond. These showings of his camera-projector, now called the Phantoscope, became major events for Jenkins.

The first demonstration occurred on June 6, 1894, at the Jenkins and Company Jewelry Store, owned by a cousin, Charles Milton Jenkins, at 726 Main Street.[24] For the occasion, Jenkins decided to make his visit a little unusual. Under the sponsorship of the League of American Wheelmen, he took a bicycle trip from Washington to Richmond. Bicycle enthusiasts in Jenkins' day were known as "wheelmen," and they "were challenged by rutted roads of gravel and dirt and faced antagonism from horsemen, wagon drivers, and pedestrians."[25] So it was for Jenkins.

He reportedly shipped his projector ahead of his ride and then set out on two wheels, to see the countryside. He managed to run into the Chesapeake and Ohio Canal. He was distracted, watching mules pull canal boats down the canal waterway, when he lost control of his bike and ran into the water.[26] The boats were likely piled with coal, or held cargoes of flour and grain being shipped from Cumberland to Georgetown. The crews used mule teams to pull the boats from the side of the canal. Jenkins, too, was an apparent distraction, as the boatmen celebrated his dunking. However, his bike was "fished out of the canal with a tree branch and [he] went on his way."[27]

Jenkins was a lover of people and nature. In Virginia, he met a mountaineer and his family and later recalled, "[T]he center of their world . . . was that log cabin on the mountainside."[28] He telegraphed his family each day, updating them on his progress. The *Richmond Telegraph* reported, "[Jenkins] will arrive home this evening, having ridden the entire distance on his wheel," and noted that he was the inventor of a machine for photographic motion. The bold headlines declared, "A Genius, Charles Francis Jenkins, Formerly a Resident of This City, His New Invention, the Phantoscope, the Most Wonderful Ever Invented by a Wayne County Man," and the article described Jenkins almost humorously: "He is fuller of inventions than a dog is of fleas, but his most wonderful piece of work is a photographic instrument for depicting actual motion. With this he takes a number of pictures in an instant of some object in motion. Then he can reproduce these pictures *life size* (by means of a stereopticon arrangement) on a curtain in the same length of time it took to take the pictures, thereby reproducing the motion."[29] It is puzzling that the newspaper reported that it was Jenkins' brother, Atwood, who had conducted this particular showing of the photographic work that "seems[ed] miraculous."[30]

The second Richmond demonstration took place four months later, on September 14, 1894. The particulars are sketchy, but the *Telegram* again noted Jenkins as a hometown boy, the inventor of a "wonderful machine for photographing motion, as described in the Telegram several weeks ago," a reference to the June 6 demonstration.[31]

The third and most notorious demonstration was reported as taking place on October 30, 1895:

> Francis Jenkins was in the city last night with this "Phantoscope," and delighted a number of his friends with an exhibition that was most marvelous. ... [P]ositive photographs that have been taken at the rate of 1,600 a minute and is [sic] shown at the same rate. ... The pictures presented upon the canvas at Jenkins' jewelry story were all life size, and those fortunate enough to see them were enraptured at the wonderful and beautiful effects seen. This is the invention of Mr. Jenkins, and he has a fortune in it.[32]

This showing inside Jenkins and Company Jewelry was for a small group of family, friends, and an editor from the local press.[33] The film, "Annabella the Dancing Girl," Jenkins had apparently hand-colored and acquired rights to use through Edward F. Murphy, who was working with Columbia Phonograph.[34] The film depicted a beautiful young lady, dressed in a butterfly costume, dancing across the screen, waving her arms in jerky movements. This was the showing where the performer slowly lifted her skirt and revealed her naked ankle: "[The] women left, but the men stayed on to see the show."[35]

Demonstrations Grow

In Washington, D.C., following the first two Richmond demonstrations, Jenkins sought to illustrate the defense possibilities of the new Phantoscope camera at the Indian Head Naval Proving Grounds. He was attempting to photograph the flight of an artillery shell when the camera was "split into kindling" from the concussion of the first shot fired from the fourteen-inch experimental gun called the Peacemaker.[36] Although unsuccessful, this would be the first of many experiments Jenkins would conduct with the military throughout his career.

Friends and family were the primary witnesses to his works, as they were persuaded to be the talent performing before the Jenkins camera. They jumped, ran, danced, and executed other athletic triumphs. Jenkins said that his first motion picture "star" was Arthur J. McElhone, a friend and fellow service worker.[37] In November 1894, Jenkins conducted a demonstration at the Pure Food Exposition in Washington with the cabinet-form projector.[38] This time, life-sized images were projected onto a twenty-foot-wide screen, and there was visible motion.[39]

Jenkins was anxious to advance his ideas. Interest was growing, people were curious, and he wanted to show off his accomplishments. However, he did not have any significant investors, and his work on motion pictures remained only a part-time labor, so he kept looking for those who would support his efforts. He used private and public exhibitions to interest prospective investors. He demonstrated each new development to pique an interest as well as test improvements in the technology. Most of what is known about these earliest experiments comes from those who witnessed them. They were conducted in a building on the north side of 1325 F Street, NW. Grace and family said that the pictures were "extra good,"[40] "[considering] of course [that we] were not being accustomed to seeing anything of the kind."[41] Rebecca L. Love, Grace's sister, remembered the fall 1894 pictures as "exceedingly clear . . . life-size . . . and they seemed to imitate a very good likeness . . . much like life." Charles H. McLellan, a coworker, described Jenkins' operations as "throwing pictures on the screen."[42] Witnesses spoke of demonstrations as free of charge, with audiences at times including as many as thirty.

Jenkins had a favorite in his portfolio of stars: the seven-year-old Annie Morrison, Jenkins' niece and the daughter of Grace's sister, Mary E. Love-Morrison. He loved taking pictures of Annie blowing bubbles or dancing.[43] He would film her, and then a few days later come to her home, pick her up, and take her to look at the pictures on the screen. Annie's mother said that Jenkins treated her daughter "like a queen." Annie thought that the pictures were for her, and this made her feel special. "He talked to her like a grown-up," even though she had just started grade school. One evening, her mother Mary heard a noise in the attic bedroom and thought she should investigate. She found Annie making her own little theater. She had ripped the pillow case open, hung it on the wall, and was making shadow pictures with her fingers.⁴⁴ Annie's mother was perhaps the first to express concern that her daughter was watching too much film. Annie was constantly trying to imitate the dancing girl. Mary would "put her to bed, and would go to see the pictures myself for fear it would influence her afterlife [or], for fear she would be a fast girl in the afterlife." Annie promised that she would be a "good girl if I would let her dance all the way to Grandma's [house]."⁴⁵ In one of the "saddest hours" of their family life, little Annie caught diphtheria and died on January 24, 1895.⁴⁶

Phantoscope Camera-Projector Experiments

Jenkins refers to several of his inventions as Phantoscopes. These were both patented and nonpatented improvements—and confusing (see the patent list in appendix A).⁴⁷ They were labeled for keeping attention focused on Jenkins and branding "the Phantoscope." He needed name recognition.

The first Phantoscope, of 1894, was a moving-picture machine not unlike Edison's Kinetoscope.⁴⁸ Jenkins' second, most important kinetographic camera patent, filed a month later, combined recording and projecting motion-picture film. This too was labeled a Phantoscope and was the instrument utilized in Jenkins' earliest experiments. It was a rotary-lens camera. The truly unique attribute of this machine was that "it was the first one that Jenkins or anyone else ever saw that embodied the principle that underlies all successful motion picture projecting machines, which is that of giving the film an intermittent motion."⁴⁹ The third Phantoscope was a joint instrument created by Jenkins and his partner Thomas Armat. Historians argue as to whether it was the second or third system that revolutionized the motion-picture industry from peephole boxes into motion-picture theaters, but the results were evident. The Phantoscope revolutionized the industry.

The Jenkins Phantoscope camera moved the film in front of a disk with rotating lenses, thus moving images and lenses together. The machine projected films of any length, as they were pulled through with a "sudden jerk on the film," and he utilized sprocket holes with a wheel feeding the film through the projector.⁵⁰ Jenkins' 1898 book *Animated Pictures* illustrated his work on these projectors, as well as film developers and printers.

Recognition of Jenkins Grows

Jenkins resigned from the Life Saving Service and became a full-time inventor at the end of 1895; "Inventing was now too fixed a habit, in the grip of which he was helpless."⁵¹ He was still courting Grace and living at a boarding house with other government workers. His Cabinet Phantoscope appeared, for a short time, along the boardwalk in Atlantic City. A patron inserted a nickel into the coin-activated mechanism, and the film ran for forty seconds. The cabinet was large, eighteen by twelve feet, and projected an image on a six-by-nine-foot screen.[52] It was this "5 cent theater" ($1.36) that, according to Jenkins, opened the "golden flood" and made "motion picture[s] a synonym for easy money fortunes."[53] Unfortunately, the gold would be distributed among the larger competitive corporations, with Jenkins reaping little for the foundations he had established. By April 1896, he had sold Armat the rights to his patent.

Jenkins was a promoter from the beginning, but getting started was challenging. His files are full of newspaper and magazine publicity generated by the various entities involved with each invention. In the beginning, the problem was not merely technology but acquiring recognition and funding. Jenkins networked among his acquaintances throughout government, press, and industry representatives to acquire financial support. He created a diversity of patents throughout his career, which he would sell off to support his primary interests in motion pictures. He established his inventive enterprises from the smallest of beginnings and would eventually be awarded the Franklin Institute Elliott Cresson Gold Medal and the John Scott Medal for his Phantoscope.

3

A Lifetime of Struggle

It is easy to romanticize a man like Charles Francis Jenkins—born into an age of discovery, traveling out West, seeing the creation of inventions that are taken for granted today. He was acquainted with the likes of Alexander Graham Bell in his time. Nevertheless, Jenkins' life was not without controversy. He was challenged by his competitors and critics. The most significant of these was Thomas J. Armat, who partnered with him for a short period of time and then moved on to create the Armat Moving-Picture Company, making millions and giving Jenkins a lifetime of aggravation. The controversy cannot be ignored, as there are reflections of Jenkins' life and invention in the conflict.

The film historian Charles Musser described the emergence of cinema in the United States as a "history of greed, dishonesty, and ineptitude."[1] This could be said of the industrial battlefields of many early inventors, in motion pictures as well as radio and television. It was a battle not only between competing laboratories and ideas. It was competition to be first—in receiving the credit, in patent rights, in the press, and in earning potentially huge profits. Battles cycled out from the small laboratory conflicts into corporate boardrooms, the media, and the courts. Lawsuits, litigation, and patent-interference charges were all just part of doing business. Film, radio, and television inventors were constantly in contention. Thomas A. Edison was involved in many infringement suits.[2] The inventors Guglielmo Marconi, Reginald Fessenden (who once worked for Edison), Lee de Forest, and Philo Farnsworth all had incorporated organizations surrounding their inventions, and all were involved in patent-infringement challenges. Perhaps the most infamous were the De Forest versus Armstrong, Armstrong versus Sarnoff/RCA, and Farnsworth versus Sarnoff/RCA cases.[3] Some sued at the drop of a hat. Lawsuits separated friendships, partnerships, inventors, and corporations. Such was the context in the development of film and television during Jenkins' lifetime. So it was for the Jenkins and Armat controversy.

Armat: Opportunity, Friends, and Adversaries

Thomas J. Armat (1866–1948) was born on October 26, 1866, in Fredericksburg, Virginia, to a well-to-do family. His father practiced law and operated a lumber mill. Thomas attended private schools and was an accomplished oarsman. He grew up and worked in Washington, D.C. In his teenage years, he clerked at a local hardware store. At the age of eighteen he labored at a railroad machine shop. Later he would become a bookkeeper for the Richmond and Danville railway.

In motion-picture history, Armat's name is inextricably connected first with Jenkins, then with Edison, and later with his own motion-picture company.[4] Like Jenkins, Armat's inventive interests were varied. He enrolled in courses at the Mechanics Institute in Richmond, Virginia, and he was developing a small portfolio of inventions. He created a rowlock for boats and a mechanical coupler for the railway cars. In 1893, he left the railroad for full-time employment in the family real estate business of Daniel and Armat in Washington, D.C.[5]

Armat's interest in photography grew out of attending the Chicago World's Columbian Exposition in 1893, where he saw Ottomar Anchutz's "tachyscope" exhibited.[6] The tachyscope utilized "photography plates on a wheel," which were illuminated by light flashing on each plate. It was a peephole device where the audience, one at a time, could see the motion, most often a nature scene, but it was not large-screen projection.[7] Armat was initially discouraged from seeking the hopeless possibilities of projection by H. A. Tabb, a family friend. Tabb was an associate of Raff and Gammon, the sales-representative firm in Washington, D.C., for the Edison Kinetoscope.[8] Tabb told him that "he did not believe it was possible to project such pictures successfully." He was aware of the failed work at the Edison Company, which had been promoted by Raff and Gammon.[9] This did not dissuade Armat but rather led to his enrollment in the Bliss School of Electricity, believing that perhaps large-screen projection could be accomplished another way. It was at the Bliss School, in the winter of 1894, that he met Jenkins.[10] At the time, Armat was working full-time in D.C. real estate, and Jenkins was still at the Treasury

Department. Successful in real estate, Armat had networked with people from Edison's organization who were developing financial underwriting. H. A. Tabb, Norman C. Raff, and Frank R. Gammon had visited the Armat family in the summer of 1894, seeking investments in behalf of Edison.[11] The seniors of the family turned them down, but the younger Armat's interest was growing, and Jenkins was to be his stepping stone to a new career. Jenkins described Armat as a "junior member" of a real estate company, "a man who possessed that great lubricant for invention, money."[12] Part-time inventing was their mutual hobby and would soon be their careers.

Opportunity, Friends, and Adversaries

At the time the Jenkins-Armat struggles began in 1895, both were young men. Armat was two years Jenkins' senior. Jenkins was single and just beginning to contemplate leaving the security of full-time government service to dedicate his attention to inventing. He was a hard-working, "aspiring, but impoverished inventor," always exploring new ideas and avenues to finance his work.[13] It was Jenkins' search for capital that interested him in Armat.

Jenkins was well into his experimental work with the camera and projector long before he met Armat. Numerous family members, friends, coworkers, and associates were subjects of his experiments and attested to his drawings, the excitement of the actual motion-picture camera, its life-sized image reproduction, and its inventor. Hugh Stuart was among the first of these witnesses. Stuart lived at the same boarding house as Jenkins and worked as a doorman at the House of Representatives. He was shown a device that was "intended to reproduce subjects in action" in the fall of 1890.[14] J. D. Boyce, proprietor of the J. D. Boyce Studio, was shown a camera projector in September 1891.[15] He described the jerky, intermittent motion of the film. John J. Hayden, a fellow worker from the Treasury Department, examined some original pencil drawings during the Christmas season in 1891.[16] In 1892, Mrs. Philo L. Bush gave two "boys a nickel each to turn a somersault while [Jenkins] photographed them with this camera."[17]

In 1892, Jenkins approached Alexander Graham Bell for funding. Typifying his strategy, he described his photographic instrument for "the recording and reproduction of action" and requested financial support. He called the apparatus he offered to demonstrate for Bell a "phenakistoscope."[18] He described it as capable of recording parades, marching columns, inaugural balls, speeches of great men, and the songs of famous artists—all of which could be played back later in real time. "What I desire," he wrote Bell, "is to interest enough capital to make and put them upon the market. The right to use the machine as an exhibition machine should result in a large income." He admitted that the machine was crude, but he offered to demonstrate it. He represented himself as "a young man with no capital, but plenty of energy and a willingness to work." Bell and Jenkins exchanged communication over the next several years. The proposal interested Bell because of its capacity to produce photographs of "moving lips," which complemented Bell's work with the deaf.[19] Bell later described the Jenkins Phantoscope as producing pictures of "remarkable fidelity to nature."[20] He congratulated Jenkins on his accomplishments and thanked him for the opportunity of seeing "the simply ingenious mechanism by which . . . an intermittent motion of the photographic film has enabled you to employ a continuous light of great intensity and I must congratulate you upon your success in throwing upon the screen life-sized moving photographs."[21] They would later conduct a lip-reading experiment for which Bell provided Jenkins limited funding and, more importantly, his endorsement.

One of Jenkins' Life Saving Service colleagues, Arthur J. McElhone, took part in several experiments in the summer of 1893, which included the high jump, running, diving, and swimming.[22] These recordings took place at the Chevy Chase Country Club in Maryland, just outside of Washington. McElhone indicated that Jenkins had used two cameras, "taking pictures by jerking the film along between exposures."[23]

Jenkins found his first investor in an engineer, James P. Freeman, and they signed a joint contract on January 17, 1893. Freeman advanced fifty dollars ($1,258 in 2012) toward the project, and when Freeman became ill, Jenkins repaid him in installments.[24] They worked together on the 1893–94 Phantoscope.[25] Their contract identified Jenkins as "the inventor and sole owner of certain new methods of apparatus for the recording and reproduction of action." Freeman was responsible for creating the drawings and providing "any sum of money that may be necessary to construct the working model" to further demonstrate its possibilities, and after successful demonstration "he will immediately interest a party of means with large business experience who will . . . successfully place the invention on the market."[26] Freeman was an investor, an assistant, and charged with marketing.

For the construction of the cameras and projectors, Jenkins and Freeman used a Washington, D.C., model maker named Emil Wellauer. Wellauer chronicled his work for them as crude and indicated that he was constructing "different machines . . . known by him as phantoscopes."[27] The work requested by Freeman and Jenkins was "always [needed] in a hurry." He was sure the machine was made prior to Jenkins' association with Armat because after Armat joined with Jenkins, Wellauer commented, their machines "were put together in a much more workman like manner."[28]

A friend of the Jenkins family, Edwin Lee, was invited in 1894 to see the pictures. He portrayed them as "remarkable ... life-size subjects [that] were projected from the back of the room onto the opposite wall. ... [The projector] made a snapping sound and had a small, very intensive light."[29] The attorney Wallace Greene was also shown the Phantoscope in the "summer of 1894," and the projector was exhibited again at the Washington, D.C., Pure Food Exhibition.[30] Jenkins' family members witnessed and participated in experiments continually. Perhaps the most poignant confirmation of these early showings comes from the diaries of Mary E. Morrison, Jenkins' sister-in-law, and her daughter Annie. In an ironic twist of history, Annie's death and resulting entries in Mary's diary marked several specific dates of Jenkins' experiments as October 7, 12, 23, and 29, 1894. Morrison would later be called as a court witness on behalf of Jenkins, who was defending the reality of these earliest experiments.[31]

Jenkins' experiments were gaining publicity. A description of the Phantoscope appeared in the *Photographic Times* in 1894, showing the combined projector and camera device.[32] The photos and narrative clearly illustrated that "Mr Jenkins' instrument is somewhat similar in its working to Edison's, except that it is small, portable and cheap." The pictures were reproduced with "an optical lantern upon any size screen, so rapidly that the eye does not see the picture except as one continuous picture with the objects apparently in motion."[33] Jenkins compared his Phantoscope to Edison's Kinetoscope and Anchetz's Tachyscope, not only as cheaper but also "constructed [so] that there [was] no limit to the number of pictures which may be made."[34]

Eighteen ninety-four was a prolific year for Jenkins. His first Phantoscope camera was effectively producing motion pictures.[35] The Phantoscope was shown in Richmond, Indiana, in June. The next month, he theorized that pictures could be electronically transmitted over varying distances. He diagrammed what was again called the Jenkins Phantoscope for "transmitting pictures by electricity." He welcomed anyone who wanted to experiment with this idea, as "I cannot at present test it myself" for lack of funds.[36] In the fall of 1894, Jenkins enrolled at the Bliss School. Here, he expanded his showings for faculty and fellow students, still in hopes of finding financial backing. One such Bliss School showing included pictures of an unidentified D.C. parade.[37]

By November 24, 1894, Jenkins had filed for a patent on this first Phantoscope; it would become patent number 536,569.[38] By this time, he had conducted exhibitions at the Food Show and the multiple demonstrations for family, friends, and work associates. He applied for his second patent on December 12, 1894, for the "Kinetographic camera." His work was all supported by court-obtained affidavits from eyewitnesses. He thought that he was ready to exhibit for profit and put his apparatus on the market, but that would take financing he had not yet acquired. It was at this time that Jenkins met Armat.

Jenkins and Armat were strangers before the fall of 1894. They met as Jenkins was working on his rotary-lens camera. He had already conducted multiple showings and was seeking financial support. In comparison, Armat's inventive portfolio to this point did not reflect any substantive technological film work. He had just begun working in film, and that was his reason for enrolling at the Bliss School and eventually partnering with Jenkins. Armat was financially well connected, and his first inventions related to railroad work. They included an automatic railroad-car coupler and an electric railway system. His photographic reputation was not yet founded.

Armat and Jenkins were both interested in motion-picture cameras and projectors. Jenkins was at Bliss, continuing his exploration of electrical options relative to his existing inventions. Armat sought to learn more about photography. Three decades later, Armat would recall that he had enrolled "largely for the purpose of acquiring practical information as to handling an arc light."[39] However, his portfolio contained no inventions, demonstrations, or activity with any photographic apparatus. At this time, he was working on an electrical propulsion system.[40] It must have been a surprise to Jenkins and Armat when the professor, Louis D. Bliss, introduced the two to one another as having similar interests.

Partnership in Investments or Invention

Professor Bliss knew of Jenkins' previous work and his need for finances, so he introduced Jenkins to Armat. Armat had just left his railroad employer and was working in Washington, D.C., as a real estate dealer with the Armat and Daniel firm. Bliss introduced Armat as a "relative of the wealthy Senator John W. Daniel, of Virginia."[41] This immediately caught Jenkins' attention. Unknown to him, Armat had apparently already conducted an investigation of Jenkins' work and "was so favorably impressed . . . that he joined Mr. Jenkins."[42] Jenkins saw investment potential, and Armat saw motion-picture ideas that he liked.

Jenkins and Armat began working on experiments that would result in the successful projection of a motion picture. Their classes at the Bliss School met three times a week in the evening, and Jenkins, now with Armat, was working feverishly.[43] Working together in and after school, they became contractual partners on March 25, 1895.[44] The agreement was a financial one that assigned "one-half interest" in the stereopticon Phantoscope to Armat, distinguishing it from the cabinet form of the projector, which Jenkins was also working on and had used at the Pure Food Exhibition. Their agreement assigned Armat the rights to "make films" and limited the number of cameras constructed to five; it swore Armat to secrecy; and it gave him the primary responsibility for "all expenses . . . incurred in the construction and exhibition."[45] The agreement left opportunities for Armat, stipulating that any apparatus manufactured and further indebtedness acquired was to be joint property, and that it would hold Armat "harmless against any suit for damages for infringement" relative to Jenkins' patent for the Phantoscope, which had just been granted on March 26, 1895. The agreement, "unless terminated," was to be in force for fifteen years.[46] Armat would later report that the contract provided him security "in the control of any invention that might result."[47] His comment reveals his mindset, but the language in the contract suggests that his control of the invention did not exist. The agreement basically had Jenkins contributing his former work on the now-patented Phantoscope to the partnership, while Armat financed the construction and developed promotion and public exhibition of the projector. There is no mention in the agreement of Armat as a contributing inventor, nor was it apparently contemplated by the agreement that he would contribute any inventions to the partnership.[48] Nevertheless, it would soon be apparent that they were working together on the machine.

When Jenkins and Armat entered into their contract, Jenkins saw it as a part of his continued efforts to acquire financing. However, Armat read it differently. He, too, saw himself as inventor. He had ideas to improve Jenkins' projector, as well as to provide himself the opportunity to enter the flurry of ongoing motion-picture development. In other words, Jenkins and Armat saw each other as a stepping stone—one into investment dollars, the other into film apparatus with significant profit potential.

This was a critical time in the development of film. The leader in the field, Edison, had made successful motion pictures, but his profits came from the nickelodeons. He was exhibiting his films on peephole devices, while Jenkins, and now Armat, were projecting the moving pictures onto a large screen.[49] As the contest between Armat and Jenkins accelerated, both strongly affirmed their positions as the sole inventor of the device.

The Cotton States Exhibition, 1895

By September 1895, a Jenkins-Armat film-projection machine was readied for the Cotton States Exposition in Atlanta. Armat borrowed $1,500 ($40,744) from his brothers and secured an additional $2,000 ($54,325) to underwrite the demonstration.[50] At the Exposition, the plan was to construct two projection theaters. A "life size screen" of twenty by twenty square feet would provide film showings in one room, where the audiences were seated, while in the adjacent room they waited their turn for the film to be shown. There was a projector in each room, and a third projector had been shipped to Atlanta as a backup. Jenkins later said that these exhibit machines were duplicates of earlier ones he had developed. The film exhibit nearly failed the first day, as few attended. People did not know what was meant by the sign "Moving Pictures," and they were reluctant to sit in a dark room full of strangers. The sign was quickly changed in an attempt to attract viewers: "Come Inside and Rest—and Look at Moving Pictures, Free of Charge."[51] The small audiences who braved the dark saw life-sized images, as the film literally tore through the projector.[52] The inventors worked each night repairing the ripped film-sprocket holes that had been shredded by the projector in preparation for the next day's showing.[53] The showings were mysterious attractions: "[W]ithout [any] warning [other] than a soft whir [of the projectors] in the darkness, a figure leaps out into the lighted spaces. With gliding steps it moves forward to the ends of the stage and gathers its flowing draperies in either hand with a sweeping curtsey. It throws back its head with a saucy jerk, flipping the point of the black lace mantilla away from the dark Spanish face and then with a familiar flash and smile of those dark Spanish eyes, whirls away in graceful, passionate, but withal stately masses of 'Santiago.'"[54] The *Baltimore Sun* declared the exhibit a "remarkable invention of a Washington stenographer"; "Mr. Edison Outdone," was the headline. "There was no Edison in this. . . . [T]he wizard of Menlo Park has been beaten at his own game by two young Washingtonians."[55] The *Atlanta Journal* labeled the exposition as "unquestionably the most wonderful electronic invention of the age . . . everyone should see it."[56]

Despite the publicity, the crowds never materialized. The showings were not as successful as the two inventors had hoped. Musser suggests that the problem was "primarily promotion" and that the significance of Jenkins and Armat's work was lost amid other photographic displays at the exhibition. Gray Latham showed his Eidoloscope, and Frank Harrison exhibited the

Edison Kinetoscope for the Raff and Gammon Kinetoscope Company.[57] The audience that came to the Jenkins and Armat showings was suspicious of having the lights turned down low; they "had the notion that expositions were dangerous places where pickpockets might be expected on every side."[58] Jenkins felt that the audience was skeptical: "people would listen to the barker [trying to convince them to see the exhibit], smile incredulously, and pass on."[59] After the admission fee was dropped, people were asked to donate as they left the theater area. However, even with free admission the attendance remained low.

As Armat and Jenkins were discussing changes needed in the projector to preserve the film, Armat's brother, J. Hunter Armat, appeared at the exhibition. Hunter had assisted the two in constructing the projectors and had shipped them to Atlanta. He brought with him the bills for the machines. At this same time, Jenkins left for Richmond to attend his brother's wedding and took one camera with him. Armat's version of the events was that he expected that following the wedding, Jenkins would return with the machine.

Then tragedy struck. All was lost when a fire started in a neighboring exhibit and destroyed the motion-picture-theater concession area. A significant investment went up in smoke. "We were tired and discouraged," reported Jenkins, and "the result [was] that I had a falling out with Mr. Armat."[60] Jenkins charged that Armat, whom he called a "promoter," had saved the third machine and taken it to New York without Jenkins' knowledge for a demonstration before Edison-company representatives. To Jenkins, this was treachery. Disheartened after the fire, he returned to work at the Life Saving Service from which he been granted a leave for the exhibition. "So far as I was concerned, the only exhibiting machine I had left was one I had where I lived on East Capitol Street."[61] However, two machines had not been destroyed, and they were both in Armat's possession. Jenkins did not return to Atlanta, and the partnership soon ended.

The Columbia Phonograph Company

By 1895, Jenkins was also courting an independent relationship with Edward F. Murphy of the Columbia Phonograph Company and the American Gramophone Company, which had supplied films for his tests. It was an effort to avoid conflict with Edison, as these were not Edison-friendly firms but aggressive competitors. Correspondence between Jenkins and Murphy through the summer indicates that Jenkins and Armat were working on the

Jenkins' cabinet nickelodeon machines displayed with the Columbia Phonography Company, Atlantic City, N.J. Courtesy Virginia Roach Family.

projector's intermittent gears as they anticipated the upcoming Atlanta Exhibition. In September, Jenkins also reported that "the lantern is simplicity itself.... It is the grandest success you could imagine. I am blessed, glad it's finished too, for I'm dreadfully tired living with it day and night."[62] Columbia Phonograph was a delicate balance for Jenkins' financing and exhibitions. Columbia was a competitor with Edison, and its agreement with Jenkins would provide Columbia a competitive edge. Jenkins was to be awarded 10 percent of parlor receipts. As a result, they created their first showroom in Atlantic City and constructed automatic projectors for the viewing audience.

While Jenkins and Armat must have been discouraged after the Atlanta fire, they were still anticipating continued work.[63] They hoped to form a stock company, and Jenkins assured Murphy that they would have "things booming again." Even though the Armat brothers' investment was lost in the exhibit fire, it seemed "only to stimulate greater effort.... We have a half dozen machines ready for use and will in all probability put them out soon. We have a camera constructed and will be trying it tomorrow."[64] Note the plurality of Jenkins' forward thinking: he was working to maintain Murphy's support and ensure the possibility of working with him in the future. He thought he had every right to do this as "sole inventor." Slight reservations were reflected, as Jenkins noted, "Armat doubts my right to let you [Murphy] have any interest in the cabinet [Phantoscope] and has refused to consider you." Nevertheless, Jenkins moved forward as the cabinet Phantoscope had been eliminated from Armat's reach in their contract. However, the Jenkins–Columbia Phonograph agreement would fall apart due to the confusion around Jenkins' patent, the Jenkins-Armat joint patent, and Jenkins' agreements with Armat. Armat would prevail.

Armat was not comfortable with the Columbia arrangement because of his Edison connections, so he sought a court injunction against the Jenkins-Columbia agreement. Although the injunction was denied, the Jenkins-Columbia agreement fell apart when Armat, with his new associates, paid Jenkins $2,500 ($67,906) to "withdraw his exclusive claim to the phantoscope patent."[65] This would later prove to be a costly deal for Jenkins and a highly profitable one for Armat.

In December 1895, after the Atlanta Exhibition and the Armat dispute, Jenkins left government service, where he had worked since 1890.[66] He had enjoyed his position and was praised for his dedication, but he wanted time to experiment. Writing his cousin Charles M. Jenkins in Richmond, Jenkins reported that he was called into his supervisor's office and told, "You are doing work in this office which no one else seems able to do so well and so rapidly and we may possibly have been piling the work on you. I wish you to know that we appreciate it."[67] However, inventing was Jenkins' first love, and work at the Life Saving Service had become only a means of supporting his growing hobby, the development of which he would now pursue as a full-time inventor.

The Sole Patent Holder

On November 25, 1895, Jenkins, attempting to go his own way, had filed for a patent as sole inventor of the Phantoscope with improvements made before the exhibition in Atlanta. He consulted with Armat's and his joint attorney and was advised that as he "was the sole inventor," he should make the application "in [his] own name, and afterward secure to Armat his interest by assignment." Jenkins later recalled that the attorneys were surprised and chagrined at Armat's negative reaction, because it left the patent in interference. Unfortunately, as he later put it, "Armat would not consent to an adjustment of our differences," so interference proceedings were triggered as the new filing was in conflict with the joint patent.[68] Armat charged Jenkins with treachery and dishonesty, claiming that he had seen the Armat improvements and stolen ideas included in the patent application. He claimed that Jenkins had created a drawing of the improvements and represented them as his own when he filed for the patent. This resulted in a patent-interference case (No. 18,032), which eventually resulted in Armat's victory, and patent number 586,953 was reissued under both inventors. This would be the only patent held jointly by both men, which was described as providing "new and useful improvements in the Phantoscope."[69] According to Armat, Jenkins all but admitted his seemingly deceptive actions, "to which he simply grinned and informed me that it would be a battle royal."[70] This seems an unlikely comment, given Jenkins' weak financial position prior to Armat's buyout. Jenkins called his first filing a mistake, thinking that this was the proper way to assure Armat of his contractual interests as he had been advised.[71] However, Jenkins underestimated Armat's determination. Armat had made improvements in the apparatus and refused to withdraw the joint filing, and the interference case moved forward. In testimony before a Patent Office examiner, Jenkins affirmed his invention, and Armat, for some reason, declined to "maintain his claim as an inventor." Yet he had purchased all of Jenkins' interests under the condition that Jenkins withdraw his claims.[72] Litigation continued as Armat sought an injunction from the District of Columbia Supreme Court, restraining Jenkins from any further use of the Phantoscope specifically with Columbia. The injunction was denied, and the joint patent stood. In the end, Armat offered cash, and Jenkins sold and withdrew.

In this nine-month period of Jenkins and Armat's association, mid-March though mid-December 1895, Armat had catapulted himself into the center of motion-picture invention, and Jenkins had sold his rights to a historic patent.

Going Their Own Way

The Jenkins-Armat partnership had collapsed, but Jenkins still needed stable financing. So he went to Philadelphia, fostering the Franklin Institute. Meanwhile, Armat moved forward and, ignoring his contractual promise of secrecy to Jenkins, approached Edison. This was made possible through Armat's longtime family friend, H. A. Tabb. Tabb knew Norman C. Raff and Frank R. Gammon, who worked for Edison and had also demonstrated at the Atlanta Exhibition. They were Edison Company distributors backed by the organization, financing, and promotion capabilities of Edison's powerful monopoly.

Armat conducted his demonstration for the representatives of the Edison company at Daniel and Armat's real estate office. "I took him [Gammon] into the basement ... threw a picture upon the screen, [and] his attitude underwent a complete transformation. His excitement and interest were most apparent."[73] Armat was hoping to get himself into a management position with the Raff and Gammon distribution company. Unfortunately for Armat, Raff and Gammon were not interested in taking on as a new partner a newcomer to the motion-picture field. They suggested a contract providing Armat with

25 percent of the sale of exhibition rights and 50 percent of the gross receipts, "up to seventy-five hundred dollars [$203,718]."[74] Before this contract was presented to Armat, Raff and Gammon approached Edison to acquire the support of his company and thus make the device more marketable.[75] Through these associates, Armat convinced the Edison Company that he had purchased the rights to the projection machine they sought. He had taken the improved Phantoscope, for which he shared a joint patent with Jenkins, to Edison, where it would eventually be marketed as Edison's creation under the name Vitascope. Armat proved himself a shrewd businessperson as he parlayed his patent rights into large profits, for the Vitascope was "really Jenkins' machine with Armat's crucial modifications."[76] By March 1896, Raff and Gammon had heard about Jenkins' involvement and insisted that Armat end his activities with him, "even if it meant making Jenkins a good stiff payment." Rather than pay Jenkins, Armat again took legal action to reclaim the remaining camera, which was at Jenkins' home.[77] Jenkins' version of the story was that Armat had literally broken into his residence and that the "machine was taken to New York by the promoter [Armat] without the knowledge of the inventor [Jenkins]."[78] Armat contended, in turn, that he was taking action to recover the device. He again accused Jenkins of stealing his ideas, and when confronted, Jenkins offered no excuse "for his treachery and dishonesty[;] he admitted that his purpose was to appropriate the whole thing to his own uses."[79]

So, while Armat's discussions were progressing with Edison's representatives, Jenkins' arrangement with Columbia came apart, and he demonstrated at the Franklin Institute. He wanted out from under threats of litigation, as he had no money to continue a legal battle, so he let go of his rights and moved forward.[80] He still insisted that he was demonstrating *his* ideas and *his* machine at the Franklin Institute in Philadelphia.

Franklin Institute Medals

Jenkins and Armat went their separate ways, but they continued their differences. Their patent litigation ended, and Armat teamed with Edison as Jenkins was demonstrating the device at the Franklin Institute, trying to advance his reputation within the scientific community and thus raise funding. The initial presentation on December 18, 1895, was a report read before members of the institute, recounting the development of photography and motion pictures. It was accompanied by a demonstration that produced "the movements and actions of dancers, gymnasts, etc., with remarkable fidelity to nature."[81]

The Franklin Institute's Committee on Sciences and Arts was intended to encourage and reward invention.[82] This focus obviously complemented Jenkins' goals—to acquire financial underwriting, he had to promote his work. The institute's Elliott Cresson Medal would be Jenkins' springboard into fame, as it was their highest award.[83] It would give him the recognition necessary for more successful fund-raising. Other recipients of Jenkins' era included some significant names: W. O. Atwater, Emile Berliner, Thomas Edison, and Alexander Graham Bell. When Jenkins received the award, it put him in good company.

The committee reviewing Jenkins' application was a prestigious group.[84] The chairman's report on the application recommended Jenkins for the Cresson Medal and provided a brief synopsis of the contextual history of motion-picture photography and Jenkins' Phantoscope, describing it as a "marvel of simplicity and perfection."[85]

The Franklin Institute's investigation of the Phantoscope and its history described "a perfected apparatus for projection upon a screen [of] a series of photographs of moving objects.... [T]he scenes and movements are in effect realistic."[86] Their report set out the scene, with pictures changing so rapidly that the "effect upon the eye ... was to give the picture the appearance of being fixed and yet showing progressive motions." There was no flickering, as previous demonstrations had produced, because Jenkins had remodeled the Phantoscope so that no shutter was required.[87] The committee concluded that "the Phantoscope of today has become ... an instrument that has been so long desired to enable us to reproduce the movements of life for analysis, profit, and pleasure."[88]

Armat protested, claiming the invention as his own. He offered to appear before the committee, and this produced a flurry of exchanges. He claimed that he had become a partner with Jenkins, had contributed largely to the invention—indeed, he had produced and "perfected the phantoscope"—and the joint application was proof of his claims.[89] In examining Armat's accusations, the committee reviewed the contract and concluded that Armat had joined Jenkins "to aid him with money and business capacity," but there was "no allusion made [in the agreement] to any inventions contemplated by Mr. Armat."[90] Insisting that he contributed significantly to the original Phantoscope, Armat pointed to the joint patent. The committee responded that the "joint patent is only evidence of ... participation" and improvements, but not the *original invention*. Although he had offered to appear before the committee, when the invitation was extended, Armat declined to provide personal testimony in support of his claims. His protest file was "made up chiefly of aspersions upon the character of Mr. Jenkins, which matters [were] not relevant to the question."[91] Jenkins responded with testimonial affidavits and substantiation of his earlier years of experimentation, in which he affirmed that the Phantoscope was already working and had been demonstrated before Armat had joined him. The committee concluded that in Armat's case against Jenkins, there was an "entire absence of proof," and in the lack of "effort to support his [Armat's] claims," the protest was dismissed. "The evidence filed by both parties, proves as far as it proves anything that Jenkins is the sole [original] inventor."[92] They moved to award the medal to Jenkins. The Elliott Cresson Award was given "to C. Francis Jenkins for his Phantoscope" in 1898.[93]

Jenkins' second Franklin Institute award was the John Scott Medal. It again recognized the value of his contributions in motion-picture art.[94] This medal was awarded on December 17, 1913, accompanied with a certificate and a check for twenty dollars ($458). Jenkins now held both Franklin Institute medals.[95] The John Scott Award was specifically for improvements made to the Phantoscope, which the institute this time called a motion-picture apparatus. Jenkins was elated, as he held the Franklin Institute medals "in higher esteem than any other recognition that I have ever received."[96] Armat would again protest with essentially the same arguments and again fail to change the committee's decision.

The Smithsonian Photography Exhibit

Shortly after the John Scott Medal was awarded, Ramsaye reported that "a projection machine resembling the projectors used in Atlanta, appeared in the National Museum of the Smithsonian." The inscription on the museum display label read, "Intermittent film projector Invented by C. Francis Jenkins."[97] This time, Armat protested the labeling. Interestingly, however, his protests surrounding the Smithsonian exhibit do not seem to have been made until the early 1920s. It is odd that Armat would wait almost a decade to launch this protest. His career had been moving forward successfully. He was receiving considerable profits "in royalties from his projection patents," derived from the Jenkins-Armat patent.[98] At the turn of the twentieth century, small storefront theaters were springing up across the nation, and Armat joined in that development. The Armat Moving-Picture Company opened its first theater in Washington, D.C., on September 10, 1902.[99] He was also involved in other litigation apart from his dispute with Jenkins.[100] Yet, Armat's first letter protesting the Smithsonian's exhibit of "Jenkins'" equipment was written on May 10, 1922. This confrontation between Armat and Jenkins reveals damning evidence, the heightened emotions of the combatants, and it gives Armat his last audience. It can be best analyzed in point-counterpoint fashion.

* * *

JENKINS-ARMAT CHRONOLOGY

Jenkins Motion-Picture Chronology, 1890 to 1894

1890–91

Jenkins produces rough sketches for his first camera.
Fall: Jenkins shows Phantoscope device to his colleague, Hugh Stuart.
September: Jenkins shows camera projector to friends and family.
Christmas: Treasury Department colleague is shown drawings of camera.

1892

Jenkins uses oil lantern to project film onto small silk handkerchief.
I. D. Boyce develops film roll for Jenkins.
Life Saving Service colleague Arthur J. McElhone performs for recordings.
Mrs. Philo T. Bush assists in photographing activities of two small boys.
September 1: Jenkins solicits support from Alexander Graham Bell.

1893

Jenkins uses carbon arc as a light source and conducts private demonstrations.
January 17: Jenkins contracts with James P. Freeman in Phantoscope construction.
Summer: McElhone participates in filmed athletics at the Chevy Chase Country Club.

1894

Summer: Attorney Wallace Greene is shown the Phantoscope.
July 6: First description of the Phantoscope appears in *Photographic Times*.
November: Jenkins exhibits at the Pure Food Exhibition and the Capital Camera Club.[101]
Jenkins enrolls at the Bliss School, conducts a demonstration, and meets Armat.
October 1984: Jenkins' sister-in-law and daughter Annie see numerous demonstrations.
November 24: Jenkins files for first Phantoscope (Patent 536,569).
December 12: Jenkins files for Kinetographic Camera (Patent 560,800).

Armat's Inventive Chronology, 1887 to 1894

1887

January: Armat files for his first patent (Patent 361,664), an automatic railroad-car coupler.

1893

Armat files for his second patent (Patent 521,562), covering an electrical railway-system conduit.
Armat ventures into the real estate business with a cousin, T. C. Daniel.

Fall: Armat attends the Chicago World's Fair with his brothers Thomas and John and sees Anschutz's film demonstrations, which sparks his interest in photography.

1894

Armat visits the Washington, D.C., Kinetoscope parlor.
Fall: Armat sees Jenkins' work at the Pure Food Exhibition. He enrolls in the Bliss School and meets Jenkins.

Jenkins-Armat Partnership

1894

Jenkins and Armat enroll at the Bliss School.
Summer: Armat family declines invitation to invest in Edison.
March 25, 1895: Jenkins and Armat sign contract.

1895

March 26: Jenkins' first Phantoscope patent is issued.
August 28: Joint patent is filed for second Phantoscope.
September: Working together, Jenkins and Armat develop a third machine for the Atlanta Cotton Exhibition.
September/October: Jenkins and Armat exhibit as partners at the Cotton States Exhibition with funding from Armat's brothers.
November 8: Fire destroys their exhibit; they consider stock company. Jenkins communicates with Columbia Phonograph Company.
November 24: The partners split.
November 25: Jenkins files for sole patent application on instrument with improvements.
December 8: Armat gives demonstration for Edison agents.
December 18: Jenkins demonstrates machine at the Franklin Institute.

International Cotton State Exhibition

September 14–December 31, 1895—The exhibit is open for one hundred days and reportedly attracts eight hundred thousand visitors.
September 14: Exhibition opens with Jenkins and Armat's display and with the equipment destroying the film.
October 16: Jenkins reports that he had left Atlanta, "compelled to come to Washington to appear before the Commissioner of Patents," afterward leaving for Indiana for his brother's wedding and to conduct the demonstration in Richmond. Atwood L. Jenkins marries Mary E. Test.
November 2 and 3: Jenkins receives a letter and telegraph from Armat noting that he and his brothers are returning to D.C. as a result of the fire, "which devoured a goodly part of the Phantoscoped building."
November 8: Letter to Edward L. Murphy from Jenkins describes discouragement, embarrassment, and optimism.

1896

February 19: Armat files for new Vitascope (Patent 673, 992).
March: Phantoscope in Jenkins' possession is taken from his home in his absence.
March: Armat demonstrates for Edison; decides to market the device as though Edison was the creator.
April 14: Armat sells patent rights to cousin and brothers.
April: Jenkins offers another demonstration for Alexander Graham Bell.
April 23: Edison premiers the new Vitascope, primarily the work of Jenkins and with Armat's improvements.
May 14: Jenkins grants license to Columbia-American Graphoscope to exhibit.
October 4: Jenkins successfully sues Armat over sale to brothers.
November 9: Armat successfully sues Jenkins over sole patent application. Joint application remains.
- Treasury watermark appears on paper of Jenkins' drawing.
- Richmond, June 6, 1894, demonstration challenged.

December 7: Jenkins sells his patent rights to Armat.
December 15: Jenkins loses case; joint patent remains. He abandons his efforts and concedes to joint application.
December 18: Jenkins applies for Franklin Institute medal.

1897

July 20: Armat sues to prevent Jenkins from using the Phantoscope. Jenkins wins, joint patent upheld.

1898

January 14: Armat enters protest with Franklin Institute.
Jenkins receives his first Franklin Institute medal.

1913

September 12: Smithsonian exhibits Jenkins machine. Armat's protest leads to labeling changes to reflect both inventors.
December 17: Franklin Institute bestows second medal on Jenkins.

1922

Armat drafts his account of the Jenkins partnership of 1894–95.

1925

Edison writes in protest of Smithsonian and Franklin medals honoring Jenkins.

1929

Edison and Armat consider lawsuit to deter Jenkins' television experimentation.

1935

March: Armat gets the last word, speaking before the Society of Motion Picture Engineers after Jenkins' death.

* * *

Point-Counterpoint: Issues and Evidence

The dispute between Jenkins and Armat began soon after their short partnership. While their contract lasted less than a year, their conflict lasted over the next three decades, with Armat's assertions appearing even after Jenkins' death.[102] The historical record is full of their point-counterpoint accusations, fading memories, some fraudulent evidence produced by Jenkins, and rambling, emotionally charged, inflammatory testimony denouncing Jenkins' character produced by Armat, much of it many years after the fact.

The primary issue was, Who should get credit for the invention of the film projector? Jenkins claimed that he was the original inventor of the first film-projector device. Armat claimed that the Jenkins machine did not work and that only his improvements made it operable. Jenkins would eventually, and unfortunately for himself, sell his patent rights to Armat. He lacked resources to embark on an endless confrontation and struggled to move past this conflict. The conflict persisted, however, with each side defending his own contributions. Armat challenged Jenkins' Franklin Institute medals and the Smithsonian exhibit, which had all praised Jenkins' inventions. Each time, Jenkins was acknowledged as the inventor of the Phantoscope, but Armat firmly claimed all the rights to it.[103] The results of these battles upheld the joint patent rights of both inventors and failed to offset Jenkins' momentum. So, who was the victor in this long war? Who made the stronger case?

A *Harper's Monthly* article in August 1917 extolled Jenkins' virtues.[104] This heated up the argument again, and Armat wrote to correct the author, Homer Croy, challenging his research and claiming that "that creature Jenkins [is] a liar . . . having stolen his [Armat's] invention" and declaring that the contract they had between them was fraudulent.[105] Jenkins responded dispassionately and seemed to delight a bit: "I think it will be enough for me to confine myself to facts, for calling one's opponent a liar a dozen times in a single letter only indicates a weak case." He reiterated his acceptance by the scientific community and his experience in exhibiting and suggested in regard to the contract that Armat, as a real estate broker, should have been familiar with contracts. The contract itself implied existence of Jenkins' machines prior to the partnership. Jenkins listed his press accolades and concluded with what was really a rhetorical response, complimenting Croy on his "painstaking research and conscientious care" as being "historically accurate."[106] Armat's credibility was eroded by time and his increasingly emotional arguments. The claims in each protest were repetitive and melodramatic, and the evidence was aging.

The technological details as to who contributed what to the Phantoscope were most effectively addressed by John Eustis and Gene G. Kelkres, who carefully substantiated Armat's contributions.[107] They argued that Armat contributed to the engineering of the Phantoscope. Both inventors made contributions—this is clear. The joint patent was strong evidence of Armat's participation and contributions, but not his claim to be the sole originating inventor. Jenkins was the first to conceive, originate, and demonstrate the basic principles—that too is clear. The argument centers on originality. Who was first with the concept and a working model of the projector? Three primary points of contention transcend almost three decades of this debate: (1) When did Jenkins start work in motion-picture photography? (2) Did the Richmond, Indiana, demonstration really occur? And (3) Which party has more credibility?

ISSUE 1: WHEN DID JENKINS START WORK IN MOTION-PICTURE PHOTOGRAPHY?

Jenkins' Phantoscopes and Kinetographic camera were developing long before he met Armat. Indeed, his first Phantoscope patent was issued one day after he had signed the contract with Armat. Testimonies, exhibits, and sworn affidavits from people participating in Jenkins' experiments are abundant. Yet Armat claimed that the Jenkins apparatus was a mechanical failure until he joined the team and made improvements. In contrast, Jenkins simply called it a grand success.[108]

Jenkins' early interest in motion-picture photography was sparked during his youthful travel in the West.[109] He claimed that he later drew his first sketches for a camera, which were seen by a coworker from the Treasury Department in December 1891.[110] In 1892, Jenkins approached Alexander Graham Bell for funding and volunteered to demonstrate a working camera for Bell.[111] In 1893, he signed a contract with James P. Freeman that identified Jenkins as the "inventor and *sole* owner of certain new methods of apparatus for, the *recording* and *reproduction* of action" (emphasis added).[112] This evidence again indicates that an apparatus existed and Jenkins was its inventor.

Armat claimed that the Freeman contract was fraudulent and challenged whether Jenkins' other demonstrations actually occurred. Jenkins affirmed those demonstrations with affidavits from seventeen people attesting that they had witnessed them and offered to produce even more for the Smithsonian.[113] These testimonies supported Jenkins' claim that he was well into his work prior to meeting Armat. The Franklin Institute awards investigation gave the testimonials further credibility. By the 1920s, however, the Smithsonian investigation seemed to ignore this evidence, dismissing it as coming from family and friends who produced journals and diaries that could have been fabricated. But Jenkins also offered affidavits from professional people who had worked on the machine, including copies from an original time book of a workman, Emil Wellauer, who had helped build cabinetry for different machines, one called the Phantoscope. In the Smithsonian dispute, Jenkins offered to return for additional testimony.[114] This evidence would again have supported Jenkins' claims. While family and friends' testimonials might be dismissed, it would seem puzzling to dismiss all seventeen sworn affidavits, especially those of independent contractors and nonrelated participants. When all evidence is considered, it is clear that Jenkins had started and was well into his work before he partnered with Armat.

ISSUE 2: DID A RICHMOND DEMONSTRATION REALLY OCCUR?

Armat argued that Jenkins was lying about the demonstration in Richmond, Indiana, and evidence Jenkins provided in support was confusing. If this demonstration occurred as reported, it would provide strong evidence of Jenkins' work predating Armat, thus making Jenkins the clear winner in the debate.

Jenkins claimed that he demonstrated the projector for family and friends at his uncle's jewelry store in Richmond on June 6, 1894. He asserts that this could be proven in a local newspaper report on that date, but he did not provide a copy. Armat countered, declaring that the Richmond presentation never occurred, and his proof came from the Treasury Department personnel records. Armat showed that in June 1894, according to Jenkins' own work records, he had taken only two and a half days off, and this was insufficient time to have completed the journey and demonstration.[115] Since Jenkins continued to insist on the experiment and the exact date, without producing actual evidence, Armat's challenge went unanswered, leaving the question open and in his favor. While several articles did appear in the local *Richmond Telegram* affirming this date, they were printed so far after the event that their accuracy can be questioned.[116] However, there remain three time-sensitive *Telegram* reports.

The critical first article, published on the date in question, June 6, 1894, was titled "A Genius." It reported that Atwood L. Jenkins was showing his brother's work on that day.[117] The Phantoscope was the apparatus being demonstrated. A second article appeared three months later, headlined, "Charles Jenkins

Has Arrived Home from Washington, D.C." Published September 14, 1894, it links Jenkins' bicycling trip and mentions only indirectly the photographic presentation of that "wonderful machine... described in the *Telegram* several weeks ago."[118] Jenkins reported that he had shipped the projector ahead of his trip, which would imply that he gave a demonstration, but the newspaper was apparently more interested in the cycling activity. The third article, headlined "The Phantoscope," appeared on October 30, 1895. This report specifically described Jenkins in Richmond, showing "his Phantoscope" to a number of family and friends. "The pictures presented last night at Jenkins' [uncle's] Jewelry Store were all life size, and those fortunate enough to see them were enraptured at the wonderful and beautiful effects seen."[119]

The first newspaper report documents that there was indeed a successful demonstration of Jenkins' equipment on the date in question, but it says that it was conducted by his brother Atwood. The second article offers only a reference to a preceding article and possible demonstration, which suggests that it was planned in some form, as Jenkins had shipped the camera ahead, likely intending to display it. However, there was no more evidence to support a demonstration having taken place. The third article documents the date of what would appear to be Jenkins' presentation, but it comes a year after the date Jenkins insisted upon. This later date is supported in Jenkins' Treasury Department work records. In October, he took thirty-one days of leave, which would have been ample time for the the Cotton States and the Richmond demonstrations.[120] It came only two months before his partnership with Armat would end.

In short, giving Jenkins the benefit of the doubt, it appears that he had his dates confused. The evidence surrounding this demonstration places the date of Jenkins' personal demonstration as October 30, 1895, after his brother's wedding and following Jenkins' time with Armat at the Atlanta Cotton States Exhibition. This was eight months following the signing of the Jenkins-Armat contract. However, Armat's charge was that a demonstration had never occurred at all, when demonstrations clearly were conducted with Jenkins' Phantoscope. One was conducted by his brother Atwood on June 6, 1894; another by Jenkins himself following his October 1894 bicycle expedition; and a third, again by Jenkins himself, on October 30, 1895, following Atwood's wedding. The first predates the Armat-Jenkins contract by ten months, and the last came at the end of their partnership. Jenkins' insistence on the exact date of June 6, 1894, was unfortunate, as it would seem to have been easily clarified with the Atwood showing or another local newspaper article, if one existed, along with Jenkins' evidence of a demonstration at the 1894 Pure Food Exhibition, thus again providing verification of substantial activity prior to his meeting Armat.

Related to this same demonstration, Jenkins specified that the film shown was "Annabelle the Dancer." Annabelle Whitford Moore Buchan was a popular dancer who appeared on stage and in early film shorts.[121] Armat claimed that she had never appeared for Jenkins' demonstrations. He answered by indicating that the film had come through Edward F. Murphy of Columbia Phonograph, with whom he had exhibited.

ISSUE 3: WHICH PARTY HAS THE MOST CREDIBILITY?

The real damage to Jenkins' credibility results from the charge of fraud and evidence tampering. In his interference case filed against Jenkins, Armat sought an injunction to prevent Jenkins' further work with the Phantoscope because such work would interfere with the joint patent. The injunction was denied, leaving both inventors' rights in place as provided in the joint patent. However, during that case, Jenkins provided two drawings that he said proved the existence of his work as early as 1890 and 1891. Unfortunately, either Jenkins made a serious error in judgment, or he had lost the original sketches and thus, believing them to be his own, thought nothing of redrawing them from memory.[122] The fact was that "the paper upon which the drawings were made or redrawn contained a watermark and from this watermark it was discovered that the paper was not made until after 1894."[123] Jenkins admitted his error and offered an apology with no explanation, which inauspiciously left the charge of fraud unanswered. It is odd that he never reproduced the Hayden affidavit, testifying to the pencil drawings in 1891. The omission appears to be an admission leading to Jenkins' abandonment of the sole-ownership application; and thus, the joint Jenkins-Armat patent was agreed. The error cost him his credibility.

Without a response from Jenkins, the charge of fraud seemed all too clear. The accusation of evidence tampering is less clear, but one dominoed to the other. This charge related to the actual date of the Jenkins-Freeman contract. Armat claimed that the contract date had been purposefully smudged, making it appear to read 1893 when the date could have been "Mar 30/9? 3? 4? 5? or 7?"[124] The contract was not found in the Smithsonian. However, the contracts in the Accession Files of the National Museum of American History, the Franklin Institute Scrapbooks, and the Jenkins Scrapbooks are clearly dated January 17, 1893, which is corroborated by the

Wellauer affidavit indicating that he had made cabinets for the Phantoscopes in 1894. The evidence of tampering reported by Armat on the Smithsonian-held Freeman contract was, in other words, not evident on the other copies of the same agreement.[125] It is difficult to assess these point-counterpoint arguments, as they came a decade after the fact. In all likelihood the dispute will remain unresolved.

The Last Defense

Years later, Jenkins did have an influential friend come to his defense, resulting in an interesting opinion from the Smithsonian. Henry D. Hubbard, who was among those earlier interested in film standards and Society of Motion Picture Engineers, wrote the Smithsonian in 1922 declaring his impartial interest and defending Jenkins against those he called "interested parties . . . selfish persons [who have] attacked Mr. Jenkins." Hubbard worked for the U.S. Bureau of Standards and said that he "had found that the first practicable motion picture machine was the phantoscope invented by Jenkins." He urged the Smithsonian to retain its original display labels.[126] This letter produced an unusual personal visit from R. P. Tolman, the assistant curator of the Smithsonian, in which he and Hubbard discussed the issues and Tolman suggested "that if Jenkins would produce a complete genuine copy of the *Telegram* of Richmond, Indiana, June 6, 1894, and that it had the claimed exhibition, Jenkins would win."[127] Nevertheless, the articles from neither the Atwood nor Jenkins' demonstrations were produced, because by this time Jenkins was totally involved in RadioVision and had moved far beyond the debate.

Armat's character was also damaged from his emotional approach throughout the debate. He appeared to be rather devious himself. He knew that Jenkins was interested in financing when they first met in 1894, and he knew the potential Columbia and Jenkins represented together. However, Armat had provided more than financing. He demonstrated the Phantoscope to Edison's representatives despite his promise of confidentiality to Jenkins. Again, when Raff and Gammon wanted to buy out Jenkins, Armat instead took possession of a camera at Jenkins' home. Armat responded by declaring that he had a warrant and thus a right to enter. His retorts became increasingly long and overly dramatic. They provided more of an attempt at character assassination than new evidence. As opposed to letting the evidence speak for itself, Armat embellished his rhetoric in the language of anger. His protests were at times difficult to follow and led many to dismiss him as well as Jenkins for lack of a substantive response. However, as Jenkins pointed out, continually labeling one's opponent a "liar" does not prove a case.

Jenkins' character also suffered. The challenge of deception went unanswered and seemed apparent from both parties, with some charges seemingly proven and others implied. Jenkins had submitted two drawings in support of his argument. They were proven fraudulent. Why did he submit the documents and fail to support them? The action seems incongruent not only with other evidence but with Jenkins' roots, personality, and work over the years. He had and would again successfully partner with others, but this error at the time called his motives into question. Similarly, Armat had moved on to become successful. In the long term, what was really to be gained by either party?

In 1925–29—again, years after the fact—Edison wrote to protest the Smithsonian crediting Jenkins. His intrusion came more to discredit Jenkins at a time when Edison was nominated for a congressional award, which Jenkins had publicly opposed. There was no new evidence provided from Edison: he had controlled the film industry in its formative years and was not about to share credit. The Smithsonian response was simply, "Mr. Thomas Edison was not in a position to know the facts before December 1895 and no weight can be given to anything he may say or infer."[128] On January 4, 1926, ever trying to get the last word, Edison declared a gross injustice.[129] And in the continued exchange, more scheming surfaced. As late as 1929, Edison threatened a lawsuit against Jenkins. The suit would have had nothing to do with motion pictures, but Edison wanted to know "if a law suit on television would help to expose the mendacity of Jenkins."[130] No suit was ever filed.

Armat's objections claimed that the Smithsonian exhibits were "decidedly not what they are proposed to be," as the film projector was not invented until "about 1896."[131] He was challenging the museum's labeling of the Jenkins Phantoscope, declaring, "This machine was not invented by C. Francis Jenkins." He castigated the Franklin Institute, implying that it had shown a strange disposition to ignore facts in this case. The invention was his, he said, and the "information furnished by Jenkins had nothing more substantial . . . than limitless nerve and brazen effrontery."[132]

Finally, R. P. Tolman, assistant curator at the Smithsonian, responded to Armat, condemning both parties. He admitted that the Smithsonian's "indiscriminate use of the words 'first, and invented' were wrong," and he had the error in the displays corrected. He refuted Armat's overall claims, however, noting that Jenkins had filed two patents before he ever met Armat, and it was around the time of the application for the second that the two men met. He noted that at this same time, in 1894, Edison had a camera but "had failed to develop a projector." "Your idea that Jenkins did nothing and that you did it all is bunk. You reaped the harvest and Jenkins proved himself unreliable, you yourself tried to sell Jenkins out the early part of 1906, which the court did not allow."[133] On December 12, 1925, Tolman reported that the case was closed.

The Final Results

The Jenkins-Armat contest was described aptly by Ramsaye, writing in 1925: "Armat had [purchased] the patent and Jenkins had the medal."[134] However, there was more to it than medals. The Jenkins-Armat patent rights (Patent 586,953) were passed to Armat in settlement of the case. Armat paid Jenkins $2,500 ($32,385). Charles Roach, writing in 1920, indicated that the settlement was $5,700 ($64,597) and contrasted the sale to Armat's profits, which emanated in "the motion picture industry based on Jenkins' patents [and] amounted to nearly $500,000,000 [$5,666,371,971] annually."[135] For Jenkins the question remained one of his having conceived the original invention despite Armat's later improvements and the sale of his patent. Jenkins remained first with the idea, a working apparatus, demonstrations, and thus the sole original 1894 inventor.[136] He clearly undersold his rights, and Armat profited handsomely. For Armat, the question of sole ownership meant *inventive contributions,* even though he had moved beyond inventing into movie-theater entrepreneurship.

So why is this debate so historically significant? Why these repeated contentions over so many years? The answer lies in the concept of intermittent motion. This development was critical in the history of motion-picture projection and even film projected for television broadcast. Jenkins and Armat's joint work resulted in the first projector that froze each individual picture frame for a fraction of a second, providing for illumination and the sense of motion. Basically, as successive frames passed rapidly through the projector, motion was created in the human eye thanks to persistence of vision. Born in conflict, this principle remains at the foundation of modern motion-picture camera and projection systems. Looking back over the conflict from a modern perspective, Kelkres summarizes, "Jenkins made a historic mistake: He sold out to Armat, thereby virtually eliminating his name from its rightful identification with the eccentric cam projector. Thus, history books talk primarily about the 'Edison Vitascope'; only rarely do we see 'Armat Vitascope';

and practically never does there appear a reference to the name which expressed the true origin of the apparatus: The 'Armat-Jenkins Phantoscope.' Yet all three names refer to the same machine."[137]

More than a century later, it is difficult to separate fact from emotion in the case. The charges of Jenkins being dishonest, treacherous, and underhanded appear in complete contrast to the man's overall life and contributions. So, which of these two inventors gets the credit is an argument that will likely continue, as current evidence does not yet provide the full, unbiased answer.

Jenkins' first Phantoscopes featured intermittent motion of the film. They were visionary, but crude and requiring improvement. Jenkins and Armat worked together for a short time, making technical changes and improvements. Henry V. Hopwood described "a new principle introduced by Jenkins in 1894," the Phantoscope camera that "employed a continuously moving film in front of which revolved a disk bearing a number of lenses." It was an apparatus he described as "standing in a manner on the borderland of history."[138] Although Terry Ramsaye was a proponent of Armat, even he credits Armat only with important adaptations of the Jenkins Phantoscope, using intermittent as opposed to continuous film motion.[139] Homer Croy boldly declared, "To C. Francis Jenkins we owe the motion picture."[140] Later writers similarly gave credit to Jenkins. Hollenback calls Armat a "dabbler in the field of electronics" and credits Jenkins with the invention.[141] Musser declares, "[T]he first machine (with an intermittent) [that was] sold outright on the American market was Jenkins' phantoscope."[142] He defines Armat's contributions as only slight revisions to the overall earlier Phantoscope and implies that Armat and Edison eventually "reaped ample financial rewards from the machine," while the "naive inventor [Jenkins] lost substantial amounts of cash."[143] Cohn describes Edison and Armat's actions as "unethical, but typical, business practice of the day."[144] The Smithsonian provides perhaps the best conclusion: "After carefully studying the evidence submitted and evidence from other sources, it is evident that neither side can be entirely proud of the steps which they have taken. . . . Both sides have shown themselves to be selfish, and have acted unlawfully to obtain their ends."[145]

Armat clearly made technical improvements to Jenkins' invention. He also gained the financial rewards from his eventual ownership of the patent rights. In the end, Armat would be rewarded for pioneering motion-picture theaters and remembered as an entrepreneur and inventor. In 1948, he posthumously received an Oscar from the film academy.[146] Jenkins should be credited for originating and patenting the initial working product, plus first demonstrations and exhibits. He will be remembered for his two Franklin Institute awards, the creation of what today is Society of Motion Picture and Television Engineers, and most importantly as a visionary inventor who pioneered motion-picture technology that led the film industry from the nickelodeon to big screens.

4

Jenkins' Motion Pictures

C. Francis Jenkins was a forward thinker. He loved travel and photography, which were constants throughout his life. He was a skilled writer and promoter, which helped him attract financing. Despite losing in the early battle with Armat, he boldly declared himself the inventor of the motion-picture projector. It was a line that regularly caught press attention and helped attract potential financial backers. He marketed his own ideas and financed them, often with little money, and at times with only hope and speculation. He was a multitasking individual open to new lines of endeavor. "Failure," he told Alexander Graham Bell, "only acts as an incentive to charge my determination."[1]

Much of Jenkins' motion-picture activity has been dwarfed in history as a result of the Armat conflicts. It was Jenkins' Phantoscope that catapulted him into motion-picture film invention as well as controversy, but he branched out into home and educational use. He would travel west, filming a Hopi Snake Dance, teach photographic expression, invent fireproof protections for the projector, and by the early 1920s, develop a high-speed camera. Milton Wright described the inventors in Jenkins' era as interested only in reaping a profit—the first millionaires of American industry. Jenkins rebuked the writer, declaring, "I know of no pioneer inventor who is actuated primarily by a desire for money." Jenkins described his own motivation simply: "I invent because inventing is what I like to do."[2]

The First Phantoscopes

Jenkins' first patented invention was the Phantoscope.[3] Technologically, the Phantoscope was the first large-screen motion-picture tool, a "useful device for exhibiting a series of pictures." It exposed the pictures to a strong light source for a fraction of a second as they passed beneath, giving "the observer the impression of real action."[4] The concept was simply a series of individual still pictures exposed in rapid succession. This is the patent that launched Jenkins' career. His second, closely related patent was for his Kinetographic Camera, filed just over a month after the Phantoscope patent.[5] This camera had multiple purposes. It was "an instrument for recording any active scene and afterward reproducing the same [scene] upon a screen."[6] These first two patents, the Phantoscope and Kinetograpahic Camera, were those that Armat had challenged. What Jenkins basically did with these patents was direct the motion-picture industry out of the peep-show business. Instead of a single viewer peering into a box and cranking animation cards or a film roll over the lens, this new bare-bones apparatus was a first in the coming world of large-screen projection. Eventually, Jenkins' ideas combined with Armat's, and Edison's finances would open the commercial world of motion pictures, but large-screen projection was conceived from these first two of Jenkins' patents.

Jenkins' third patent was the joint one with Armat. It was a modified Phantoscope, integrating Armat's and Jenkins' ideas and resulting in "new and useful improvements in phantoscopes."[7] The significance of this joint patent is not only in settling matters between the inventors, which it did only legally, but in allowing both inventors rights to the technology of the Phantoscope for future development. In other words, as they went their separate ways, they each could continue building photographic equipment based on this patent without fear of intervention, even from each other. And they did so.

The Franklin Institute Elliott Cresson Award coincided with publication of Jenkins' book, *Animated Pictures,* which sold for $2.50 ($67) and provided a chronology of the history and development of motion pictures. More importantly, at the time, it illustrated Jenkins' future vision for the Phantoscope's commercial application.[8] *Animated Pictures* was really a public relations tool, but today it serves as a valuable historical document, establishing positively Jenkins' "phantoscope [as] the first successful moving picture projection apparatus."[9] It advanced an understanding of Jenkins' technology and his developing ideas for entertainment for group audiences, ribbon photography, film boxes detached from the camera, telephoto lenses, animation with high-speed cameras, and coding that included subject, time, day, season, and distance.[10]

Jenkins refers little to Edison's Vitascope, "which is not an Edison invention at all": "[T]he so-called Edison Vitascope . . . exhibited at the Cotton States and International Exposition, Atlanta, Ga., 1895, and on March 17th,

1896, [was] surreptitiously taken from the writer's residence [by Armat]."[11] He commented that even the Armat-Edison agreement eventually ended because "the poor deluded purchaser learned to his chagrin that the sale was based upon no monopoly and that he had paid his good money for a little patch of blue sky."[12] Most of the *Animated Pictures* text chronicles the history of the Phantoscope and the commercial applications of animated photography. The patents, award, and book provided the recognition Jenkins desired. He had underscored the importance of his invention and validated his work within the scientific community. From this foundation he progressed.

Motion Pictures at Home and in Science

While promoting his photographic inventions and moving them forward, Jenkins ventured into similar directions as others of the time. Fundamentally, he followed the trends of Edison and those seeking a way into the anticipated profits of the commercial entertainment nickelodeon. His niche within this peep-show world would be found in the possibilities of motion pictures in the home, as opposed to storefront parlors. His systems featured scientific and educational content, utilizing a device where the viewer could learn from film at home. This niche also distanced him from Edison.

Why couldn't the home serve as a science classroom? The use of motion pictures in science was not new; birds, horses, and animals of all sorts had been the subjects of study. Jenkins took his ideas to Major John Wesley Powell, Gifford Pinchot, and Alexander Graham Bell. These scientists shared Jenkins' interests in nature, the West, service, and education, and he was now able to get their attention.

Powell gave Jenkins' educational-film idea an endorsement. Powell was a western explorer who is best known for his exploration of the Grand Canyon areas of southern Utah and northern Arizona. No doubt Jenkins' affinity for Powell came from his own travels and their mutual love for the American West. They were also both government employees in the District of Columbia between 1892 and 1894, when Powell headed the U.S. Geological Survey and Jenkins worked in the U.S. Life Saving Service. The result of Jenkins' inquiry to Powell was simply that Jenkins received affirmation of his educational directions: "I see a wide field for use of your Phantoscope in the teaching of science to children of the schools of our country," Powell said.[13] That was it, but coming from Powell it was still an endorsement.

Gifford Pinchot was a government employee also, and this contact produced a short-lived experiment with potential. Pinchot was the creator of

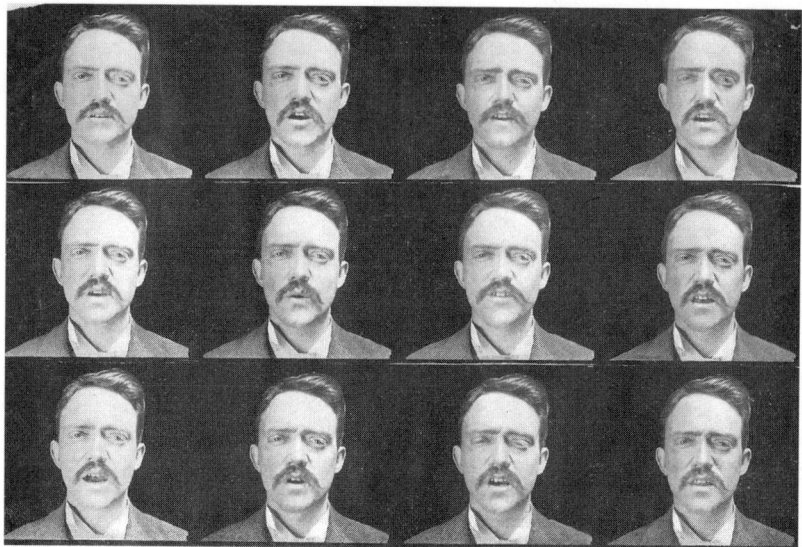

Jenkins filmed himself lip reading the Lord's Prayer. Courtesy Wayne County Historical Museum.

the National Forest Service and had procured a budget for filming along with his new duties as chief of the Forest Service. Jenkins was commissioned by Pinchot to use film to record plant growth. Jenkins set up a stationary camera, but unfortunately a rain storm destroyed it. The experiment was a failure, but Jenkins summed it all up positively: "[E]ven a physical bit of the experience [was], perhaps, enough to indicate a usefulness in educational work."[14] Why Jenkins did not get another camera and continue the work is puzzling.

The most meaningful demonstration before a person of science was performed for Alexander Graham Bell, after which Jenkins would acquire his endorsement and a little funding. Bell had created the telephone and a recording and playback device, and he worked extensively with the deaf.[15] So his support of Jenkins was extremely significant and would be quoted in Jenkins' marketing materials over the next two decades. His pitch to Bell was that motion-picture technology could be advantageous in teaching the deaf. Jenkins suggested that his camera could enhance communication by teaching lip reading.

Bell's wife, Mabel Gardiner Hubbart Bell, had contracted scarlet fever as a child, and the illness had rendered her without the ability to hear.[16] Bell was interested in Jenkins' experimentation, and the two men set up a test. Jenkins produced a film, reciting the Lord's Prayer, and presented it full-screen to Mabel. Without knowing the content of the message, she was able to understand by reading the lips of the presenter.[17] The film contained two thousand images of the "lips and tongue position in recitation of the Lord's Prayer."[18] The demonstration illustrated possibilities for lip reading. The idea was helping an "oft-times timid people. . . . [Liberating] them is a gracious act."[19] It appealed to Bell and gained his support. He provided limited funding, covering expenses, and perhaps more importantly gave permission for Jenkins to use his name in the promotion of his work. Jenkins was pleased to receive the endorsement and financial support, with Bell's "kind offer to assist in bearing the incidental expenses of the first experiments."[20]

The utilization of film as a teaching tool would progressively gain a foothold. Dr. Arthur G. Bretz heralded chronophotography in the medical sciences.[21] Chronophotography—a term used before *cinematography*—was the scientific study of motion. Bretz quoted Jenkins as emphasizing that the power of the photograph was beyond entertainment, having "a more serious mission." Bretz underscored the advantages of the camera in the medical clinic and the laboratory.[22] Within the next decade, Jenkins would be producing projectors for the classroom.

In the midst of Jenkins' moves to create a niche in science and education for his work, in August 1899, almost a year after receiving the first Franklin medal, he took a short hiatus from his laboratory activities and made a trip to the western United States, this time to film the legendary native Hopi Snake Dance.

Filming the Hopi Snake Dance

Jenkins wanted a film of his own to use in promotion, and the Hopi tribe in northeastern Arizona provided the opportunity.[23] He took his camera back to the high desert north of Canyon Diablo, Arizona, where he filmed a Hopi Snake Dance.[24] He hand-cranked his motion-picture camera during the two-week Hopi celebration, capturing the expression and understanding of a unique culture and people.[25] For Jenkins, the trip had twin purposes: he could use the photos to teach and promote, and he loved to travel in the western states.

The travel was rustic and rough. Jenkins described his arrival at the Arizona Canyon Diablo train stop as a departure from its historically rough Wild West image. The city of Canyon Diablo was said to have been more lawless than Tombstone, but Jenkins said, "[I]t wasn't a very exciting place . . . with just a box-like [railroad] station, the trading post store, and a corral."[26] Getting to Canyon Diablo via the Santa Fe railroad was only half the journey and the easy part. Jenkins still had to traverse a rocky sand-filled desert in a two-day journey north into the dry mesas of the Hopi Nation. The trail only went halfway, after which local directions referred to merely geographical landmarks. The Canyon Diablo storekeeper rented him two mules and a wagon and told him details of the route: "[K]eep a little to the east of north, and before sundown you should reach the middle trading post. . . . [T]here is no trail from here on . . . all sand; but if you bear a little east of north you should be in sight of the mesa by morning."[27] His wagon was loaded with provisions for the animals and his equipment. He arrived at the midway station, where fresh mules were provided, and he journeyed into the night. As the sun set over the desert, he was afraid he was getting lost and was "stricken with a little terror." He drove his animals hard into the night. Finally stopping, he fed the mule team, rolled himself "up in a blanket, wriggled down into the sand, and slept."[28] He awoke at sunrise and was relieved to see the high-topped mesa of the Hopi. He described the desert beautifully.

> There is a feeling of unequal peace and contentment [that] steals over one in the deathlike quiet of the desert. No city noise or other clamor of civilization disturbs the great quiet. I can almost hear my pulse; not a sound except an occasional squeak from the harness as the mules stand eating oats out of the feed box of the wagon.
>
> Nor is there any interruption to the view as far as the eye can see across that great sandy waste, except the welcoming sight of the mesa ahead. It reminds me of the purpose of my errand. I must get going now without delay.[29]

He met the Hopi, who guided him up the steep slopes of the mesa into their village. Two young men carried his camera equipment. The village crier barked his arrival to the people of the area. There was an air of excitement as everyone was anxiously awaiting the dance celebrations, which lasted a few weeks and were held only every two years.[30] Jenkins set up his camera and described the Native people lining the walls and sitting atop both walls and their pueblos. "A hush falls on the spectators. . . . And here they come, the snake priests, in single file, stately procession, their white-stripped bodies undulating in unison and in time with their weird chanting."[31] The rattlesnakes were taken by the neck, resisting their capture, then dropped and immediately recaptured by the priest as the parade passed through the village. The sacred ceremony concluded with the priests grabbing as many snakes as they could carry and running to the desert floor, where they were released to carry the Hopi prayers for rain "to those above."[32]

Jenkins stayed throughout the celebration. As he was leaving, he was "presented with a saddle blanket of native weave and dye."[33] The way back to the train station included a detour around a new lake—the prayer for rain had been answered. Indeed, he reported, "it poured."[34] Jenkins now had a film of his own and still photographs that would appear in promotional materials and coming experiments.

The Jenkins Phantoscope Company created a nickelodeon for home sales and education, calling it again the Phantoscope. Note the horn speaker and the record mounted on the side for sound to accompany the visual images. Courtesy Wayne Country Historical Museum.

The Jenkins Phantoscope Company in Education

The Jenkins Phantoscope Company evolved following the contract with the Columbia Phonograph Company to exhibit the Phantoscope. Columbia was in competition with Edison, so for Columbia's use of Jenkins' Phantoscopes, Jenkins was to receive a percentage of sales and exhibit revenues. The Phantoscope did go on exhibition in Atlantic City, Philadelphia, and Chicago, but in Chicago the demonstration was sabotaged by a vaudeville-theater manager, John D. Hopkins, who was negotiating a more lucrative deal with Edison representatives, the familiar Raff and Gammon. Armat filed a legal injunction against Columbia's use of the Phantoscope. Even though the injunction was denied, the end of the Columbia contract was clear.[35] Thus, Jenkins moved forward into home and educational markets.

The Jenkins Phantoscope Company, a Washington, D.C., corporation, was Jenkins' first formal business, organized in 1904 to manufacture another Phantoscope, this one another peephole system.[36] The goal was exhibiting, "preparing pictures" for sale, and leasing photo equipment.[37] This company, however, should not be confused with the earlier Phantoscope camera-projector. This Phantoscope was similar to Edison's Kinetoscope and Herman Casler's Mutoscope, peephole devices already on the market.[38] In these turn-of-the-century parlors or arcades, individual viewers paid from a penny to a nickel (twenty-five cents to $6.79) to see the motion film.[39] The viewer watched travel films, parades, exotic animals, and sporting activities. The Edison and Casler companies had served the mainstream marketplace for almost a decade with their own nickelodeons.

Jenkins modified this peephole Phantoscope for educational household use. When his arcade-exhibit partnership with Columbia derailed, he began marketing directly to consumers with prices of $2.50 to $3.50 ($63-$89). His device could be shipped directly to the home, came with two reels of pictures, and additional reels could be purchased at fifty cents ($12.58) per pair. "Designed for entertainment and education," this Home Phantoscope was described as "a real educator."[40]

The Phantoscope Company sought capital stock of one hundred thousand dollars ($2,516,154) for "manufacturing, marketing, exhibiting and otherwise dealing in Animated Picture Apparatus."[41] As of September 30, 1904, twenty-five thousand dollars ($629,938) had been raised.[42]

Jenkins purposely kept using the term "phantoscope" for a growing number of devices, promoting and connecting them with the now-famous name.

Even though the connection was slight in terms of technology, the names *Phantoscope* and *Jenkins* had become marketable. Pamphlets promoting home sales made reference to the 1890s machine used to record and project motion pictures. They suggested to the consumer, "You've seen Life-motion Pictures Theaters, Church, etc. doubtless. With the phantoscope you can now have them in your home."[43] However, this home Phantoscope could neither record nor project film like earlier apparatus with the same name. The home Phantoscopes were clearly a different card-animation system with educational content. Hundreds of such reels were available from various catalogs. The Phantoscope advertised bringing to "your fireside living scenes from all over the world."[44] A 1904 *Ladies' Home Journal* advertisement said that this Phantoscope would be ready for Christmas, and Jenkins was hiring temporary Christmas help. "Wanted: Six Girls until After Christmas. Light work."[45] Shortly after the holidays, an inventory of picture reels was available for the public, and by 1905 machines were sold successfully enough to include musical accompaniments. A record player was mounted on the side of the Phantoscope and turned as the handle was cranked, producing both the visual and the audio. On the newer models, a lens could be mounted over the viewfinder to magnify the image.

Jenkins filed for another patent in 1904 for a "Moving Picture Apparatus." It was actually an improvement on the existing machine, but the patent was granted not to Jenkins but to his wife Grace,[46] though its author was clearly Francis. This apparatus was "designed for synthetically exhibiting . . . an animated scene." The model utilized "picture carriers [that were] elastic cards attached to a [moving] band."[47] It was Jenkins' first photographic patent following his wedding, and when the couple were married, Grace had given Francis her entire savings, which funded the establishment of a new laboratory, pulling Jenkins out of bankruptcy. "In return he promised her that one-half of everything he made on inventions would be put into a special account . . . to be hers alone."[48]

Marveloscope versus Mutoscope

Jenkins continued to work within the mainstream even while he was working on an educational home market. The Mutoscope, created by Herman Casler, was the paramount peephole projector of the day.[49] However, it was not without competitors, and among them was Jenkins, who had created a machine he called the Marveloscope. This 1905 "Moving Picture Apparatus" belonged to the "special class [of motion pictures] in which the several pictures [in a series were] printed on cards." It basically provided for more flexible movement of the photos.[50]

Jenkins and Jeremiah Edward Cahill (1873–1943) filed a joint patent, and within the year they would form a competing enterprise, the Marveloscope Company, in which Jenkins was the secretary-treasurer.[51] The Marveloscope Company filed for incorporation on November 4, 1905, with a capital investment of five thousand dollars ($125,808). The investors were C. C. Dieudonne, J. E. Cahill, John Kramme, and Jenkins.[52] This was the Christmas season, so while the Marveloscope machine was being developed for marketing under the Marveloscope Company, at the same time the Jenkins Phantoscope Company was still actively pushing "live motion pictures . . . [at your] fireside, scenes from all over the world with the vividness of actual life. . . . Simple and strong . . . a child can operate it," in what appears to have been a national campaign.[53]

The Marvelosope was another peephole system. It had an ornate cabinet design, with a circus elephant on top carrying a sign that read, "Live Motion Pictures, One Cent." The laborious viewer's handle was replaced by an automatic crank, which drew the photos through the machine so that there was "no [hand] cranking, no winding, no lost key" to inhibit the viewer.[54] The Cahill-Jenkins patent for the "Motion-Picture Machine" provided for a greater number of picture-animation cards, an attractive casing, and an inexpensive device for home, and the exhibit machines were interchangeable.[55] A second patent by Cahill and Jenkins grouped the cards closer together, providing uniform production.[56] Improving even this nickelodeon film technology was a continually evolving challenge. What had been exhibited and distributed for animated-card scenes was transitioning into film rolls. In the nickelodeon industry, Jenkins followed the trends, each time working to improve his systems and earn his share of the financial action. The only significant differences in his approach were his educational-content orientation and home-sales target. The Marveloscope Company added a "Photographic Printing Apparatus," which printed a positive film-strip transparency from the original negative, which Jenkins recommended be stored separately for preservation and protection.[57] Adapting it for the Jenkins-Cahill system, it could utilize film or printed cards and rotated the pictures, "giving the appearance of objects in motion."[58] Jenkins' Marveloscope and the Stereoscopic Mutoscope were in direct competition with Casler's American Mutoscope, the Biography Company's Mutoscope, and Edison's Kinetoscope.[59] None projected on screen for a large audience; they were all still coin-operated individual viewing machines using printed cards or thirty-five-millimeter film, basically wrapped around a drum. The turning of the handle, flipping the cards, or rotating the film created the appearance of motion.[60]

Between 1896 and 1912, Edison would reap windfall profits from the Edison Vitascope. He had begun his battle to maintain control over the motion-picture-distribution and film industry, while Jenkins looked to education and home markets for his success. The motion-picture film *The Great American Train Robbery* would debut as the industry transitioned from nickelodeon box to large-screen, group-audience theatrical and feature-length film.

In August 1904, doctors suggested that Jenkins needed to slow down. He was a workaholic; pressure from the competition motivated his inventive activities as well as introducing fatigue. He worked at a pace that raised health issues. So Francis and Grace would relax, taking a trip home to Richmond, Indiana, to attend the wedding of Jenkins' sister Alice. It was a small Quaker wedding, attended by the immediate family, friends, and a few out-of-town guests. Jenkins was very much a family man.[61] The trip served as a respite from his Washington, D.C., life, his work, and routine. But nothing seemed to slow him down. Over the next several years, the diversity of Jenkins' inventions widened, with varying applications mixed into his photographic patent portfolio. It would appear that the variety of his work was stretching the workload and increasing the stress.

Motion-Picture Classroom Safety

Motion pictures for the home segued slowly into classroom education. For Jenkins, the move from the home to the classroom was a natural progression. He had participated in science film and created home instruments, so it was a logical segue into the schoolhouse, as well as a way of staying clear of Edison. However, moving the projector into the classroom created a new challenge: namely, fire.[62] The use of the film projector in the classroom was hampered by the fire hazard it presented. There was no "projection booth" in the classroom, as had been originated for theaters to protect the audience from the fire potential. However, during the years before World War I, the projector had been used for defense training in public buildings. With the permission of the War Department, it was used in "training camps, schools, public buildings, and aboard transport," and "even the [ship] *George Washington* had four such machines when it carried the President [Wilson] to and from France."[63]

Jenkins' response to the challenge was to establish another business, the Graphoscope Corporation, with himself as president. The Graphoscope projector, created by Jenkins, was manufactured by the A. S. Campbell Company; it was a self-contained thirty-five-millimeter projector.[64] This projector and related patents restrained any hazard within a fireproof chamber and at the same time reduced noise.[65] Jenkins reported the corporation's success: "ten times more business was done in the territory covered by all the competitors combined."[66] With this success, he set up the new corporation and acquired a "skilled manufacturer."[67]

Over the next several years, Jenkins worked on issues surrounding the heat generated by the projector lamps and the arc lights with the objective of controlling temperature. He wanted an apparatus that even "children or other unskilled persons" could operate without danger to themselves.[68]

Teaching Photographic Expression

It was at this point that Jenkins turned his attention to promoting visual literacy as a means of marketing. Throughout the 1800s, still and motion pictures had been utilized in scientific studies of birds and animals, but it

Photo illustrations by Jenkins' own work illustrate his attention to visual literacy in a photograph. Courtesy Virginia Roach Family.

took a few more decades for photography to become a popular hobby. By the turn of the twentieth century, when consumers were popping nickels into nickelodeons, still photography had become a growing consumer hobby. Jenkins contributed by writing and illustrating visual-composition instructions and ideas for education to the growing photo-hobby industry. He was a talented photographer and writer.

Expression in Photography directed the amateur photographer's attention to the significance of framing and composition in the communication emanating from a photo. Jenkins addressed the communication values of photography as having meaning beyond just the production of a photograph, a document, or a scientific record of study. He provided instructions relating to facial expressions and the various positions of the head and body as "a sort of pictorial resumé of human expression."[69] The photographer was encouraged to note the importance of facial features. Jenkins was critical of those before him who had simply said, "the eyes are the windows of the soul," and those who made the costume the heart of the portrait. He argued "that quality of expression . . . [was] in the delicate muscles of the human face."[70] These gave the photograph its human communication value. The pamphlet featured small diagrams examining different poses and the expressions created. It illustrated camera angles for expressions of fear, interest, superiority, bashfulness, objection, humility, supplication, and dejection. It blocked the different body positions, the muscles of the face, and their relevance to composition as well as communication. This was a publication before its time, but it was published during a time of transition. It addressed the issues of communication as well as the art form. Still photography was a growing hobby, the peephole devices were at their climax, and the large-screen audience was on the horizon.

At the turn of the century, Jenkins was showing himself to be a skilled photographer. His visit to the Hopi Nation produced still and motion pictures, and he was "one of the most prominent amateur photographers in this city [Washington, D.C.]."[71] This reputation offered an opportunity to promote himself and support the nonprofessional photographers at the same time. His personal photos graced *Expression in Photography* and every one of his publications. In 1901, as director of the Capital City Camera Club, he organized the annual photographic exhibit. He encouraged the hobby of photography in high schools, clubs, and civic meetings, where he would be the featured speaker, judge exhibits, or loan his own works for display.

Jenkins believed that a child should be able to operate a camera and that the impact of film was solely dependent upon the individual's ability to visualize, then interpret the visualization. "Visualization" was the "only universal language.... We have the laboratory method in chemistry; the geography class has the field; and engineering students visit nearby industrial plants."[72] Yet, little in his time existed to help the photographer produce and analyze the visual as the universal communications tool Jenkins believed it would become. Students learned to read and write, but in film and photography, the interpretation of communications, persuasion, feelings, and impressions were primarily based on personal experience.[73]

A second pamphlet, *Motion Pictures in Teaching* (1916), promoted the wide use of cameras in the classroom. "Nothing else stimulates the imagination like the motion picture," Jenkins suggested, as he described the camera's advantages in teaching geography, history, botany, zoology, chemistry, mineralogy, physics, engineering, literature, geology, astronomy, agriculture, manual operations, mathematics, and drawing.[74] The use of photography in education, according to Jenkins, was endless.

Producers: Change, Challenge, and Opportunity

Jenkins testified in the trial between the Motion Picture Patent Company and the independent film producers at the turn of the century. The Motion Picture Patent Company was a collaborative alliance of companies with the Edison Manufacturing Company in control. In that alignment, continued patent litigation was a means of commercial control; new practices were adopted into place, and old ones were defended. It created a period of industry-wide "commercial warfare . . . ferment and rapid change."[75] The controversy, described as bizarre by Ramsaye and Jenkins, was about the efforts of the Motion Picture Patent Company's attempt to "monopolize the technology of the motion picture."[76] Edison and other controlling corporations were suing the Independent Motion Picture Company to maintain technical and creative control.[77] It was a long and arduous battle for dominance from which the independent filmmaker would assume creative control, and Jenkins supported the independent film producers.[78] He provided independents with cameras, and his court testimony assisted in a decision that favored the independents, setting "the infant industry free [from the Edison monopoly] to grow as its merit as a public entertainer might warrant."[79]

The motion-picture camera had come a long way since 1894. By the early 1900s, film producers were creating content for the nickelodeon parlors, in-home use, and traveling exhibitors. The time would soon arrive when picture theaters would spring up in towns across the nation, as producers began taking creative control away from the technical creators. A new production-content business was growing. This transition was fraught with challenges for the industry: inventive technology, equipment, film-size standardization, and patent and copyrights created "cutthroat competition"—and an audience that had growing expectations for what could be provided.[80] The Motion Picture Company corporations and Edison were frustrated in the wake of change, annoyed by the independents and the increasing competition along with the "invasion of numerous other imported and homemade" Vitascopes.[81] The "homemade" reference was an apparent allusion to Jenkins marketing systems and content directly to the home. However chaotic the history of motion pictures, it moved forward into the commercial framework of the cinema that we see today.

Meeting the needs of the independent film producers, Jenkins teamed with Oscar B. Depue of Chicago on a "Motion Picture Camera" patent, filed March 24, 1908. Depue was a projectionist who traveled with the famous Burton Holmes, an international travel-film lecturer and filmmaker, who had purchased his equipment from Jenkins.[82] As a producer, Holmes integrated elements of still and motion photography into his Chautauqua-style theatrical presentations.[83] By 1910, Holmes was drawing "fashionable people to Carnegie Hall at prices of $2 and $2.50 [$49 and $61]," and the established nickelodeon parlors were complaining that they were losing money while they showed "the very same pictures, for five cents."[84] Depue, as Holmes' projectionist, brought new production skills and a technical orientation to Jenkins' coming inventions. Holmes described Depue as devoted to "perfecting the art of cinematography."[85] Thus, he made an excellent inventive partner who brought an operational perspective to the inventions. Many projectionists in those days were not simply operators but skilled technicians as well as artisans.[86] The Depue-Jenkins camera saved film, eliminated static, and prevented fogging in projection during the presentation.[87] Authorities were not agreed as to the cause of the static and fog, but DePue and Jenkins suggested that it was due to friction as the film moved through the film box and that this could be curtailed with changes in the construction of the camera.

Jenkins also produced a device for traveling lecturers that coordinated the presentation of photographs, slides, and film strips.[88] Independent film producers who were moving into a field that was heretofore limited to technical inventors were now potential equipment-leasing customers. They were producing public content for devices rented from Jenkins. While the inventors struggled to become producers, a new creative role of the producers was emerging.

Among the successful independent producers to whom Jenkins provided technology was Siegmund Lubin. He created a large production-and-distribution organization, the Philadelphia Lubin Manufacturing Company. He too was a thorn in the side of Edison and the larger power brokers of the time, basically bootlegging some of their films for his own exhibits. For example, his 1903 production of *Uncle Tom's Cabin* came only one week after Edison's release of the film. He was sued by Edison for patent violations.[89] The Lubin studios were equipped with Jenkins cameras and technology.[90] Lubin had purchased the Jenkins camera after seeing it in Philadelphia.[91]

Herbert J. Miles was another producer who was using Jenkins technology. He accompanied miners on the trail of the Alaskan gold rush.[92] Although he used Jenkins' camera, his films along the West Coast were supplied to Edison and the Mutoscope Company for distribution.

In 1911, Jenkins made his most significant sale of equipment, to the film producer Carl Laemmle. Laemmle had worked in Chicago's nickelodeon theater and in 1906 had founded the Independent Motion Picture Company.[93] At this time, Chicago was the growing center of the motion picture and particularly the nickelodeon industry.[94] Laemmle was a prolific producer, and Edison, true to form, tried unsuccessfully to shut him down. Jenkins' cameras were again a part of this independent producer's success. Using Jenkins' equipment, Laemmle avoided a shutdown by multiple attacks from Edison, including charges of patent infringement. This was no small achievement. Jenkins had sold a camera to Laemmle for ten thousand dollars ($242,718) and "twenty-five percent of stock of [a] company to be incorporated by said Carl Laemmle for exploiting said camera."[95] One year later, the Independent Motion Picture Company was reorganized into Universal Studios.[96] Jenkins was networking with people who would create one of the world's largest motion-picture production and distribution firms.

Jenkins never ventured into commercial film production or distribution other than films he used for demonstration purposes. His love was for invention. In 1911, he developed practical devices for in-camera film movement[97] and another camera that provided "the desired exposure" with a continuously successive movement of the film.[98] His film-editing-marking mechanism accelerated the process of editing by creating permanent marks on the film, identifying pictures within a "common class."[99] The camera had a rotary rack utilized for storing exposed film until it was needed in processing and editing. Over the next few months, several inventions would appear to guide the film movement within the camera, relieving and maintaining the tension on the film as it was pulled through the projector and then rewound, thus decreasing film wear.[100] By 1919, Jenkins' editing system provided a method for film-sequence identification that marked the film at various intervals and helped establish an editing order. He noted that "some of the artists [editors] employed get as high as $10,000 [$131,233] per week" while making films, and his invention helped to speed the process and help control those costs for the company.[101]

High-Speed Camera

The last of Jenkins' photographic inventions was a high-speed camera. He had started this project in 1921, producing a cylindrical mirror that increased speed of film movement and exposure.[102] This work came at the same time as RadioVision, the electronic transmission of a still photograph. The high-speed camera produced pictures at the rate of one thousand per second, with a "possible 2,000 having been attained" in the lab.[103] This camera allowed for the analysis of motion by filming the human body or any movements at high speed and playing the film back slowly to examine the delicacies of movement.

The camera proved useful for the study of industrial machinery near the end of the 1920s. The film was shot through the camera, according to press reports, "at a rate of three miles a minute to slow down rapidly moving machinery."[104] It debuted at the annual meeting of the Society of Automotive Engineers on January 26, 1928, where Jenkins was invited to speak regarding the advantages of the camera in industry.[105] Calling it "The Chronoteine Camera . . . a time stretching camera," Jenkins extolled its virtues in the study of guns, airplanes, and any mechanism where high-speed action could not be seen by the human eye. In fact, he declared, "anything that moves too fast for the eye to follow can be shown slowed down and . . . examined in detail at leisure and repeatedly."[106]

Three years later, the camera had found its way into sports, where Robert "Bobby" Tyre Jones Jr. enlisted the assistance of the Jenkins Laboratory's ultra high-speed camera to improve his golf swing. Jones was an amateur golfer, writer, promoter, and lawyer whose British and American Open amateur championships made him a legend. He used the camera over several years. "It is always interesting to see one's own swing as others see it . . . [but] now that we have years of slow motion study almost every first-class golfer knows well enough how he hits the ball. The guess has been removed," Jones wrote.[107]

Jenkins' life and work never ventured far from his cameras. He had created the foundations for full theatrical-size screen presentations, taking the industry from the nickelodeon into the movie house, then into the home and the classroom. He would continue with the camera in creating RadioVision and the coming of television technology.

5

Founding the Society of Motion Picture and Television Engineers

There is little doubt that among the most significant and lasting contributions C. Francis Jenkins made in the film and television industries was the formation of the Society of Motion Picture Engineers (SMPE). In 1950, "television" was added, making the organization the Society of Motion Picture and Television Engineers (SMPTE). In less than four decades, the SMPTE would become an international association with industry, technological, and creative influence around the globe.

The foundation of the SMPE comes straight out of the weaning years of the Industrial and Gilded Ages. It helped guide film from an awkward novelty into a technology with scientific and entertainment purposes. It fostered technological development and the spirit of cooperation. One major problem among pre–World War I competitors was that each utilized different systems. The answer was standardization.

Over the years, each inventor, manufacturer, and filmmaker had developed their own systems. Each sought to outflank the other with improved technology. By 1915, the plethora of apparatus was creating confusion and slowing overall industry growth. The manufacturers had tried to agree on technical standards but failed "principally because of a lack of mutual confidence in the integrity of . . . competing interests."[1] Jenkins was disappointed in the inability of the Motion Picture Board of Trade's Committee on Standards to establish at least some uniformity. Terry Ramsaye notes that the Board of Trades, organized in 1915, had a short, uneventful life, likely due to the hotter issue of censorship.[2] It was followed by the National Association of the Motion

Picture Industry, which too was of no help in standardization. Censorship had surfaced as early as 1909, when the newly created Motion Picture Patents Company organized to freeze out the small independent operators in favor of a group called the National Board of Censorship of Motion Pictures.[3] While the group had no intention of promoting censorship, its unfortunate choice of a name stimulated the topic and dwarfed discussion of standardization. The industry of the early 1900s, as now, was strongly opposed to censorship, and consideration of standards was simply diverted by the issue and "a lack of confidence and common interest."[4]

It was within this atmosphere that Jenkins took the leadership reins and responsibilities to organize a fledgling industry association whose sole purpose would be a discussion of technology and its standards. Jenkins' own inventions, his photographic works, his scientific awards, and his competitive involvement positioned him as an influential proponent. In 1915, he had written about the possibility of creating an industry-wide organization.[5] Informal conversation about another possible organization to work on standards first took place on the boardwalk in Atlantic City with his friends E. K. Gillett and Nat I. Brown.[6] Jenkins was a bit unnerved by the earlier failures, as no one seemed willing to step forward and provide leadership. The hesitation remained as the prevailing concerns were self-interest, self-preservation, and simply disinterest. Jenkins reported his own frustration at putting "my personal standing in the industry to the risky test."[7] Would anyone show up if an organizational meeting were called?

The First SMPE Meeting

The primary task Jenkins faced was bringing together highly differing competitive interests. They were simply afraid to provide information to competitors. World War I proved a catalyst.[8] If the film industry could not get itself together and create its own consistent standards, the government would step in and mandate them. Effective defense communication was essential in the military conflict, and this was inhibited by a lack of standards. The world faced a growing crisis, and the film industry was divided. It was at this time that Jenkins' leadership made a difference. On July 14, 1916, he invited a few engineers to an initial meeting to discuss the formation of a motion-picture engineering association. He sent telegrams to friends, manufacturers, and technicians whom he thought might participate and invited them to a meeting at the Raleigh Hotel in Washington, D.C.[9] Eleven people attended, representing the industry in Washington, Cleveland, Chicago, Boston, and New York.[10] Jenkins was elected the temporary chair. The group immediately drafted a Certificate of Incorporation, and Jenkins was charged to formally "incorporate the society."[11] The petition for incorporation was signed by Jenkins, Paul Brockett, and Herbert Miles.[12] Brockett was from Washington, D.C., and would serve as the treasurer of the newly formed association. Miles was a San Francisco film-exchange entrepreneur and would serve on the first SMPE membership committee. Incorporation papers were filed in the District of Columbia on August 11, 1916, making the SMPE a formal organization.[13] In addition to incorporation business, the group heard a presentation by Henry D. Hubbard, the secretary of the National Bureau of Standards. Hubbard's remarks pushed the necessity of standardization within the industry and how it could be applied throughout the nation.[14] At this meeting, by-laws were adopted, a membership fee was established at twenty-five dollars ($520) per year, and recruiting for membership began. The SMPE declared its objectives to be the "advancement in the theory and practice [in motion-picture engineering,] . . . the standardization of the mechanisms and practices employed therein, and the maintenance of a high professional standing" among membership.[15] The group's next meeting was set for October 1916 in New York City.

SMPE Influence Grows

Starting the new group was not simple. Missed communication almost derailed the second meeting, but after a little floundering, corrected communication was sent via telegram, and the meeting was held in early October 1916. The group now more than doubled in size.[16] This time, Jenkins was elected the society's first president. He set the tone of the organization in his first address: "[The] motion picture is making the whole world kin. . . . [It is] the only universal language. . . . [I]t is making the world one great family. . . . Every new industry standardizes sooner or later. . . . [O]ur duty . . . is to wisely direct this standardization, to secure the best standards of equipment, quality, performance, nomenclature, and, unconsciously, perhaps, a code of ethics. [Our purpose] is an unselfish exchange of views."[17] Jenkins' appeal focused on standardization. He and Donald J. Bell, one of the founders of the Bell and Howell corporation, presented papers at this second meeting.[18] Jenkins' first SMPE paper was "Condensers, Their Contour, Size, Location, and Support," and Bell's was on "Motion Picture Film Perforation."[19] In addition to speeches and papers, new committees were established: the paper, auditing, and membership committees, and several technical committees on camera and perforation issues, electrical devices, theater equipment, and optics.

The next meeting of the SMPE was in early April 1917 in Atlantic City. Jenkins would later describe this as a pivotal meeting marking "the beginning of our real purpose, namely the dissemination of specialized data relating to our art."[20] The theme, established in his president's address, was again for the "Establishment of Standards." "We came to further our standardization," he said. "There are many looking to us to establish these standards . . . [in fact] they are anxious [for us] to do so as they have no other authoritative body to consult."[21] Discussion continued, a paper was presented by the cameraman Carl Louis Gregory on the "Motion Picture Camera," and a report was heard from the Committee on Electrical Devices.[22] The SMPE's membership had now tripled in the nine months since its organization.[23]

The pressing crisis of World War I—"the war to end all wars," as it was commonly called—had the world nervous. The government was impatient with the motion-picture industry. Hubbard called on Jenkins, issuing an ultimatum: "[W]rite the specifications for the war service motion picture camera for the Army and Navy," or they would mandate them.[24] Jenkins' earlier employment with the Life Saving Service and his continued association with the government military establishment no doubt kept him in their minds. They wanted motion pictures that were similar to one another in quality and in frames per second. He had forty-eight hours to draft these standards.[25] To accomplish the task, he called upon the new members of SMPE, who were studying cameras and perforation. Together, they drafted the first standards.

It is noteworthy of Jenkins' character that, despite the critical nature of these standards, there is little written about them or his leadership as a major contributor; "The credit therefore [was] given [by Jenkins in his president's address] to our Society for the added prestige."[26] He did not use the association's success to bolster his individual contributions or promote his self-interests but to establish the SMPE as the authoritative organization within the industry.

Jenkins' president's address at the 1917 Chicago meeting took the discussion of standardization beyond the camera and film-perforation issues. He appealed for "standards with due regard for the rights of all," and in this tone noted the importance of the inception of the *Transactions of the Society of Motion Picture Engineers*.[27] The standardization issue was pressing, and thus there were no papers presented at this meeting. Three reports all reflected standardization themes. The report of the Committee on Optics focused on lenses, the aperture, and angles of projection.[28] The Motion Picture Standards and Nomenclature Committees were working on terms and definitions.[29] The standards adopted at this 1917 meeting related to film speed, size, frame lines, and perforation, along with projection angles and lenses and the aperture of the picture.

The fall meeting took place in New York in October 1917. Here again standards were refined, and Jenkins continued his strong personal role. The SMPE was "not trying to legislate . . . [but] simply reflect the consensus of opinion of engineers of our industry" before the government decided to legislate.[30] He personalized his appeal, talking about two separate photographers who went to Yellowstone National Park to film the wonders of the mountains and the geysers. At a private viewing of their films, to which Jenkins was invited, he described "the photography of both men [was] so good that the work of each was spliced into that of the other without discordant tone." However, despite complementary visual quality and consistency in tone, when the projectionist tried to edit the two film stocks together, there were immediate problems, "misframes. . . . A beautiful picture ruined, and much time and money wasted, all because two cameras with different frame lines were used on the same work."[31] It was a convincing argument that moved the debate forward.

Reports at this meeting came from the Committees on Standards, Nomenclature, and Electrical Devices. Papers were presented on the effects of the projection angle, the use of various incandescent lamps, and screen brightness. Jenkins discussed the motion picture booth, a topic arising from the ongoing issue of fire hazards. His appeal this time was for safety.[32] The lamps of projectors burned at high temperatures and presented a fire hazard that needed attention.

Membership had grown to forty-four people. Interest in the SMPE's work was spreading within the industry. However, that success did not mean that standardization was widespread. Progress would come with continued effort in the late 1920s, when further issues were resolved.[33]

Closing Jenkins' SMPE Leadership

Rochester, New York, was site of the April 1918 meeting. Jenkins' president's address followed up on his earlier presentation related to safety and fire protection.[34] The number of papers increased to eight, including one by Jenkins on condensers.[35] Before the meeting concluded, Jenkins announced that he would not be a candidate for another term as SMPE president. He would remain active within the association throughout his life, but he felt that he had established sufficient foundations and was anxious to get back to his own work. The SMPE had been consuming; time that could have been spent inventing "had been lost in the office of the President."[36] He expressed no regrets.

The SMPE's final meeting under Jenkins as president was in Cleveland in November 1918. It was almost two and a half years and twenty-seven technical papers later, following the first meeting in Washington, D.C., that Jenkins would return to his own work of inventing. He had provided unselfish leadership, bringing technical people together for mutual benefit. He led in the formulation of standards on a range of motion-picture technologies. His leadership averted government intervention by bringing together industry-wide professional representation and organizing the SMPE. In less than three years, membership had grown to over fifty. He saw to it that papers and actions of the membership were recorded in the SMPE *Transactions*, which later became the *Journal*, an open forum; he formalized committees; and he diverted time from his own work for the betterment of the industry.

Jenkins' final address as president was a combined report and a brief society history. He was still pushing for refined standardization. He looked forward to "a stable future of still greater usefulness and an enlargement of our society into an international organization."[37] He warned the membership not to fall back into competitive and conflicting self-interests or to discriminate against one another. He concluded with a statement supporting continued sharing and publication of technical papers: "I myself have found the most interest in our meetings has come from the valuable papers read and printed, and I don't believe the limited time of our meetings can be spent in a more worthwhile manner."[38]

Jenkins would remain an influence on and active participant in the SMPE throughout his life. Between 1919 and 1930 he would deliver nineteen additional papers contributing to motion pictures, RadioVision, and television. In 1926, he would organize the national spring meeting in Washington, D.C., where members met with President Calvin Coolidge and other government officials. Pictures from the meeting taken at the White House were widely circulated within the industry press.

The SMPE would continue expanding, gaining a reputation as an active research, publication, and service organization. Glen Frank, president of the University of Wisconsin, characterized it as providing a part in "the service of humanity." It had "nothing to sell, and nothing to ask" for its work in the scientific and technological communities. From the beginning, Jenkins had a keen awareness "of its relationship to other departments of the motion picture industry" as it directed and linked the science and the art. It was primarily a technical organization with a broad approach to the free exchange of ideas. Although the original purpose of the SMPE organization was engineering standards, it rapidly grew into an important forum. Among those early presenters have been historical names of significance—George Eastman, Thomas Alva Edison, Lee De Forest, Herbert E. Ives, Terry Ramsaye, Vladmir Zworykin, and Philo T. Farnsworth. And even Jenkins' arch rival Thomas Armat was invited without a single protest. All had a forum for presentation and discussion, which continues today. The SMPE became the premier association of the industries of motion pictures and later television engineering.[39]

The accolades and foundations established for the SMPE were not forgotten. Herford Tynes Cowling, a member of the SMPE Board of Directors, noted Jenkins' absence from the meeting in August 1931 and commented, "[W]henever you are with us [at SMPE] it makes us feel more like a large family than a scientific Society. . . . [Y]ou can be assured that the Society . . . is going to remain the outstanding technical Society . . . and we intend to stay in the lead. You can well be proud of the movement you inaugurated."[40]

Today the worldwide organization of the SMPTE has more than eight thousand members and stands as a monument to its founder, C. Francis Jenkins. He remained an honorary lifetime member—the SMPTE's highest honor—until his death in 1934.[41]

6

Visionary Entrepreneur

The motto on Jenkins' laboratory desk for years read, "If a thing is very difficult, it is as good as accomplished; if it is impossible, it will take a little more time."[1] Jenkins was forever optimistic. He took his inspiration from needs that surrounded him, merging new ideas with technology. He branched out whenever the need put his mind to work. He responded to the demands of the time, and from the profits of his labors he was able to support his primary interests in film and television technology. Until the late 1920s, he had no significant outside corporate sponsors, so he supported himself with a variety of inventions that he would create and sell. Some were dramatically different from each other as well as from motion pictures and television, and successful, while some were not. Nevertheless, his inventions reflect his broad interests. The American humorist Strickland Gillilan visited Jenkins' home in 1932 and commented that afterward, he was going to "take a blue pencil and mark the word, 'impossible' out of every dictionary I can lay my hands on."[2]

Today Jenkins might be labeled a multitasker, but inventing was always his focus. While working in the field of photography and television, he contributed to the development of the automobile, aeronautics, and even a new concept for a milk bottle—this little-known patent likely brought him greater financial return than anything else during his career. These and a multitude of novelty industries kept Jenkins busy (see the time line in the preface). He was a workaholic and received repeated stern warnings from doctors to slow down. These warnings produced little more than a few vacations.

The Jenkins Horseless Carriages

George Baldwin filed the first American patent for the gasoline engine in 1885. The Stanley Steamer automobile was first produced in 1896 and lasted until 1924. The New England states were the center of automotive production at the time. In 1899, there were eighty-four manufacturers, but half of these, including Jenkins, would be out of business by 1902. In these years, three power sources existed: steam, electricity, and gasoline. The steam engine burned kerosene or gasoline, which heated water, creating steam pressure to push the crank shaft. The electric engine was quiet but heavy and slow, and the driver had to stop too often and recharge batteries. The gasoline-powered engine would take over the market simply because of the abundance of cheap oil discovered in Texas. It was within this context that the Jenkins Automobile Corporation was formed—at a time the Jenkins Phantoscope Company was selling its version of peephole nickelodeons.

The Jenkins Automobile Company was incorporated under the laws of Delaware in 1900. Jenkins hoped to raise one hundred thousand dollars ($2,716,239), with shares in the company selling for ten dollars ($272), and for a "limited time and limited amount" at five dollars ($136) per share. "No investment could be safer where the profits are so large," he said as he sought to raise capital.[3] The company produced four different classes of motorized vehicles. The first were short-run personal passenger cars; the second, freight trucks; the third, a tourist bus; and finally, the world's smallest electric car. There were no Jenkins automotive manufacturing plants. He would design the vehicle, then produce it for demonstration, promotion, and sale.

In 1898, Jenkins introduced his first car onto the streets of Washington, D.C. It was steam-driven, carried two to four passengers, and rolled on wooden wheels. These he called "Trap," or "Steam Trap." They traveled twenty miles before they had to stop for water. Selling for $1,200 [$32,595] each, Jenkins happily drove prospective buyers through "the greater part of Washington."[4] He touted his cars as nonpollutant and described the steam trap as a "queer-looking-contraption without springs, steered with a tiller [stick], but with very fancy upholstery.... The steam generator was hardly larger than a large bucket.... [A]bout the most that can be said of this early horseless carriage is that it had a wheel under each corner."[5] Top speed was eight miles an hour.

Among personal passenger vehicles, Jenkins designed one of the first cars with the engine up front. He also built what was referred to as a "runabout." These too were steam-propelled.[6] The Jenkins locomotives were promoted

The Jenkins Automobile Company produced several steam-propelled vehicles, including cars, a bus, trucks, and a little novelty vehicle used at the Pan American Exposition in 1901. Courtesy Virginia Roach Family.

as cleaning up the city of Washington by replacing the horse-drawn wagons with clean steam-powered vehicles. Comparing them to the horse-drawn carriage of the day, these Jenkins machines did "not get sick . . . are always ready . . . [had no need] of hay or straw, no grain bins, no refuse [horse droppings], no offensive exhalations, no flies and no stomping horses. Oil and water [were] the only articles required."[7]

The freight trucks were also steam-propelled. Jenkins manufactured the trucks by converting horse-drawn-wagon base frames—minus the horse—into motorized trucks. The cost of the conversion depended upon the size of the wagon but started at "$350 [$9,507] upward." He predicted that converting twenty-five wagons per month would result in 100 percent profit,[8] a calculation that would seem to suggest that Jenkins was working to sell a business as well as a vehicle. In addition to wagon conversions, he produced a three-ton steam truck. This was sixteen feet long and six feet wide, with a flatbed of approximately twelve by six feet. The top speed was six miles per hour if the truck was loaded and ten when it was empty.[9]

The Jenkins tour bus, dubbed the "Observation Automobile," was painted dark green and black with a gold leaf stripe. It was sixteen feet long and six feet wide. The interior was hardwood and contained twenty-two rotary chairs for

The Jenkins Steam Trap Automobile, 1912. Courtesy Wayne County Historical Museum.

tourists. Large windows surrounded the coach; open in the summer for fresh air, they were closed in the winter when the compartment was heated with steam. Top speed was eighteen miles per hour on good roads. The twenty-horsepower engine made no noise.[10] Sightseers toured around Washington in comfort. Jenkins even provided a tour itinerary for prospective buyers: the Observation Automobile circled the sights of Washington every two hours.[11] He promoted the sale of this vehicle by providing potential owners with some speculative statistics. With twenty-two passenger seats and five to six trips daily, an owner's potential income was "$50 to $60 per day [$1,358 to $1,630]. The estimated costs for operating the sightseer included: salary for the driver, fuel, and a tour guide all for $12 to $15 per day [$326 to $407]." Jenkins presented the vehicle "on good authority that the gentleman who originated the 'Seeing Washington' railway cars made $60,000 [$1,629,743] in two years."[12]

Unfortunately, Jenkins' financiers did not see the same vision for the Observation Automobile. They refused to put up more cash, and the one that had been completed for demonstration was sold at auction. *Scientific American* carried a simple classified-style advertisement: "For Sale: Handsome 24-passenger automobile coach: also a 2-ton steam freight wagon. Both new. C. Francis Jenkins, 1103 H St. Washington, D.C."[13] The touring vehicle was sold to a New York investor, who modified it "for sightseeing service in Central Park and Riverside Drive."[14] Interestingly, today's horse-drawn touring vehicles in Washington, D.C., look a bit like the Jenkins Observation Automobile.

The small electric car, dubbed the "littlest automobile ever built," was likely Jenkins' best-known vehicle and received the most publicity.[15] It premiered at the 1901 Pan-American Exposition in Buffalo, New York. It was constructed as a gift for Chiquita, a thirty-one-year-old Cuban entertainer; it was a special car for a special person. Chiquita was "a living doll" at just over two feet

The Observation Automobile Coach was a tour bus that looks a little like a tour bus of today. Courtesy Wayne County Historical Museum.

Luxury auto. Courtesy Virginia Roach Family.

tall, weighing nearly nineteen pounds. The vehicle weighed just a little less than one hundred pounds, with a two-foot wheelbase. It was a Gilded Age promotional novelty item featuring a unique Victorian top, luxurious leather interior, electric lights, an electric gong, nickel-plated wheels, leather fenders, and compressed-air tires.[16] The miniature car was street-ready. Chiquita drove it from exhibit to exhibit as a part of her performances, also advertising for the Jenkins Automobile Corporation.[17] The exposition celebrations were unfortunately all overshadowed by a dramatically more memorable event: President William McKinley was assassinated at the Pan-American Exhibition on September 6.[18]

Jenkins reported resistance to his automotive work from horse-drawn-vehicle operators, railroads, their employees, and regulators, who proposed restrictions, licensing, and new laws. He lobbied hard against new regulations. "Why should motorists be singled out and numbered like dogs and tagged [requiring license plates] like convicts?" He felt that existing laws were sufficient and all that was needed was enforcement.[19] He was obviously unsuccessful, and auto regulation proceeded apace.

Aeronautic Conceptions

The airplane was a means of enjoyment, respite, and work throughout Jenkins' life. He conducted aerial photographic and television experiments, as well as developing new ideas for flight. He described his aeronautical patents as primarily improvements on existing technology and processes.

Jenkins marveled at the fact that the "lift of an airplane [came] from the air, which passes over the top of the wings."[20] Five years after Orville and Wilbur Wright made their first sustained, powered flight in 1903, he was filming the marvel. The Wrights at this time were working on a stretch of farmland near College Park, Maryland, training Signal Corps officers from Fort Myer, Virginia.[21] Both locations were not far from Jenkins' home and lab. On September 17, 1908, he filmed the Wright Flyer, piloted by Orville Wright, a model the Wright brothers were trying to sell to the army. The flight circled the Fort Myer parade ground and the bordering Arlington Cemetery several times before a crack in the propeller sent the plane, pilot, and passenger, Lt. Thomas Selfridge, crashing into Arlington. The accident cost the lieutenant his life, and Wright was seriously injured.[22] Two decades later, on December 16, 1928, Wright and the members of the District of Columbia National Aeronautic Association were invited to a luncheon where they saw Jenkins' film of the crash, along with a demonstration film of how "pigeons fly like planes."[23] The film was taken with Jenkins' new high-speed camera, reportedly at 3,200 images per second and projected at sixteen frames per second, so the demonstration was spectacular.

The popular histories of aviation do not mention Jenkins as a primary contributor to flight history, and he would likely agree with this assessment.[24] However, his film work and patent portfolio reflect significant interests. His first aeronautical patent was for an "aeroplane or flying machine." The drawings appear much like the earliest of the Wrights' planes. The primary objective was in its "easy and manageable construction," as well as "providing greater stability in the air."[25] Still others titled "Flying Machine" focused on maintaining the automatic equilibrium of the plane, accomplished by utilizing the principles of the gyroscope,[26] and then by a related patent to "maintain lateral equilibrium without affecting the steering in any manner."[27] A gyroscope today stabilizes ships and planes, and by 1914 it would provide the foundation for the future automatic pilot. Jenkins' familiarity with the combustion engine led him to seek improvements on the two-stroke aeroplane engine.[28] Still seeking greater stability, he turned his attention to the wing. He suggested that manipulating the back edges of the wing would create stability and drag. Jenkins' wing design minimized drag and kept the plane on course.[29] Anyone who has sat in a window seat in a modern airplane has seen derivatives of this wing's extended downward movement.

As early as 1912, the United States began to feel the political tensions from Europe. Jenkins wrote that this startled Americans into preparing, "in case we should be drawn into the conflict in defense of our own security."[30] In 1915 he patented his first device directly related to the war—a "Device for Aerial Warfare,"[31] used to attach and release a bomb from an airplane. However, most of his direct contributions to the war effort came through his work in aerial photography. Few patents were granted during the war (1914–18).[32] Secrecy was a part of national defense, and Jenkins worked along with all others under this umbrella of silence. His first in-flight motion-picture apparatus served to stabilize the camera in the plane. Flights were so bumpy that he considered the photos as the better alternative to motion pictures. However, in the long run, motion pictures were the defense preference. So he developed a camera mount used by the photographer, who stood in the "gunner's cockpit of the plane."[33] Another method suspended the camera from the plane. Both alternatives resulted in improved motion pictures. Jenkins' camera adaptations produced a wide, panoramic view of the flight paths and activity below.

Jenkins purchased his first airplane—a flying boat from navy surplus—on Armistice Day, the formal end of World War I, November 11, 1918. The navy was reluctant to sell to Jenkins, because they had no knowledge of his experience, so he wrote a letter to the assistant secretary of the navy, Franklin D. Roosevelt, requesting permission to buy the plane.[34] Apparently, "after some banter about the danger of flying," Roosevelt gave his permission, and Jenkins' wife Grace commented, "I hope we will not regret having bought it."[35] The plane arrived in several thirty-foot crates. Jenkins solicited the help of a navy test pilot, Lt. Edward W. Rounds, to help him assemble it and got a little free advice along with the assistance. Rounds strongly recommended twenty flying hours of training before Jenkins' first solo, but Jenkins was given his license after only twelve.[36] Grace and a group of onlookers watched as he skimmed along the water during the takeoff on his first solo flight. As he gained confidence, friends were taken with him. One lady offered "to wash his clothes for a year if only Mr. Jenkins would take her up."[37] Despite the attractive offer, for some reason he refused. Jenkins' second plane, an enclosed land "cabin plane," carried a few more passengers. His planes were the objects of research and pleasure. In research, he tested his photographic apparatus. In pleasure, he transported friends and just relaxed. "A recreational activity . . . and a delightful sport," he dubbed it later.[38]

Jenkins loved to fly.[39] His flight logs record the dangers of his new fascination—two accidents almost cost him his life. On August 15, 1929, a forced landing was really a crash. "A wheel collapsed—plane flopped over on its back, probably because I bumped my head on a brace, lost consciousness, and let the stick go forward. The plane skid[ded] 30 ft before turning over."[40] He wrote on the next page of the log that he believed "that someone had cut the safety wiring on the cylinder bolts." Again, on September 26, 1929, he claimed that "gas has been syphoned out of the plane—theft," and as a result, on a trip from New York to Boston, he was forced to land in a rocky field outside of New London, Connecticut.[41] With his fuel running low, he headed for a beach to make his landing, but it was apparently too far, as "he glided between two houses and struck rocky ground. The engine was partly ripped from its base and the fuselage torn."[42] Luckily, no one was injured. He did not mention any motivation behind the vandalism, but the dangers did not deter him. He took up most anyone who was willing to go—family, friends, associates, and potential investors. In the top right-hand corner of each log-book page was printed a legal waiver: "In taking this aeroplane flight with C. Francis Jenkins, I, the undersigned, understand and agree that I do so at my own risk, and I hereby release him from any liability or otherwise, that may result therefrom."[43]

One of Jenkins' most fascinating aeronautical ideas was an airplane catapult. Long before skyscrapers had their first helicopter-landing platforms, he was working a system that could land and launch airplanes from the top of a building. This idea was not totally new; the navy had used a catapult to propel planes from battleships, and there were proposals from many other inventors for short landings and catapults on flat roofs. "Roofs of many business buildings, and all business blocks in downtown Boston and other New England cities, [were] potential practical airports" for the smaller planes of the day.[44]

Jenkins' ideas for takeoff and landing were to make both possible "in as little space as possible." His first landing mechanisms, in 1919, maneuvered the top wing to provide a slight lift of the nose and thus slow the plane's momentum without affecting the engine or causing any shift away from the landing path.[45] Another alternative was reversing the propeller. Jenkins' goal was to "stop an airplane traveling one hundred miles an hour within one or two lengths of the machine."[46] Jenkins' takeoff and landing catapults looked something like a large, old-fashioned child's playground slide or a rollercoaster.[47] He envisioned doing away with multiple regional airports developing at the time.[48] The whole mechanism would fit atop a building, was set on a large turntable, and could rotate, allowing the planes to utilize existing wind direction to their advantage. The cost of this space-saving device was "less than $10,000 [$131,233]."[49] According to Jenkins, it would now be possible to establish landing and takeoff sites on commercial buildings, to slow the rising costs of mail, and to get passengers closer to their real destinations.[50] The reverse propeller and the landing apparatus "together . . . remove the problem of expensive and inaccessible airports. . . . Big airports are unnecessary," he said.[51]

An important invention in regard to flying and landing was Jenkins' altimeter, which allowed the pilot to know the height of the plane before touchdown. It helped determine the time and pitch of the propeller blades changing to slow the plane. It also controlled the noise and lessened interference of radio communication with the plane, all in preparation for landing.[52]

Jenkins' final aeronautical invention came during a time of ill health, only two years before his death: it introduced fresh air into the cabin of the plane. The challenge was not only fresh air but the elimination of exhaust fumes from the engine.[53] Jenkins flew from 1910 to the end of his life. He flew in his own planes as well as in others at the invitation of navy and U.S. Post Office officials. His ideas for commercial aircraft were left behind with the growing regional airports and his own increased activity in television. However, flying was good for his health and relaxation—dangerous at times, but as always, his work was central in all that he did, seeking solutions and progress.

Milk Bottles

Ironically, perhaps Jenkins' most profitable invention was the one in which he was least interested—a mere milk bottle.[54] It was a "spiral wound paper container for liquids that [was] used in every grocery, drug, candy, or produce store in the country."[55] Between 1906 and 1913, Jenkins produced almost thirty patents related to this singular endeavor.

Milk had always been a home staple. In Jenkins' day, fresh unpasteurized milk was delivered to doorsteps by a milkman. Milk came directly from the farmer's milk bucket, sold to the milkman, or the farmer himself delivered it. At the home, a hand-held dipper scooped the milk from the can into a pitcher or milk-pan, which the homeowner had placed on the doorstep. The milk was then retrieved and stored in the house icebox or a cellar.[56] Glass bottles replaced the milk scoop, as glass seemed more sanitary. Homemakers placed empty glass milk bottles out on their porches, and those empty bottles were replaced with bottles full of milk. The empty glass containers were retrieved, washed, and refilled for the next morning's delivery.

Reusing glass bottles drew the attention of health-conscious government officials, who were seriously concerned that the washed and refilled bottles were not thoroughly sanitized and were spreading disease. In 1907, the Surgeon General of the U.S. Marine Hospital Service was charged with finding the cause of the spreading typhoid fever.[57] He organized a committee that visited farms and dairies across the country and concluded that "the conditions surrounding the handling and delivery of milk were about as bad as it is possible to imagine." The committee recommended the use of disposable "paper bottles in the delivery of milk," to curtail the spread of this illness.[58]

Jenkins had already patented the answer before the report was filed: a sanitary "cylindrical box of paraffined [waxed] paper, with fixed bottom and slipcover top."[59] This waxed-paper container is comparable to today's ice-cream containers or waxed milk cartons. The new "bottle" had no effect on the milk, and it was sanitary and cheaper to produce than glass bottles.

The Jenkins Paper Milk Bottle improved sanitation and prevented disease. Courtesy Wayne County Historical Museum.

Glass was expensive, at five dollars ($126) per hundred bottles. Jenkins' paper bottles cost five dollars per thousand. Health professionals strongly endorsed use of the new bottle.[60] What began as milk cartons would eventually hold dairy products and seafoods of all kinds.[61]

Jenkins' first "Paper Receptacle" patent was filed in 1905. It was described as a vessel used for the delivery of "small quantities of milk, thus taking the place of the glass milk-bottle... which is repeatedly filled under notoriously unsanitary conditions."[62] He sold the rights to this patent for ten thousand dollars ($251,615) and was swindled out of the payment by someone he called a "crooked" attorney. The lesson, he said, was to "select honest civil attorneys as well as honest patent attorneys. Both are available."[63]

Jenkins forged ahead. Another company was formed to manage coming paper-bottle patents. The Jenkins Paper Milk Bottle Company was organized under the laws of the District of Columbia and Virginia. It appears that Jenkins did not have as much difficulty acquiring investors on this venture as he had with previous business startups. He asked for two thousand dollars ($50,323) to create the first manufacturing machines for a demonstration. However, the company was capitalized with more than was needed as the Columbia Paper Milk-Bottle Company.[64]

Jenkins' waxed-paper cartons resulted in patents for many different kinds and shapes of receptacles and the machines required for manufacturing them. If there was to be a milk carton, then there had to be a means of filling it.[65] Caps, too, were necessary for the older glass bottles as well as the top and bottom of the cartons.[66] Spiral tubes the size of modern ice-cream cartons could also be manufactured in any size for any product.[67] Folding boxes and boxes for the bottles all fulfilled different packaging needs, along with the machinery to manufacture these products. Notably, these patents have been referenced thirty times by modern inventors.[68]

What Jenkins lost in his first milk-bottle patent was retrieved in royalties from new variations of this first product and the manufacturing equipment necessary to produce them. It all revolved around the sanitary paper bottle, the growing number of differing paper products, their specific use, and the machinery to produce them. By 1907, the last Jenkins bottle corporation was called the Liquid Paper Package Company. Jenkins was a member of the board of directors, vice president, and general manager of the new company.[69]

The paper-bottle companies, like other Jenkins operations, evolved into making paper products for different needs: folding boxes for storing and moving merchandise, and direct-mail folders used in advertising and promotions, which closely resembled mass-mailers seen today. In addition, he developed the machinery necessary to fold, glue, and assemble these paper products.⁷⁰ Jenkins would eventually hold more than eighty patents related to this paper "bottle" before he sold his rights.⁷¹

Business and Home Accessories

Following each of the automotive, aeroplane, and milk bottle inventions were patents for accessories and novelty items. Jenkins' manufacturing of automobiles ended, but his involvement in the industry did not. He would adapt his approach to the manufacturing of automotive parts. The Jenkins company provided any automotive accessory needed in the steam or the emerging gasoline engine. The intuitive mechanic could order anything from the Jenkins catalog. A Brooklyn doctor testified that the Jenkins Kerosene Gas-Burning automotive unit was the best. "I used it in my carriage (mobile) in my professional calls and it steams very much better than it did when I used gasoline as a fuel and I now feel at all times a degree of safety that I never felt when using gasoline."⁷² Under the auspices of the Jenkins Automobile Burner, advertised in the *Automotive Review* and the *Horseless Age*, the Jenkins converter drove the car "three times as far on a dollar's worth of fuel."⁷³

Jenkins developed and patented many automotive parts. A small steel storage apparatus provided for safe acetylene storage.⁷⁴ A "Wheel Rim" was related to durable and easily repaired wheels.⁷⁵ Two patents simply titled "Valve" were for starting the internal-combustion engine. Starting one of these early cars was laborious,⁷⁶ so Jenkins invented a starter that could be used from the driver's seat, eliminating the danger and effort of starting it with a hand crank from the front. Jenkins patented a "Repair Device for Pneumatic Tires," a kit for repairing punctures and cuts. In 1918, Jenkins filed for his last automobile-related patent: a flexible transparent sheet utilized as an inexpensive window pane or windshield in the cars. As it was reinforced and lightweight, it was easily adapted to the various models and windowing needs.⁷⁷ All of these automobile products were featured in the Jenkins catalogs distributed to the public.

At the same time that Jenkins was working to improve automotive transportation, he sought to transform the coal-burning home stove into a liquid-fueled appliance. His early steam automotive burners that used oil and kerosene to heat water and turn it into steam were easily converted for household use as cooking stoves and furnaces.⁷⁸ The Jenkins Kerosene-Gas Burner company converted the burners for just about any size "apartments, houses, restaurants, lunchrooms, bakeries, breweries, florists, china kilns . . . yachts, steamboats and a thousand other industries."⁷⁹ The burner used oil instead of coal. Oil was advertised as twenty-five cents to two dollars ($6 to $48) per barrel (fifty gallons), and two barrels of oil were equal to one ton of coal. By 1910, he was still improving the furnace appliances and working on what he called the blast furnace. This innovation increased the furnace's output by "enriching the content of the oxygen" so that the fire burned more easily, therefore increasing the efficiency of the furnace.[80] Today's wood-burning-stove enthusiasts will recognize the value of the damper in the stove pipe. Developed during the period of Northeast regional coal strikes, these Jenkins heating implements controlled the air supply to the fire by opening and closing a valve in the stove pipe. With the fire going full force, closing the damper cycled oxygen in the stove for a more efficient burning of the fuel.[81]

From 1895 to 1910, coal strikes troubled the national landscape. Coal was the fuel of the era for home, business, and industry. Even after the discovery of oil, the railroads still thrived on coal. So Jenkins put his mind to work on the challenge of transporting coal. His idea was a gravity-railway device that would ease the process for the railroads. Jenkins adapted roller-coaster rail-gravity designs by utilizing the electrical power generated by the fully loaded cars moving downhill to supply energy to return empty cars back to the higher elevations.[82] He used a storage battery to save the excess power created going downhill and redirected it back into the power needed for the uphill grades. But the device suffered power loss at various stages; it simply stored insufficient energy to move the train on longer runs, and additional power sources were necessary.[83] Jenkins' gravity device has nevertheless been referenced in ten patents between 1990 and 2007, so apparently the idea had merit.

Novelty Inventions

It is easy to see Jenkins' motivations in automobiles and home-improvement products as following the national trends of innovation in transportation, industry, and oil production. He changed as needs changed. The novelty items in his portfolio seem to reflect the inventor's mind, fulfilling smaller, more immediate, and sometimes humorous needs. At the same time that he was working on cars and home heating, he developed a mathematical pocket calculator and instructions for multiplying on one's fingers, a Christmas-tree holder, and talking signs. The Jenkins pocket calculator sold for thirty-five cents ($9) and could be ordered directly from his lab or the Scranton Novelty Company of Scranton, Pennsylvania.[84] His Christmas-tree holder required no nails to affix the base of the tree to a holder, and "after the holiday . . . made a splendid workbasket, Jardiniere holder, or stool."[85] His talking signs were created with Edward Clifton Thomas; they provided illuminated lettering for advertising.[86] Akin to today's news-ticker building signs, "the strange changing of the sign [was] of almost irresistible attraction to the average observer."[87] The "Jenkins-Thomas Changing Signs" were marketed to merchants at a "fraction of [the cost of] the good electric sign, and this one 'talks.'"[88]

Other diverse projects developed during 1908 and 1909 included keys and a needle sharpener. The key made it possible for a single key to unlock a series of locks, provided they were approached in a set order. This patent was referenced eight times in patents from 1979 to 2004.[89] The needle sharpener was created to sharpen needles, razors, and a chisel with the aid of electrolytic action.[90] In 1920, Jenkins created a power lawn mower. The first mower engines mirrored Jenkins' interests in the small engines used in electric cars. His patent for his version of the machine was referenced twelve times between July 1996 and July 2008.

Several of Jenkins' novelty items reflect wartime issues: an electrical meter, depth meter, and gaiters. His electrical meter was used to measure the strength of an electrical current by the heat created as it passed down the wire.[91] What Jenkins specifically intended for the utilization of his meter is not apparent, likely due to wartime secrecy.[92] His depth meter was for navigating ships in shallow water. Jenkins utilized sound waves bounding off icebergs and objects in the water to measure depth, similar to modern sonar devices. The gaiter is akin to a shoe covering.[93] Horseback riders utilize gaiters to protect their legs. During World War I, gaiters were worn by soldiers to protect their ankles. Jenkins' gaiter was also worn for protection. Drawings indicate a unique design to also protect the arch of the foot. How Jenkins intended its use remains a mystery. However, the patent was cited once in 1993 by someone developing a high-top insert.

Another interesting novelty, which he called a "plaything . . . a mere toy," sat on Jenkins' desk in the laboratory. It was a static motor. This small motor, developed in 1928, took electricity, apparently from the air, and made fan blades turn. There were no batteries or external power sources, but when Jenkins hooked it up to a radio antenna, the blades of the motor would spin. "It slow[ed] down in pleasant weather," Jenkins commented, "but it runs like the dickens at other times."[94] *Radio News* described the motor as "resembling a spider web" that connected to the antenna.[95] Despite its being described as a toy, it attracted significant press attention as a potential independent power source—"just as a wind wheel [took] power out of moving air," like the windmill.[96] Jenkins did little to develop the motor, as his focus by 1928 was totally on television, but he did feel that there was potential behind the idea and the mystery of how it worked.[97]

The last of the patents within this eclectic grouping was a "diathermy contact pad," filed in 1933. Jenkins was ill at this time, and there would seem to be little doubt that this was meant to relieve muscle pain. In the medical practices of his day, diathermy was the practice of utilizing heat, produced by alternating electricity, introduced into the body below the skin for therapeutic purposes. The pads were localized for specific parts of the body and enhanced recovery.[98] His contact pads replaced lead plates that were painful. They created a snug fit to the body's form and "eliminated the danger of sparking between the contact pads and the flesh of the patient."[99] This invention was likely related to Jenkins' own heath issues.

The milk-bottle industry, like the automotive and photographic industries before them, started with a perceived need, and for the inventor this was always the challenge. In each case, Jenkins sought and found solutions, providing machinery, parts, information, and ideas. He was still often competing with much larger corporations. Success was feast or famine, and he was working too hard. His health was troubling him. Though he would not abandon his efforts, he was forced to consider his health and the toll exacted by his demanding work habits. His doctor had told him that without rest, his life would end prematurely. A restful vacation was again his part of his answer to the doctor's orders to relax and slow down. He would take an ocean-to-ocean trip.

Healthy Working Vacation

Little is known about the earliest of Jenkins' health issues, only that by 1910 his doctor had started telling him that he would not live another six months if he didn't slow down and rest.[100] One of his first ideas for "slowing down" was to sign on as a photographer for an "Ocean to Ocean" automobile tour in 1911,[101] a nationwide publicity stunt to sell the Premier automobile.[102] Jenkins had obvious interests in the tour—he was an automobile enthusiast, the trip would take him through his hometown of Richmond, Indiana, he loved the West, and it was supposedly a vacation at the doctor's order. Jenkins made it into something more as one of two photographers commissioned to document the trip. He "equipped his personal car with one of his 'movies' [camera] and made the entire journey."[103]

The 1911 tour started in Atlantic City on June 26 with thirty-six people packed and ready in twelve Premier cars, which ceremoniously dipped their rear tires in the Atlantic Ocean before heading off for the Pacific.[104] The trip was intended for forty-five days—twenty-nine on the road—with a planned finish in Los Angeles on August 13. New experiences would daily astonish

A crowd gathers in Los Angeles as the cross-country travelers "dip their wheels" into the Pacific Ocean, concluding their journey. Courtesy Virginia Roach Family.

Jenkins readies his camera along the journey. He was one of two photographers documenting the trek. Courtesy Virginia Roach Family.

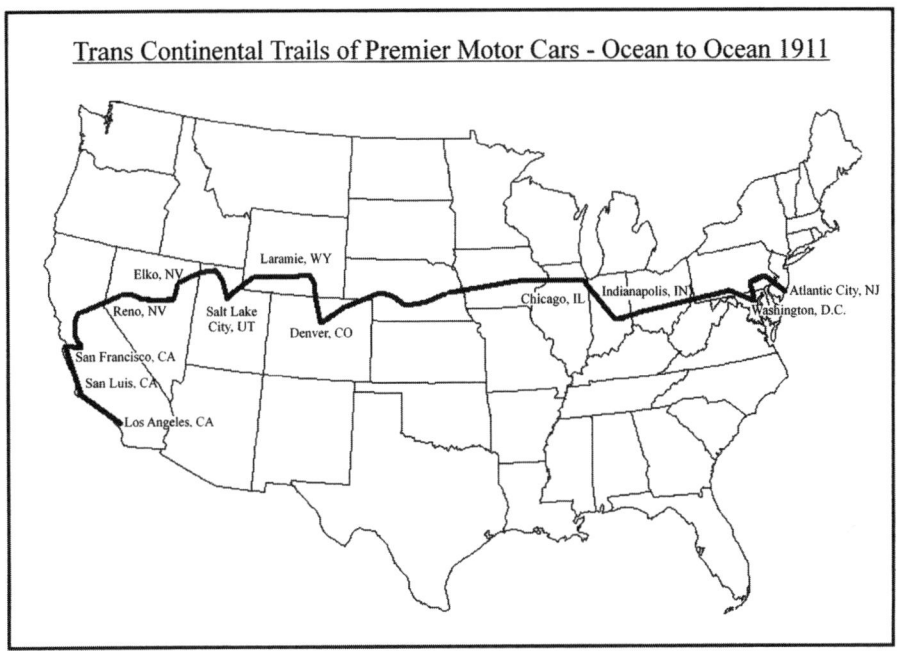

The Ocean to Ocean Tour was chronicled by Grace Jenkins in her diary, beginning June 29, 1911, through August 14, 1911. It was a rugged and picturesque journey. Courtesy Knight Graphics.

the touring party, and especially Grace Jenkins, who had never been west. According to *The Automobile,* "gasoline was as common as sage brush ... the only exception being on the run from Ogden, Utah, to Montello, Nevada, a distance of approximately 150 miles over the roughest going on the transcontinental tour."[105] The price of gasoline encountered on the journey would range from eight to fifty cents ($1.94 to $12) per gallon, and the vehicles averaged 12.5 miles per gallon.[106] The thirty-six passengers were to be hosted by municipal dignitaries along the route and scheduled to stay in many of the finest hotels. They carried camping equipment: ten seven-by-seven-foot tents, blankets, and portable toilets, ovens, and cooking burners. It must have been an interesting sight as the vehicles caravanned slowly from one city to the next along gravelly, muddy, and, in places, almost nonexistent hillside roads. Most of the vehicles were canopy-covered, four-passenger touring cars. The canopies could be pulled back on sunny days. One car looked more like a covered wagon than the modern vehicle of the day.

The tour was an adventure: it was dangerous, but it was also sometimes restful for Jenkins. Grace and Francis Jenkins missed the opening ceremonies in Atlantic City, driving from Washington, D.C. Out of Omaha they traveled along the railroad line, stopping at the "halfway post—1,733 miles from Boston and 1,733 miles to San Francisco."[107] Grace exclaimed, "Do you mean to say that we are only half way across?" Jenkins responded that it did not matter, as the journey "had become an enjoyable vacation."[108]

The group experienced treacherous roads, fine hotels, and camping. Headed north outside of Denver, Grace reported "heavy rain." They stopped at the Forks Hotel, and the "gents slept in the hay barn, ladies in the house."[109] The tour was apparently caught in a summer cloudburst that left the road "slipperier and stickier . . . than gumbo." Forks, Colorado, was the "stage station on the route to Estes Park, Colorado."[110]

The tour crossed into Utah just west of Evanston, Wyoming, from the top of the Wasatch mountain range. The elevation dropped from 8,000 feet to 4,200 feet into the Great Salt Lake Basin—a drop of 3,800 feet over the final miles into Salt Lake City.[111] They rolled over the old Mormon trail, arriving just before sunset, and were put up at the newly constructed Hotel Utah.[112] They found excellent highways until in northern Utah they turned westward toward the Promontory Summit, where in 1869 the Union Pacific and Central Pacific railroads joined, becoming the first transcontinental railway. In Elko, Grace chastised several travelers for delaying their early morning departure: "four of our boys got in trouble gambling—could not get away from there until 11 o'clock."[113]

The journey officially ended on Sunday, August 13, 1911, when all the vehicles drove to the Pacific Ocean, "where wheels were dipped" into the water. The tour concluded with a "glorious cheer" from the thousands who had gathered at the beach to greet them. Lunch was provided, and Grace was able to attend the Baptist church's Auditorium Theater.[114] Interestingly, no one mentioned how the group returned to Washington—most likely by train.

They had been on the road over a month. Hardly a restful vacation, but Jenkins enjoyed himself. He would later describe the journey pleasantly, with only a little sarcasm aimed at those who tried to take advantage of the travelers. "I have toured largely over the United States from ocean to ocean, but the pretty villages whose signs spoke to me so kindly like in my memory as cases of good-fellowship in a desert of graft on the traveler." It was a tongue-in-cheek description of times and the signs along the route as they related to the character as well as kindness of people along the way: "speed limit six miles an hour or $50 fine for driving faster than a walk through town."[115]

7

RadioVision:
The Genesis and Promotion

RadioVision, as Jenkins initially perceived it, would be as an instrument of industry and government communication through the wireless transmission of photographs and text messages. In basic form, this was an idea dating back to the turn of the century, but with primarily wire transmissions.[1] Jenkins centered his efforts on "the development of radio-transmitted pictures, an address to the eye; while others have been developing radio-transmitted speech, addressed to the ear."[2] His RadioVision evolved into a system for transmitting still pictures, maps, messages, and eventually into an early form of televised motion entertainment.

This effort materialized during the Roaring Twenties when a multitude of new devices were dramatically revolutionizing communications. This was a time and a marketplace "already glutted with more miraculous gadgets . . . than the public had ever been confronted with at once."[3] These gadgets surrounded radio, the phonograph, photography, and early television. Today, that list sounds limited, but the roots of modern electronic telecommunications can be traced to this era when wireless communication was largely theory—and tests of theory.

The idea for transmitting pictures by electrical impulse was not original to Jenkins. He joined in a global flurry of scientific innovation that began decades earlier, during the Industrial Era. In 1865, James Clerk Maxwell, a Scot, put forth his first theories of electromagnetic energy.[4] In the 1880s, Heinrich Hertz, a German academic physicist, put those theories into practice and discovered that electromagnetic energy could indeed be projected through the air. These "Hertzian waves . . . became the subject of laboratory experiments in all lands."⁵ In 1884, the German-born Paul Nipkow gave birth to the idea of using flat, spinning disks to scan visuals. A light source behind the rotating disk produced lines of a picture as it turned. This was the mechanical concept that Jenkins would use in his earliest television patents until the close of the 1920s, when he predicted the end of the disk.

Most inventors were designing systems that sent "pictures over wires as a unit."⁶ Noah Amstutz brought the elements of visual-facsimile technology together in Cleveland, Ohio, in 1891, though he never marketed the device commercially. The German physicist, mathematician, and inventor Arthur Korn is credited with "the first photo transmissions system" in the late 1890s, and he demonstrated it in 1902, using a light-sensitive scanning device.⁷ The American Nathan B. Stubblefield began working with wireless voice transmissions.⁸ In 1899, Lee de Forest, whose company would much later purchase the Jenkins Television Corporation, had just earned a Ph.D. from Yale and was entering the wireless business. He later proclaimed himself the "father of radio," the controversial claim based on his "audion" tube that facilitated voice transmissions.⁹ The Italian Guglielmo Marconi worked in conjunction with the Royal Navy on wireless ship-to-shore communication, a foundation of what would later become the Marconi Companies of Britain, Canada, Italy, and the United States.¹⁰ In 1899, using Samuel Morse's code, Marconi successfully transmitted signals across the English Channel and by 1901 sent the

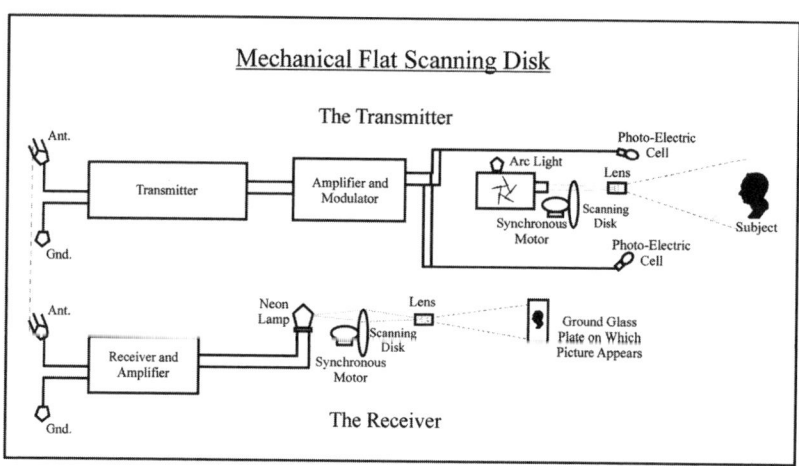

The Nipkow mechanical television, with scanning lines produced by rotating a flat metal disk with holes through which light was passed to create lines and pictures. The lines were then amplified and transmitted. The process was reversed on the receiving end, where the picture appeared on a ground-glass plate. Courtesy Knight Graphics.

letter "S" across the Atlantic between Newfoundland and England.[11] In 1900, the U.S. Weather Bureau hired the Canadian Reginald A. Fessenden, a rival to de Forest in wireless transmission,[12] to conduct "experiments in wireless telegraphy as an aid in weather forecasting and in storm warnings."[13] In 1907, Edouard Belin of France used a stylus to scan still pictures. The Belin system was a wet-chemical photographic-development process.[14] In that same year, Korn sent scanned pictures from Paris to London and three years later linked these cities with Berlin.

Jenkins filed his first patent in facsimile technology in 1908. It was a machine for copying handwriting, known as the art of telautographs.[15] Basically, the "tel-autograph" was used like the facsimile for reproducing graphic material transmitted with the electromagnetic signal controlling the movements of the pencil. However, Korn's telecopying methods dominated the experimentation until World War I, when the U.S. navy moved the wireless scene in earnest. With the American entry into the war, the navy was given control over all wireless telecommunication in the country. It remained under their control until 1920, when patent rights were returned to their prewar owners.[16]

The military, scientists, academics, and inventors from the 1800s into the 1920 were opening the doors for new devices and creating the foundations for Jenkins RadioVision and modern television. By the 1920s, wireless had become "one of the most interesting subjects before the scientific community,"[17] and Jenkins RadioVision would be in the forefront.

Telecommunication Pioneering

Much of television's earliest technological history relates more to what is now called "facsimile" (fax) than radio and television, but the roots are intermingled.[18] Jenkins published his first telecommunication designs in an 1894 issue of *Electrical Engineer*.[19] It was a crude scheme for transmitting pictures. The technical historian Albert Abramson described it not as a television system but as a device for transmitting images: "[I]t was merely a restatement of principles," Abramson said, as opposed to an original design. Nevertheless, while Jenkins continued his work on developing the camera and projector, his interest in television was taking root.[20] In 1908, his telautography apparatus for replicating handwriting was similar to his competitor's, consisting of a "transmitting station and a reproducing station." The transmitting station was not a "station" in today's broadcast sense of the word; it was instead a modest device for "introducing electricity into pairs of wires . . . to actuate the writing arms of the reproducing station" at the other end of the transmission.[21] Even at this early stage, Jenkins envisioned the telautograph's use for billboard displays and in newspapers. In 1913, he proposed the transmission of movies by wire. This was different from most other experimenters, who were working largely with still pictures and messaging machines. While Jenkins later indicated that these ideas were "impractical," they nevertheless "provoke[d] discussion" and initiated progress.[22]

The two primary pre–World War I inventors working toward a system of television were Korn and Jenkins, in Europe and the United States, respectively. In 1913, Jenkins wrote for the *Moving Picture News* describing RadioVision, but the article was really more about developing a favorable public opinion for his futuristic ideas, which included live theater-stage images received in the home. The publisher wrote of the "wonderful possibilities of motion picture progress" and thanked the enterprise of William Randolph Hearst, who had sent photographs by wire from New York to London.[23] It would be another decade before Jenkins would begin practical experimentation. His articles mark the theoretical beginning of his quest for pictorial wireless transmission and television. Large-screen television systems seemed a natural next step, paralleling his work in early motion pictures. "It seems not impossible," he said, "[that] one might with a reasonable degree of confidence expect to transmit *motion* pictures by wire or even perhaps wireless."[24] The difference between the transmission of still and moving images was only that in motion the pictures had to be created and projected more rapidly in electronic form. The transition to motion would follow still transmissions as his next challenge.[25]

In May 1920, Jenkins' presentation before the newly organized Society of Motion Picture Engineers described an important new element. It was the "first crude description of the ring prism device" that would be used in his wireless photography and the earliest motion pictures to optically produce thin scanning lines, "each about one five-hundredth of an inch thick."[26] The rings were rotating glass prisms of variable depth; they hung at an angle and manipulated the intensity of a beam of light, producing what was like "a small pencil of light coming from a fixed source . . . the lights and shadows [made] up the picture, line by line."[27] The variegating shades were then converted into corresponding variable electrical values and transmitted. At the receiving end, the process was reversed.[28] This produced a visual facsimile or faxed, wireless photograph. Jenkins would eventually shift from using the prismatic rings to other optical concepts, as his thinking was decades ahead of mainstream technology in optics and projection. He described the idea as like a child taking a nickel under a sheet of paper and "draw[ing] *straight* lines across it with a dull pencil, [until] a picture of the Indian appears."[29]

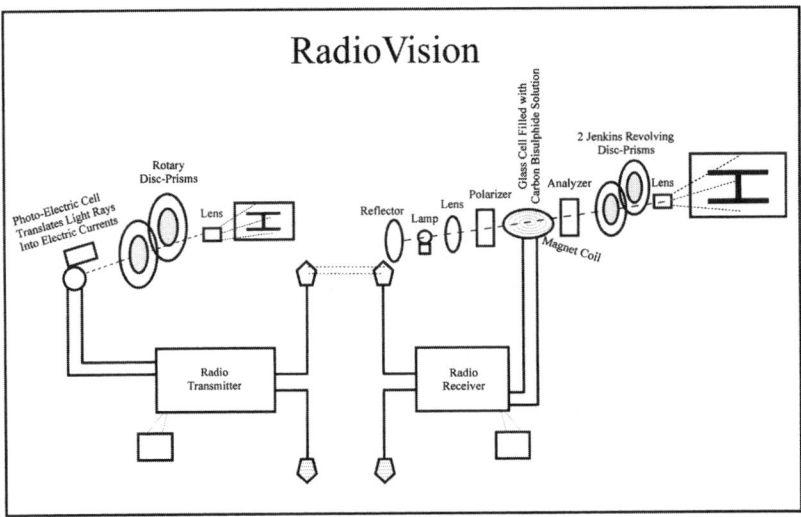

RadioVision used several designs of prismatic rings, prisms through which light was passed and then transmitted to a receiver, where pictures were rendered. Courtesy Knight Graphics.

Instead of a pencil, in Jenkins' mechanism light is drawn across the picture and transposed into optical lines and electromagnetic signals that could be transmitted and reproduced anywhere. Reading Jenkins' descriptions of this test brings to mind the children's Etch-a-Sketch toy, electronically advanced to transmit the single line as it were "sliced" over the toy's small screen. Jenkins called the process of his sending machines "zinc etching."[30] This scanning principle resembles Farnsworth's later scanning ideas, envisioned while plowing the field and gazing at the lines the plow created. Both inventors saw line-scanning visions, which would evolve from mechanics, to optics, and followed by electronic lines.[31]

Scanning lines to draw a picture are illustrated in this simple child's pencil drawing, as they create an image of the nickel by moving the pencil back and forth over the coin. Jenkins used the mechanical disks, later prismatic rings, and a drum scanner for the purpose of drawing television lines. Courtesy Knight Graphics.

Future Vision

Up to World War I, Marconi's wireless and the use of Morse code dominated radio transmissions. In Europe, Korn's methods of picture transmission were leading in still-pictorial relays.[32] However, postwar advances spurred many to work in wire and wireless communication. Jenkins and other inventors broadened the uses of wireless beyond defense to include applications in government and such industries as mail delivery, newspapers, and commercial AM-radio broadcasting.[33] The press and consumers followed the techonological progress with the enthusiasm of the Roaring Twenties.

The press was always interested in speculation about emerging technological trends. They had been reading about the technology over the years, and the new radio industry was growing rapidly. Predictions often sounded like science fiction. This was as Jenkins would have wanted it, because the radio and wireless-messaging fields were crowded with competitive inventors. In comparison, the transmission of still pictures or motion pictures was unique to Jenkins, as he enjoyed the competitive advantages gained from his years in motion-picture technology.

Jenkins acknowledged that he was among "a veritable army of engineers" working in the development of wireless radio, but he was moving in a different direction than the mainstream. Other inventors saw radio "as a service to the *ear*."[34] Jenkins described his own efforts as rather lonely, because he was "devoting his efforts in the development of radio" for the human *eye*. He stressed the "undreamed-of possibilities in radio in the unlimited frequencies above audibility . . . [in] the workable range where light instead of sound is employed" to send messages.[35] The messages sent were photographic works dispatched to the home firesides "from distant world points . . . a daily source of news . . . [an] instructional class . . . the evening's entertainment . . . [giving life to] shut-ins . . . and in far places less lonely."[36] The flair of Jenkins' vision was exciting, and he took advantage of every means to communicate publicly about his work.

In industry, education, and consumer markets, there were real possibilities for wireless photos. The concept of wireless textual transmissions had already excited the growing newspaper business.[37] Jenkins' photo-transmission systems were of interest because they were easily transmitted and remarkably clearer than those of his competitors. This wireless photo technology facilitated information processing with the rapid exchange and distribution of photos.[38] Jenkins promoted the use of the wirephoto, which he called "radio photograms," for a newspaper's visual representation of an event, whether it be "a mine explosion, train wreck . . . a street riot. . . . or a [sports] scrimmage

... in a ball game."[39] The photograph was "the most rapid means of copying anything," and when the photo was sent via radio, the combination "simply hatched the two together for a better means of communication."[40] Secretary of Commerce Herbert Hoover congratulated Jenkins on adding this "electrical channel of intercourse [to the telegraph and telephone], a third means, the radio photo letter."[41]

Jenkins proposed an interesting application of RadioVision for the U.S Postal Service. He advanced photo and message transmissions as practical for "mailing" messages "by radio instead of by steamship." It was possible, he said, for the post office to deliver photos and business letters "at the speed of light" with his new RadioVision. Such an exchange would speed up commercial enterprise because "industry can go no faster than its means of communication."[42] Addressing a large gathering of students and officials at Washington's Central High School, he gave another simplified example of how the RadioVision scanning operated. Suppose, he suggested, that an artist creates a picture made up of colorful rocks laid in rows. The artist creating the mosaic begins at the upper left-hand corner and works left to right, laying the rocks in descending lines to complete the scene. Jenkins proposed that the picture could then be taken apart, "one little stone at a time," and mailed via the post office to Boston for reassembly, or using RadioVision, "it could take the picture apart one line at a time ... [and] sen[d] it immediately to Boston where the picture would be electronically put back together again one line at a time."[43] Pushing his sales point, Jenkins hinted that in education, "instead of the country student struggling along with incompetent teachers ... [the student could] have the inspiration gained from listening to a lecture given in person by the greatest authorities in the world."[44]

Every time Jenkins spoke, he touted electronic motion pictures as soon becoming the norm. He described the future as not limited by wires or point-to-point transmissions but rather a time when every city and business would be connected across the nation. "Present metallic channels now employed for ... power lines, railroads, city lighting ... can be made a new source of revenue at a ridiculously insignificant cost." Jenkins declared that neglecting these other intercity links, "which lead into every place of business, and into every home, is an unnecessary waste."[45] Indeed, Jenkins' concept of *seeing* as well as *hearing* with radio technology helped usher in a new era of development. We see much of his vision in the internet and social media.

By 1922, Jenkins was in the forefront of the scientific community with his RadioVision, devoting his primary efforts to the transmission of pictures by radio, with radio motion pictures clearly on the horizon.[46] He pushed his advantages of leadership and enthusiasm at every opportunity. After an interview with Jenkins, the *Boston Post* reporter John Brady wrote that the image was "[s]tartling and reminiscent of the fanciful tales of Aladdin's lamp and the 'magic carpet'"; everything seemed reflective of what would soon be in the homes across America.[47] In an ironic twist of history, Jenkins' work was even used in the *Edison Sale Builder,* a publication for Edison company representatives. The article said nothing of Jenkins (an archrival in photograph technology), but it bragged about the fact that he was using Edison lamps for his successful RadioVison transmissions. Edison saw Jenkins' use of his lamps as a stamp of approval endorsing their own products. As the magazine reported, the possibilities for Jenkins' work "almost stagger the imagination."[48]

Not everyone was so optimistic. American Telephone and Telegraph, which would participate with Jenkins in his fax experiments, called the possibility of live wireless telecasts for "action pictures of ball games, riots, prize fights, parades, etc. . . . almost negligible."[49] Yet at the time Jenkins was already successful with still photographs and had transmitted simple motion films in his lab.

Jenkins' technical skills, working with the complexity of multiple photographic machines over the years, moved his vision for the future easily into electronic wireless technology. He was helping establish the foundations for combining film and wireless in what would be America's first motion-picture radio transmissions, but the still photos came first. His earliest experiments, as described by the *New York Times* radio editor Orrin Dunlap, were like "seeing around the world by radio."[50] They were wireless relays of single photographs, letters, and written messages—all with the promise of the coming motion-pictures transmissions. "In due time," Jenkins said, "folks in California and in Maine, and all those in between may see . . . inaugural ceremonies of their president . . . football games, baseball, polo, regattas, flower festivals, and baby parades."[51]

Demonstrations were progressing toward the televising of motion pictures, which Jenkins predicted were a little over a year away.[52] Arthur C. Hardy of the Massachusetts Institute of Technology credited the use of radio movies in private homes as due "largely to the efforts of C. Francis Jenkins."[53] Even though inventors would later change scanning systems from the disk to electronics, he had created the foundations from which the future evolved.

Unofficial First Demonstrations

Jenkins' first demonstrations on any new idea were conducted in the lab. Before he involved the press, Jenkins wanted a working demonstration. He referred to these tests as "unofficial." His wireless RadioVision experiments were initially conducted under the auspices of the Jenkins Laboratories, which had been incorporated in 1921. The purpose of the lab was to create the devices and experiment with them. He found a new building, at 1519 Connecticut Avenue, and hired the staff, which ranged form three to six people over the years. His first photographic wireless transmissions were sent between rooms in the lab. Continuing patent applications and experiments perfected improvements in "speed . . . shades and half tones . . . light intensity [and] picture clarity."[54] In May 1922, transmissions were dispatched the short distances between the lab to his home, where he had a small personal lab in his basement. He described it as the "first radio photograph."[55]

To understand the lab experiments of these days is to realize that there were no readily available parts. If a part or tool was needed, it was often made by the lab staff—discs and prisms cut and ground by hand, tubes blown. Jenkins had two hundred applicants for the job of cutting the prisms, and he hired a dressmaker because of her "fine touch, an expert touch."[56] The lab staff pioneered all of the original apparatus. He obviously hoped that whatever was created would be eventually placed into manufacturing and general use. Hence, numerous patents were established around the invented products.

In the early 1920s, Jenkins had no real competition. John Logie Baird in England was the only other engineer actively working on television. Vladimir Zworykin was working for Westinghouse, where the focus was on radio, and Philo Farnsworth was a teenager in Idaho. Arguably, neither Zworykin nor Farnsworth entered the television realm seriously until the late 1920s.[57] In contrast, Jenkins took out his first television patent and began demonstrating his work to the press in 1923 and 1924. So Jenkins was ahead of all three. Because Baird was carrying on his work in London, Jenkins took little note and made no mention of him in his own work; Baird's biographers mention Jenkins only as one of two contemporary competitors working on the problems of scanning.[58] These researchers were likely all aware of one another, as their work was covered in the trade press, and they quietly learned from each other, but it would be almost a decade before Zworykin and Farnsworth would garner significant influence—and the Jenkins Laboratories was already in full swing.

By 1922, Jenkins was actively promoting his wireless picture transmission. His RadioVision was potentially profitable, and electronic motion pictures in every home held even more promise. In August, he filed for four patents simultaneously, each of which he considered "[n]ew and useful improvements . . . for broadcasting of photographs and motion pictures by radio."[59] As he achieved growing success in the transmission of still photos, he was anxiously anticipating doing the same with motion. He foresaw motion pictures following the pattern of commercial radio, "transmitted by radio from a *central broadcasting station* into the homes of people. . . . [P]ictures of people or happenings on the other side of the world may be transmitted instantly into newspaper offices . . . and even notorious criminals . . . [may be] stood up in one central station and their likeness shown to the police in every city."[60] This conception marks an important shift for Jenkins. He was actively thinking about general broadcasts with receiving sets in each home and business. The central station, using RadioVision, would serve a mass audience and thus generate potential sales of the technology. While WEAF-AM had experimented with their first radio commercial on August 28, 1922, the focus for producing a profit during this period of television history was still in selling technology.[61] The selling of commercial time was yet to evolve, and the early experimental television stations were, in fact, forbidden to advertise.

Jenkins' most significant tools in the early 1920s RadioVision were his prismatic rings. He patented three prism-ring devices on the same day in 1921, improving scanning quality and RadioVision transmissions. These rings separated and directed refracted lines of light to create scanning. It was a significant improvement over the spinning Nipkow disk with holes though which light was merely passed. Jenkins' first was a circular sleeve-styled prism that sharpened the picture produced by either the camera or projector and was built upon the principles of his motion-picture machine.[62] The second was a lens that bent the light to be transmitted in the still or the motion picture.[63] The third was the "high-speed motion picture machine."[64] This used two overlapping parallel prisms spun in opposite directions, producing a still picture. The prismatic rings were optical, using a variable-depth glass disk rotating in front of a light source, converting light into electromagnetic signals to create scanning lines, which could then be transmitted to reproduce pictures. The high-speed camera provides slow motion for industrial and commercial use.

Jenkins' longtime association with the Life Saving Service networked him into military circles of influence and provided the platform for his study of RadioVision. The navy was interested in RadioVision for defense purposes: ship-to-shore messaging, weather reports, military communication, and various radio systems. With RadioVision, more than just Morse code and sound could be broadcast. Jenkins would conduct several transmission experiments with navy support. On October 3, 1922, he utilized AT&T's phone lines to send photographs from his office telephone to the U.S. navy radio station, NOF, at the Anacostia, naval base in D.C.[65] The base then relayed the photos by wireless, broadcasting them to the Navy Research Laboratories in northwestern

Washington, D.C.—5502 Sixteenth Street NW.[66] The experimenters monitored the pictorial quality and scanning-line definition at each stage of the relay, and when they were broadcast, "Jenkins' home . . . radiovisor picked the signals out of the air and changed them back into the original light values, which in turn were recorded photographically."[67] The *Rochester (N.Y.) Post Express* reported the success of the radio pictures, but the story was more about motion pictures: "the day of the [motion picture] film exchange is doomed . . . theaters will do away with projection machines and operators and use nothing but the radio receiving set."[68] The forecast was decades early, and the *Motion Picture News* would call the prediction "somewhat far-fetched."[69] The established silent-motion-picture industry would indeed survive, grow, and adjust rather than decline. But what was evident was that Jenkins' broadcast of still photographs was reflecting the road ahead as well as promoting a new industry in the relay of still photographs and messages in newspapers, industry, government, law enforcement, and entertainment.[70]

Jenkins' scrapbooks are full of pictures utilized in his wireless pictorial experiments. Portraits of people were relayed, dated, documented, and then sent to the individuals. In 1922–23, more than eighty different still photographs were used. These included individual character portraits, Washington, D.C. scenes, street maps, signs, automobiles, homes, a piano player, phone operators, Native Americans, and messages written in Japanese, Chinese, and English.[71] A photo of President Warren G. Harding was used, and a copy was sent to the president, who thanked Jenkins for the "handsomely mounted copy which will be preserved as a much prized souvenir."[72] Charles F. Marvin, head of the Weather Bureau, requested Jenkins' assistance in transmitting weather maps.[73] Different written messages, photos, and images were sent, each testing picture quality, resolution, and general readability. In the fall of 1923, transmission experiments were conducted relative to color, light-definition variances, and transmissions from paper drawings to paper prints. The tools remained the prismatic rings, while Jenkins experimented with different subject matter and picture definition. Unlike other inventors, Jenkins' wireless system replaced the need for photo processing.[74]

Ever the promoter, Jenkins used portraits of famous people of the time. Elihu Thomson, a celebrated scientist at General Electric, praised Jenkins on his prismatic rings as "the solution of a problem which [he] had often thought as impossible."[75] George A. Hoadley of the Franklin Institute complimented Jenkins, praising his "tremendous energy and [the] persistence that you have put into the development of this new art."[76] Herbert Hoover, the secretary of commerce and the governmental authority regulating radio at the time, wrote in appreciation of the photograph received, "it represents a very startling development.... I would be interested in discussing the method with you."[77] Hoover's photograph was scanned at one hundred lines per inch.[78] Williams Jennings Bryan, an influential politician and orator, thanked Jenkins for the photograph.[79] Bryan's comment reflects the exuberance of the 1920s: "[I]t is wonderful: What is there left to be discovered?"[80]

At the genesis of RadioVision, during the fall of 1923, the beginning of testing it with motion pictures, Jenkins took an unusual opportunity to relax. It was "a great treat," a brief respite from his hectic schedule.[81] Through his connections with the navy, he was invited to witness "the experimental sinking of a battleship by airplane bombs" on September 5, 1923. Two obsolete battleships were used as targets and film study. The sinking was conducted as an army and navy bombing test.[82] Jenkins was invited as a celebrity viewer, and he enjoyed it thoroughly. General John J. Pershing, Major General M. M. Patrick, and Rear Admiral W. A. Moffett were watching.[83] The event had been filmed and used as stock footage by the military for training as well as in Hollywood war pictures.[84] After the sinking of the ship, Jenkins' friend Preston Bassett, the chief engineer of the Sperry Gyroscope Company, took him to the wheelhouse of the ship in which they were sailing to show him the "iron mike [a Bassett invention] . . . the gyro-controlled automatic steersman which holds the ship on her course from port to port."[85] Jenkins and Bassett had common interests in naval navigation, aviation, night flying, and weather instruments. Bassett held patents for high-intensity carbon-arc lamps used in motion-picture projection, one of which was filed two months before the sinking of the ships.[86] So, Jenkins was among friends. The episode was a respite from his usual work, but it ended as always with Jenkins back to business.

Jenkins would file several patents to improve RadioVision over the next several years, perfecting the still-photo transmission systems. His next move was to take the equipment, demonstrate it wherever possible, generate publicity, and start selling the apparatus for a profit. He would organize two corporations to promote and sell the industrial applications and apparatus, the Discrola and Radio Pictures Corporation.

8

Radio Pictures: Going Operational

Jenkins' private demonstrations of RadioVision and the transmitted photographs of public figures produced volumes of publicity. By 1922, even *Scientific American* was on board, foreseeing a distinguished future: "it is obvious that we already have broadcasting concerts and opera [on radio], there is no reason why we should not . . . broadcast an entire theatrical or operatic performance." While the publicity produced anticipation for home theaters, available technology was not yet close to approaching that goal.[1]

The earliest "official demonstration," as Jenkins called it, took place on December 12, 1922.[2] He transmitted still pictures between his lab and the Anacostia naval radio station NOF in Washington, D.C.[3] The demostration was attended by leaders of the military and the motion-picture industry. Military representatives included Admiral S. S. Robinson from the Naval Board; Admiral Henry R. Ziegemeier, Bureau of Communications; Captain J. T. Tompkins and Commander Stanford C. Hooper, Naval Radio Section; Lieutenant Commander A. Hoyt Taylor of the NOF radio station; D. C. Edgerton, supervisor of the post office's radio activities; and John M. Joy, representing William H. Hays, who had most recently been President Warren G. Harding's postmaster general and was now reassigned by the president to "clean up Hollywood." He created the "Hays Code," or what today is known as the Motion Picture Production Code, establishing standards and a code of conduct for filmmakers. Joy and Hays both would have known of Jenkins' motion-picture inventions, as they were active members of the Society of Motion Picture Engineers.[4] As the dignitaries watched, a photograph of Edwin Denby, the new secretary of the navy, was transmitted.[5] The photo was projected "across a photoelectric cell . . . [with] only a thin 'slice' of the

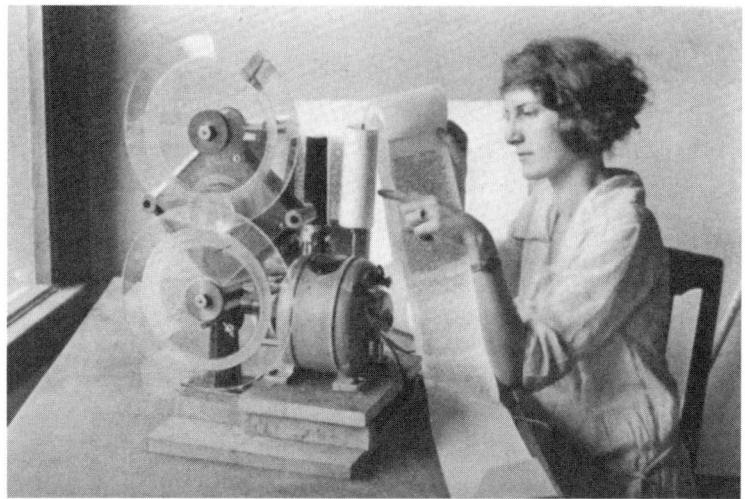

Jenkins' prismatic ring was a spinning prism though which light passed to create scanning lines. The young model here reads the manuscript being transmitted. Courtesy Wayne County Historical Museum.

image drawn across the cell" by the prismatic rings, and then transmitted. At the receiving end, "the rings automatically reversed the 'sweeping' process" to reproduce the picture line by line. It took six minutes to transmit it from the lab to the military base.

Jenkins took advantage of his military audience to point out that secrecy was a bonus in his system, as "maps, coded messages and similar secret documents may be transmitted in such a way that no one, but the person for whom they are intended may receive them."[6] The process was so simple and inexpensive, it "may be attached to the ordinary sending and receiving radio outfits already in use at a cost of $50 to $150 [$677 to $2,032]."[7] The navy experiments were more than a technological success—they were also a promotional success. The photographs reproduced so well, surprising the authorities, that the "Navy official continued their generous aid in further experiments."[8]

On March 3, 1923, under the auspices of the North American Newspaper Alliance, radio photos were publically demonstrated, this time transmitting over a distance of 130 miles from NOF in Anacostia to the *Philadelphia Evening Bulletin*'s newspaper building.[9] The *Bulletin* called it "the first time in world's history . . . [that pictures had been sent] from city to city."[10] The photographic dignitaries included here were President Harding, Secretary Hoover, Governor Pinchot of Pennsylvania, and the *Bulletin*'s managing editor Robert McLean. Navy Commander Stanford C. Hooper observed that the method was "capable of transmitting and reproducing the most delicate shading effects met within black and white photography.... It is therefore, very much superior in this respect as well as in others."[11] Lawrence C. Porter, 1922–23 SMPE president, lauded the demonstration: "while these were still photographs and took a few minutes to send, Jenkins has all the elements necessary for instantaneous vision as far as present audio radio will travel."[12]

On April 28, 1923, Jenkins was invited by an apparently skeptical National Press Club to demonstrate his pictorial transmissions. The portrait transmitted this time was of the president of the club, a *New York Tribune* reporter, Carter Field.[13] Jenkins explained to his audience the technology of scanning and relay, then demonstrated the transmission of a clear photograph. The resolution the reporters had seen previously was vastly improved, as no lines were apparent in the transmitted photo. Jenkins proposed that his system provided reporters a means "by which sport news pictures could be broadcast and picked up by every newspaper subscriber.... [Such a plan] would give the San Francisco newspaper a picture of spot news in New York as promptly as the Boston newspaper got it." In fact, he claimed that "every newspaper in the country would get its spot news pictures in the same five minutes."[14]

In that same year, Jenkins demonstrated the significance of his high-speed camera to the postmaster general and foreign officials. It was once again clearer and faster. The assistant postmaster general Paul Henderson received a transmitted communique explaining, "this is an example of our new radio photo letter, a method of transmitting messages via radio." Autographed by Jenkins, it indicated that "the time will soon come when the post office will deliver by radio."[15] Foreign dignitaries participating in similar demonstrations included ambassadors from China and Japan; that transmission was between rooms at the lab. Another transmission sent to the State Department for translation, and public-relations effect, read, "Ten thousand joys on your journey."[16] The Chinese transmission was sent to the ambassador Shi Zhaoji at the Chinese embassy.[17] A similar message was also sent from I. Yoshida in the the Japanese embassy's Charge d'Affairs office to a ship captain in Boston Harbor.[18] A later report from the *Washington Post* compared Jenkins' Oriental-lettering transmissions to the story of Alexander Graham Bell, who was apparently asked by a young Japanese student if his new invention, the telephone, could speak Japanese. The potential of transmitting Chinese characters could revolutionize international printing, as "in the future a Chinaman could write out a message which could be flashed to all parts of China by radio."[19]

The transmissions usually enlisted the technical assistance of the navy and NOF. From Anacostia, Jenkins relayed the signal to the broadcast station WGI near Boston, a distance of 450 miles.[20] During this transmission, there was apparently severe interference from a lightning storm, but Jenkins reported that a clear message was received during "the worst static conditions of the year."[21] The experiments illustrated the ability of Jenkins' system to transmit delicate Oriental lettering in complete detail. As well as being technologically successful, this experiment noted that messages could now be sent in any language, with no need for written or spoken translations.[22]

The distances and speed with which the signal was transmitted increased with each experiment. It was only a matter of time before "it would be possible to see around the world."[23] Achieving distance and speed in the transmissions was a dual challenge. There was, first, the speed it took to transmit the image; and, second, the speed at which the motion-picture images could be scanned once received. The first transmissions from the Jenkins lab to the Anacostia station had taken six minutes. Jenkins reported to newspaper groups in Washington and Ottawa that transmitting to newspapers across the country would take less than five minutes. By 1923, signals from the lab to the Anacostia station and on to Philadelphia had taken only three minutes. The Japanese and Chinese messages were transmitted in ten seconds.

Jenkins was approaching the peak of his accomplishments. "Photography being the swiftest means of recording and radio the most rapid method in travel, it seems sensible to combine the two for the speedy transmission of [any] subject matter."[24] The speed and distance of transmitting motion had increased significantly.

Publicity Growing

Jenkins' earliest RadioVision experiments had originated from his labs; then more formalized tests were conducted with the navy and sent to different and distant locations, as he took advantage of every opportunity.[25] His first tests had utilized friends, his lab secretary, and acquaintances—the "splendid young folks" who participated in lab experiments.[26] The process was simple. Point-to-point transmission of the photographs was less a matter of mechanical operation than one of photographic quality, "a matter of blending line and tone, just exactly as it is with the artist."[27] And Jenkins was already showing his capability as a photographic artisan.

While traveling, Jenkins followed up each experiment by speaking and replicating his work.[28] In Omaha, in August 1923 at the invitation of the assistant postmaster general Paul Henderson, Jenkins observed air-mail experiments.[29] In addition to continually lobbying the post office administration for support, he talked with local officials about utilizing the Omaha broadcast stations WOAW and WAAW in RadioVision as a revenue producer.[30] These demonstrations were conducted between Omaha and Iowa City, along the post office's "night flying airmail route" from New York to San Francisco. Little is known about this experiment of December 15, 1923, which was cut short by General H. M. Lord, the federal budget bureau chief, who concluded "that there was no appropriation out of which such experiments could be paid for, and they must stop." Jenkins was home by Christmas.[31]

While the vision of a "radioized" postal service seemed out of the question in 1923, it would reappear decades later. By 1926, Henry B. Hubbard of the National Bureau of Standards predicted that radio would deliver "the world's correspondence instantly. . . . A greater system is in sight . . . no mail bags, no sorting, no long trips overland! Swift as light."[32] "Do you realize," Hubbard stated, "that C. Francis Jenkins is the man who not only had the vision, but the daring to attempt to unite the two most popular subjects, radio and motion pictures?"[33]

In Ottawa, Ontario, in October 1923, Jenkins addressed a friendly audience of the Society of Motion Picture Engineers. He showed some of the photographs recently transmitted in Philadelphia and stressed that, using his system, if a fire were "raging in Ottawa . . . within five minutes a moving picture of its spread or conquest could be flashed on screens as far away as San Francisco."[34]

In early December 1923, Jenkins gave a significant private demonstration to Hugo Gernsback and Watson Davis. Gernsback was a prolific and influential writer of fiction and nonfiction, a fact obviously recognized by Jenkins. He founded the Wireless Association of America and the *Modern Electrics* and *Electrical Experimenter* magazines. He was a notable "dreamer who made us fall in love with the future."[35] This was a description applicable to Jenkins himself. Gernsback's writings in electronics and broadcasting would influence the future.[36] He had written previously about Jenkins, who was anxious for the publicity Gernsback could provide. In the demonstration, Gernsback saw his hand moving, "projected by radio and being received by radio," and he saw other lab objects moving, a key and a clamp. "It was possible to wave these small objects in the path of the light ray . . . and amuse oneself by seeing how these objects were actually being transmitted."[37] Gernsback's evaluation was that the principles of transmitting still and motion pictures, while different in complexity, are fundamentally the same. "The day will come when you will be able to sit at home and witness a baseball game as it is being played five thousand miles away or . . . actually see an opera as it is being sung and acted." Or, as he continued, it would be possible for "an Admiral to witness a naval battle and follow it with his own eyes, although his battle ship squadron may be thousands of miles away."[38] This was the kind of industrial and popular publicity Jenkins sought to achieve. Gernsback's vision mirrored Jenkins' own dreams and rhetoric. The private lab demonstration for Gernsback was not unusual for Jenkins, as he had done similar things for friends, family, employees, and acquaintances from the beginning. But demonstrating for Gernsback was a uniquely public opportunity.

Watson Davis was given a similar demonstration. Jenkins invited him to the lab and sat Davis down to watch a screen while his back was to Jenkins. Jenkins then waved at Davis. "I was seeing by radio!" Davis declared. "I could hardly refrain from hoping that [Jenkins] would form a shadowy rabbit or bird with a wing neck or some other strange animal." Davis described RadioVision as the forerunner of pantomime entertainment that would be widely available within a few years.[39]

In Lynn, Massachusetts, in December 1923, Jenkins told his audience of the American Institute of Electrical Engineers and the Thomson Radio Club of the latest development in "photographs by radio."[40] He explained in an accompanying WNAC Boston radio address that he had been working on the idea for years, but "could not work it out, so that was when he took his cross country excursion, resting and coming home with new ideas."[41]

On June 15, 1924, Jenkins conducted his first high-resolution transmission. It included one-hundred-line pictures and, most importantly, again improved the quality of the transmitted photos. He called the results "portraits of true photographic value."[42] Putting this in historical perspective, the question of picture resolution had been a constant challenge, and improvements were continual. At this time in motion-picture history, the industry was moving toward a thirty-five-millimeter film standard. For years, this remained the quality picture standard to be reached by competing media. Television's early pioneers settled with pictures scanning sixty lines in their experimentation, so Jenkins' move to one hundred lines was a significant improvement. Even Farnsworth and Zworykin's earliest electronic television experiments started at these earlier levels. By comparison, today's analog standard remains at 525 lines (625 in Europe and Asia), and high-definition television ranges near 1,080.

Jenkins new high-speed camera and projector used for taking RadioVision pictures was additionally capable of producing quality at a variety of speeds—most often reported at four thousand pictures per second and as high as two hundred thousand pictures per minute.[43] The idea of the high-speed camera was to slow the action. The film shot at two hundred thousand frames per minute, when played back at normal speed, slowed the motion by a factor of two hundred.[44] The danger was tearing the film at this speed.

However, after Jenkins conducted demonstrations for the army and navy, they purchased the high-speed camera.[45] A modern-day Hollywood film is shot on thirty-five-millimeter film at twenty-four frames per second, or ninety feet per minute; this was the standard for decades, until the movement toward digital. The modern high-speed camera can shoot "at 2,000+ frames per second."[46]

Jenkins' reputation was growing, as was seen in the notable cooperation he received from many important figures of the day. Carl E. Akeley, a naturalist photographer and a fellow recipient of the Franklin Institute's John Scott Medal, described Jenkins' wireless photos: "I think few people realize or appreciate the transmission of radio photographs and the high development to which you have brought the art. . . . [I] wish a speedy realization of your dream."[47] Even Kodak's George Eastman sent his congratulations: "[Y]our feat seems marvelous to me and I heartily congratulate you."[48] Jenkins wrote Secretary Hoover, hoping for leads to financial support, and reported that his expenditures to that point were $240,000 ($3,186,690). He asked Hoover if there were any "public spirited American who would gladly support our further work."[49] Industrial giants, too, were noticing—AT&T, the U.S. navy, General Electric, and Westinghouse among them.

Radio Pictures Corporation

By this point, Jenkins was looking beyond RadioVision toward television. In 1924, the Radio Pictures Corporation was organized to facilitate that transition. Jenkins' radio-pictures technology had progressed to the point where it was ready for sale. So, the new company was formed to promote, market, and sell the various devices he created for still-photography transmissions, including the prismatic disk machines, synchronizing mechanisms to control various motors, a picture-strip-machine receiving device, the duplex machine used for two-way service that included visual and audio transmission, a talking machine, and RadioVision machines for ham-radio-operator, industrial, and home use.[50] Jenkins charged the new corporation with selling the devices for the "complete radio picture coverage of the whole United States; [and] serving thousands of radio cameras in newspapers offices, photo studios, banks, municipal and government departments."[51] With the new firm in place, he turned toward improving resolution and television experiments.[52]

The Radio Pictures Corporation was a turning point for Jenkins in many ways—not an abrupt change in direction, but a segue from still-images transmission to work on a more concentrated effort in broadcast television and motion pictures. It provided a point from which sales of equipment might help Jenkins continue working on television. At this stage of his career, he was seen as "one of the Top Ten Greatest Figures in Motion Pictures."[53]

Reaching the Red Planet

One of Jenkins' little-known experiments was his participation in "reaching out to the Red Planet."[54] In 1924, his prior work with the navy brought him into a network of scientists from the U.S. Signal Corps, who were listening for radio signals from Mars. Being "in a neighborly mood," that planet was passing closer to Earth than it had come since 1909, some thirty-five million miles away.[55] In a special to the *New York Times*, the engineer at the Yerkes Observatory at Williams Bay, Wisconsin, Frank R. Sullivan, suggested that August 24 was the "best night for observing Mars we have had in 34 years."[56]

In preparation for the close passing of the planets, radio stations belonging to the army and navy were to "stand by from midnight tonight to 8:00 A.M. Monday to 'listen in' for possible signals from Mars." Some "seventy-eight stations around the globe [were] listening for any signs of life."[57] RCA even agreed to have its New York radio station WRC listen for five minutes each hour.[58] The army had William F. Friedman, a codebreaking expert, standing by to "translate any peculiar messages that might come by radio from Mars."[59]

Not everyone was overly excited with anticipation of this Mars-monitoring experiment. "Astronomers generally smiled over the expectations." Dr. J. H. Dillinger of the National Bureau of Standards declared that the "heavy atmosphere shield surrounding the Earth . . . would prevent any signals from reaching this globe." The Smithsonian scientist E. F. Fowle "took no stock in the idea that communications could be established."[60] Most commercial AM-radio stations refused to shut down unless directed to do so by the Department of Commerce, the regulatory body of the time, which did not so order.[61] Nevertheless, elaborate arrangements were made to see and perhaps hear the thrill of Mars passing so closely.

Jenkins was brought into the effort by a friend, Dr. David Todd, an astronomer from Amherst College. Jenkins was asked if he could record any communication received from the planet.[62] He responded affirmatively, and his RadioVision, a radio-camera at the Jenkins Lab, was set up "in a dark room so that the camera could be opened and the paper [recording] strip put in and taken out without getting light-struck."[63] The paper strip moved

The Mars experiment team included C. Francis Jenkins (left), William Friedman (center), and David Todd (right). Courtesy Virginia Roach Family.

about two inches an hour. Jenkins put a twenty-five-foot ribbon of recording paper into the camera at 1:20 P.M. on Friday, August 22, to record the "mystery messages," and he removed it Saturday afternoon.[64] The strip was to record "all incoming radio signals."[65] The system used optical recorded points of light to register any frequencies and the content of any messages.[66]

The results were inconclusive. Dr. Todd clung to his conviction that the experiment "proves beyond a shadow of doubt, that intelligence . . . exists on the planet Mars."[67] Jenkins was mystified; his conclusion was that the signals recorded were likely atmospheric interferences, and that what some thought was a "crudely drawn face [on the film's surface]" was merely a "freak which we can't explain."[68] Nothing seemed to come of this Mars-monitoring project, and Mars would not again pass the Earth so closely within Jenkins' lifetime.

American Radio Relay League

Jenkins received unexpected attention not only from the Mars-monitoring experiment but also from the growing number of radio hobbyists of the time. These radio enthusiasts, most commonly called "amateurs" or "hams," had become a valuable source of military-personnel support during World War I, training radio operators for the military as well as contributing technological advances to the war effort.[69]

In the early 1900s, thousands of ham operators were coming on the air, and in 1914 the American Radio Relay League (ARRL) was formed.[70] These pioneers competed with ships, military stations, and experimenters for the limited radio-spectrum space controlled by the government.[71] Then, as today, the hams' focus was on "just talking," usually in Morse code, to other people around the globe. The terms *amateur* or *ham* connote an image of radio operators that is inconsistent with what many of these people really contributed. They were, in fact, skilled professional radio experts. Historically, there has always been an attempt to keep amateurs separated from commercial operators. But technologically, particularly during World War I and the early 1920s, amateur operators worked alongside the military, inventors, and future commercial operators.[72] They worked in the shortwave frequencies from 15 to 60 meters (a radio wave–frequency measurement) on the spectrum, as did Jenkins, for he, too, considered himself a ham operator, as he was experimenting in the same frequencies and called on other ham operators for technical reports and ideas.[73] Unlike commercial broadcasters, their focus was distance-challenged, point-to-point personal-hobbyist-style communication and, during the war, serious military communications. As accomplished operators, they worked to improve technology and their signals. They participated with the military in improving emergency preparedness.[74] And they too were concerned with distances and the time it took to send and receive messages.[75] Initially, following World War I, the navy felt that amateurs were cluttering the airwaves with meaningless communication.[76] However, by 1927, the ARRL numbered well over ten thousand members and proved an important force in television and radio developments.[77]

Jenkins reached out to amateur radio operators for technical solutions for RadioVision and his future image of television. Through the ARRL membership magazine *QST*, Jenkins proposed adding "visible radio to our old friend audible radio."[78] Remember, Jenkins' work in the early 1920s involved what today would be called the faxing of still photographs, and his experimental television broadcasting of silhouettes was yet to come. He saw film-style motion, with its better picture quality, as only months away and declared that the scanning of motion pictures was similar to the fax or a still photo, an oversimplification (and Jenkins knew it). While sixty to one hundred scanning lines was an acceptable definition for experimentation, even audiences in the 1920s sought more as large-screen quality developed in motion-picture projection. Looking for new technological ideas and solutions, he solicited the help of the ARRL membership via *QST*. He hinted at the difficulty, to be ironed out, and promised that with their help and opinons he would soon have a new system of entertainment for the home.[79] In May 1925, the Department of Commerce had given amateur operations permission to transmit "pictures and facsimiles" under their standard amateur licenses.[80] Another significant benefit of involving the amateurs was that the Department of Commerce did not specify any wavelength for their use, as it did for experimental transmissions. Jenkins shared his facsimile system with amateurs and offered prizes for good technological ideas.[81] He followed up his first invitation, from July 1925, sending *QST* a telegram outlining the terms of his discussions with its membership and offering them a copy of his book *Radio Pictures* in addition to cash prizes.[82] Jenkins was willing to pay twenty-five, fifty, and one hundred dollars ($324, $647, and $1,295). He wanted "to bring forth ideas that had not occurred to the laboratorians."[83] He set up a panel of three judges to evaluate submissions: Kenneth B. Warner, secretary of the ARRL; Dr. A. Hoyt Taylor, a physicist from the Bellevue Naval Research Laboratories; and Major J. O. Mauborgne from the U.S. Army Signal Corps. He did attach conditions: the decision of the judges was binding; "the scheme proposed must be one not disclosed in my book, *Radio Vision* and *Radio Pictures*";[84] the ideas "must deal with mediums and mechanism for transmissions and/or reproduction; and ... they must be new and original."[85] The first award went to G. J. Shadick, a Canadian from Regina, Saskatchewan. His idea related to still-photography carbon-ink reproductions.[86] There was no indication that this contest produced specific patent applications, but related correspondence and ham QSL cards, which amateurs collected to acknowledge and confirm transmission existed, filled Jenkins' files, and their increasing numbers resulted in potential ideas and a market for equipment sales.

His company offered amateurs RadioVision equipment for transmitting their own radio pictures, which could be connected to their existing equipment. Jenkins' Radio Pictures Corporation had "four to five" facsimile-model machines, ranging from $45 to $250 ($582 to $3,239), sold by the lab. The less expensive used a stylus to mark "electro-sensitive paper," and the more expensive used prismatic rings and optically sensitive film. Each machine came with the related accessories "and a book describing not only the work of Mr. Jenkins but that of almost everyone else working on the problem."[87] Jenkins opened new opportunities for amateurs. Sending and receiving "scripts, sketches, maps . . . [even] alternately transmitting from station to station ludicrous cartoons . . . there will be an irresistible fascination in this unbroken ground of experimentation . . . because the approximately 20,000 bona fide amateurs are on the threshold of a new and fruitful period of experimentation."[88]

The ham's transmissions were of still photographs, not motion-picture cartoons, but the singular cartoon drawings sent from the "Radio-Pen Transmitter" that the receiver simply reproduced on paper opened new applications for the amateurs.[89] *QST* pronounced, "The two most popular picture transmitters now available for amateur use are the Midget and Junior, both made by the Jenkins Laboratories."[90] Jenkins' outreach to the ham-radio operator was mutually beneficial. It promoted new experimental directions and provided Jenkins with ideas for new solutions in his move to the increasingly complex issues of television technology as well as equipment sales.

Weather Maps by Radio

After World War I, Jenkins revolutionized the communication of weather information with his wireless transmission of weather maps.[91] Newspapers heralded the idea of getting weather information via the radio, as they gained access to printable maps.

In 1925, Jenkins was at the height of marketing his RadioVision. His technology for transmission of weather maps came from a familiar association—the U.S. navy. In July 1926, a group of navy officials attended a demonstration at the Jenkins Laboratories. In attendance were Captain Ridley McLean, director of naval communication; Captain S. C. Hooper, an engineer and delegate to the Fourth Annual Radio Conference; and C. F. Martin, a professor of meteorology and chief of the U.S. Weather Bureau. The meeting with Jenkins was successful, as they apparently gave him a weather map for each day, and he was given five weeks to deliver the RadioVision system he had sold them. Within a few weeks, Jenkins had two transmitters and six receivers ready. The transmitters were set up at the Jenkins Laboratories and in the "recreation room of the Arlington wireless station."[92] Eventually, the latter transmitter was moved to the "Navy Building in Washington . . . Radio Central in the Bureau of Communications." Other receivers were set up in Chicago and two aboard the USS *Trenton* and the USS *Kittery*.[93] Wireless delivery of the maps was the same as other RadioVision still photographs. A code was transmitted first to get the attention of a ship operator, who heard a voice announce, "Stand by for a weather map." Within a few minutes the map was transmitted and printed.

The system was proven significant. A month following these first tests, the USS *Kittery* was off the coast of Florida during the hurricane season. On September 11, 1926, the U.S. Weather Bureau notified ships in the area of an approaching storm. Jenkins' system aboard the *Kittery* provided the information that averted a disaster.[94] Despite the storms, the *Science News-Letter* called the location of the *Kittery* fortunate from a scientific point of view because it made it possible to track several hurricanes with the ship "artfully dodging and evading every blow."[95] In this near catastrophe, the service had proven its value.

The quality of the map reproductions continued to improve, and Jenkins received a landslide of publicity from the *Kittery* hurricane stories. The *Philadelphia Evening Bulletin* and *Philadelphia Record* printed pictorial spreads of the leading scientists in town, including Jenkins.[96] The *New York American* called it "Wonders!" while the *Philadelphia Public Ledger* called it "PROGRESS." The *Washington Post* concluded that the results were "so far very promising," and the *New York Herald* exclaimed that it was the "ultimate equipment of all ships at sea."[97] Jenkins had become a master of public relations, and the press could not seem to get enough of his work.

As the publicity over the *Kittery* subsided, Jenkins moved to expand his weather maps to assist aircraft navigation. Equipping military airfields and the growing number of civil airports was the next application for the delivery of weather information via the radio map. Transmitters and receivers were set up at airfields in Los Angeles and the Lakehurst Naval Air Station in New Jersey.[98] Their tests lasted only a few days but "proved entirely successful."[99] The weight of the apparatus had to be decreased from the one hundred pounds used in the navy experiments to a mere fifteen to twenty pounds in airplanes, but the equipment worked.[100] It took only twenty-five minutes to send and print a complete map from the Weather Bureau to the plane, which picked it up using an antenna wire hung from the aircraft.[101] The newly created Federal Radio Commission was sufficiently impressed that it was considering specific spectrum space for the broadcasts.[102]

Newspaper by RadioVision

By the late 1920s, Jenkins was again pushing newspaper applications of his technology. The idea was not merely the transmission of newspaper photos or text articles, which he had previously demonstrated, but the transmission of a full newspaper page, measuring fourteen by nineteen inches. Jenkins described it as "the largest surface ever transmitted over a radio channel" to date.[103] Eventually, pages would be transmitted thousands of miles "before the main publication would be put on the streets of its home town."[104] Despite his optimistic predictions, however, Jenkins was spending more time focusing on television, and his refinements of the newspaper-page transmissions dwindled.

Jenkins was not disappointed; he wanted to move on. So he delegated the responsibilities of sales and promotion in radio-pictures technology to the Radio Pictures Corporation. His newspaper technology would be progressed by others in the late 1920s and mid-1930s.[105]

9

Television: Seeing by Electricity

The word *television* first appeared in France in 1907. It simply meant "vision," bridging large distances. In Jenkins' writings, it does not appear until around 1925.[1] The word and the work evolved over time, with inventors borrowing from Industrial Age telegraphy and the telephone. The prefix *tele* or *tel* means "at a distance," hence *tele-vision* was "seeing over a distance," or bridging those distances.[2] Applications of the term unfolded with each new apparatus, application, and inventor. The name was often both descriptive and a market-branding tag. By the mid-1920s, Jenkins' definition was focused primarily on "wireless motion pictures in the home"—although technologically, in his earliest definition, *television* was the electrical "transmission of living images by wire." His *RadioVision*, with the center capital letter, referred to wireless transmission of still images or facsimiles. *Radiovision* with the small "v" and *Radio Vision* as separate words were his terms for the wireless broadcast "transmission of [moving] images by radio from living subjects," whether the images be from film or live action. *Radio movies* were the "broadcast records on film of these persons and scenes."[3] In 1927, Jenkins concluded that it was necessary to "coin a new word here ... we speak of 'television' as sight transmitted over wires, and of radiovision when the transmitting medium is the air."[4] But his usage of these terms was constantly intermixed and confusing, forcing the reader to establish context for each application. Nevertheless, these terms linked inventors, technology, commodities, and corporations. They began as broadly descriptive promotion and marketing tools as the inventors worked on technological solutions, providing the foundations for what has evolved into "television" as we know it.

Positioning Jenkins' Television Inventions

Historians portray Jenkins as the American inventor of *mechanical* television. They too often portray him as something of a failure because by the early 1930s, his ideas were purchased by competitive corporate powers and then overtaken with the far-better Farnsworth and Zworykin electronic scanning techniques. Thus, Jenkins' overriding contributions in television faded from the mainstream. The mechanical-only portrayals, however, are oversimplifications that tell only a small part of the Jenkins story, ignoring his fundamental contributions in television. In fact, mechanics and electronics remain critical components of modern systems.

Jenkins' experimental television had deep roots in mechanics, which came from the Industrial Age. The technology's evolution waned through the years of World War I and resurfaced during the prosperity of the Roaring Twenties, when Jenkins began his experimentation using the Nipkow disks, before moving on to optical experiments. The stock-market crash of 1929 and the Great Depression cleared many inventors and smaller corporate players from the marketplace. They were simply economically overpowered and unable to compete. Their progress was buried, thwarted by the larger corporations—Westinghouse, General Electric, AT&T, and particularly RCA.

The three television inventors who were the most influential during Jenkins' lifetime were Valdmir Kosma Zworykin, John Logie Baird, and Philo Taylor Farnsworth. In 1923, thirty-five-year-old Zworykin, a recent immigrant from Russia with a doctorate in engineering science, was working with Westinghouse, where meeting the growing demand for radio receivers was the goal. Zworykin filed his first electronic scanning television patent on December 29, 1923. He was dissuaded from spending too much of his time with television by his employer, however, and he made little attempt to test and place the patent into practice.[5] He did scant work with television until 1929, when it had already evolved to include motion pictures. He was hired by RCA to spearhead David Sarnoff's technological drive, launching television's reality.[6]

A British inventor, thirty-five-year-old John Logie Baird, "had just started on what was to be a lifelong task." After recuperating from illness and several speculative business failures, Baird produced the first *public* television exhibition in April 1925. His demonstration was for a birthday celebration of the Selfridge department store. Baird was really not concerned about being first, or even about the publicity; "it was the case that he could not refuse the weekly cheque without which it would have been difficult for him to have carried on."[7]

At this same time, seventeen-year-old Farnsworth had just drawn a schematic for an electronic television scanning system on the blackboard for his high-school teacher. It would be another four years before he would file for his patents and produce electronic scanning lines, which would later head the industry in a new direction.[8]

Each of these inventors produced crude visual demonstrations. They each shared the desire for improved picture quality. Their goals were the same—creating a picture that would equal the increasing clarity of Hollywood film. In 1923, the fifty-six-year-old Jenkins was clearly out in front.

The Jenkins and Baird systems both used mechanical rotating disks. Several different mechanical-disk scanning devices were used by the inventors of this era to create picture resolution, including flat metal disks of varying sizes, paper disks, metal drums, and Jenkins' prismatic rings. This was technology inherited from the late nineteenth century that had evolved by the 1930s. Jenkins and Baird were both trying to improve upon what had already been established. It is, however, incorrect to label their systems wholly "mechanical," because many components were primarily electronic—including vacuum tubes and the electronic circuitry for processing, transmitting, and receiving signals. These combined with the mechanized scanning disks to create lined picture resolution. Crude images? Absolutely, by modern standards. However, for the first time images were clearly seen. Perfecting them was the challenge. Farnsworth and Zworykin were more than a decade away from consummate electronic scanning. Jenkins used available mechanical technology, but his prismatic disks were prisms. He was encouraged by new ideas conceived from his photographic foundations. Between 1921 and 1934, he produced more than 120 television patents.[9] These were the genesis of the Jenkins Laboratories, the Jenkins Television Corporation, and several experimental television-station operations.

In contrast to the historically famous work of Baird, Zworykin, and Farnsworth, the name of the now fifty-nine-year-old Jenkins was not widely associated with television. His foundations were in motion pictures, photography, and photographic projection.[10] This was consistent throughout his writing and the writings of others about him. Jenkins himself declared that the time of the flat disk was over and that he had progressed beyond it as a basic scanning tool. He first moved onto his rotating drum with quartz rods. The drum was an "ingenious optical system [that] project[ed] whirling dots of light on to a magnifier screen."[11] This concept would lead to other optical scanning solutions and theater-size screens. By 1927, Jenkins had progressed to the point where he was labeled "the man who was keeping America in the front ranks in the development of radio vision, or 'seeing' by radio."[12] Of this time, he later reminisced that television was the venture in which he had never "had so much advance publicity,"[13] much of which he had helped create.

Lonely Inventor

Jenkins had written of transmitting wireless pictures as early as 1894 and about wireless motion pictures in 1913.[14] The 1894 scheme was criticized by the technical historian Albert Abramson as "not a true television system . . . [as it] used a multiplicity of short wires of either sulphur or selenium . . . inserted into a non-conducting plate." The plate thus had an image "projected on it . . . with each individual wire picking up a part of the image."[15] It might not have been a complete system by definition, but the descriptions sound much like some modern liquid-crystal and digital-projection receivers. Not a new idea with Jenkins, he called it a "telectroscope."[16] This device used photographic technology, crystalline selenium, to conduct photos and was sometimes called the electronic telescope.[17] Crystalline selenium is a crystalized nonmetallic chemical element used in photocells. It has a sensitivity to light and is used even in modern photographic systems. The 1913 apparatus using it was supposed to be explained in two *Motion Picture News* articles. The first enthusiastically endorsed Jenkins' work: "not to anticipate interesting details with which Mr. Jenkins has supplied us, we go so far as to say that, in our opinion, the time is not far off when this marvel will be made a practical accomplishment," the editor reported.[18] The method and description of the devices were to have appeared in the next issue. Without explanation, they did not appear. While both of these systems may not have been practical at the time, they certainly reflected Jenkins' theories and his futuristic vision.

In 1922, Jenkins was the sole American inventor working on transmitting television motion.[19] His first television patent was filed on October 23, 1919.[20] His work with the prismatic rings and radiovision broadcast of visuals for multiple applications positioned him to transition from still-image to motion-picture transmission. A great deal of publicity accompanied each still-image transmission and always included the dual concept of the coming "radio movies in your home."

The first Jenkins television wireless transmission of motion occurred in April 1923, in private demonstrations given to friends at the lab and a broadcast demonstration transmission from the lab to the naval radio station at Anacostia.[21] These displays were questioned by critics who suggested that at

The Jenkins Laboratory was established on the second floor of the Marion Building at 1519 Connecticut Ave., Washington, D.C. The first experimental transmissions came from this lab. Courtesy Special Collections, University of Maryland Libraries, Herbert Cooper Phi Mu Paper.

the events "there were no disinterested witnesses."[22] This was Jenkins' pattern. Throughout his life, he had conducted private demonstrations for anyone he could commandeer before he offered the more publicity-targeted public exhibitions. Little is known of the first demonstrations, except that Jenkins "transmitted by radio the outline of a woman's face illuminated by sunlight . . . five miles to his laboratory on 16th Street."[23]

Jenkins' final work in television is most appropriately portrayed as *optical* electrical technology. His systems built upon his film foundations. Improving on the older Nipkow disk from the Industrial Era, Jenkins' 1920s prismatic rings were optical prisms creating a "tiny point of light travel[ing] across a photographic plate"; used in the transmission of radio still pictures, they

Discrola was produced utilizing a paper disk for motion pictures. It was something akin to later DVD players. Courtesy Virginia Roach Family.

were now being adapted using multiple rings for motion transmission.[24] In October 1923, he disclosed his progress at the regular SMPE conference, emphasizing increased speed in producing pictures and incorporating a projector that sent an optical image signal onto the radio photo transmitter.[25] This concept used the prismatic rings and a series of lenses on the ring. He pointed out the connection between this new radio-movies device and his first patent for the Kinetoscope, "where a series of lenses ran with the film."[26] Radio had at last found its first motion for our eyes, coming full circle in Jenkins' career.

The more public and well-publicized demonstration occurred three months later in mid-June 1923. This time government officials were invited to witness radio pictures in motion. Officials of the navy, post office, and the National Bureau of Standards viewed the "movements of [Jenkins'] hand and other objects held by them in front of his 'radio eye' device."[27] These were lab experiments, viewed from an adjoining room, and unmistakable motion could be discerned. Later in 1923, Jenkins would "demonstrate the transmission of shadowy motion."[28]

The two demonstrations in December 1923 were conducted for Gernsback and Davis, illustrating the transmission of motion and producing favorable publicity. Gernsback reported that he had seen televised motion of his hand during a private Jenkins demonstration.[29] Davis described the demonstrations as shadowgraphs utilizing prismatic disks spinning at 960 rotations per minute.[30] These demonstrations are historically significant. Baird, who would become Jenkins' only real competitor at the time, would not present his first demonstration for another two years.

The Jenkins Laboratories

The Jenkins Laboratories were founded in 1921, located at 1519 Connecticut Avenue in Washington, D.C. They were a "modest group of rooms over a store . . . unpretentious. . . . [Go up] one flight of stairs to his workshop any day and you'll see a well set up man, who looks upon you with inquiring eyes."[31] Jenkins had almost three decades of photographic and electronic invention as a foundation for the laboratories. He had a working staff and significant connections with Washington's influential people.[32] The laboratories produced RadioVision success and imaginativeness with the utilization of still photographs in ship-to-shore images, weather reports to naval craft and airplanes, and newspaper applications of the new visual technology. From these activities, two corporations had developed—Discrola Inc. and the Radio Pictures Corporation. The Discrola was an early home motion-picture appliance utilizing disks. The Radio Pictures Corporation's mission was promoting military, industrial, and amateur operators' use of RadioVision's facsimile devices for sending still images.

The lab-manufactured Discrola was an attempt to develop "radio movies to be broadcast for entertainment in the home."[33] The Discrola was a "practical and safe motion picture machine for the home."[34] It had a wooden home-furnishing-styled cabinet, making it decorative. Internally, it provided movies using printing lithographed pictures on "a plurality of paper disks [that] was the equivalent of 1,000 feet of film," meaning some three minutes of entertainment. Jenkins reported that these picture recordings were easily made and could be exchanged or sold for one dollar ($13) or "rented for twenty-five cents [$3]."[35] It was a rental concept coming forward from the home film-catalogs Jenkins had earlier offered those who purchased his nickelodeon projectors—a concept that has become commonplace today. Four patents were assigned to the corporation.[36] Three were filed the same day, February 5, 1921, and the fourth was filed on April 14, 1921. They are each related to disk-motion experiments.[37] Discrola Inc. was organized for promotion and marketing of this home system. The Discrola set was only distantly related to television and today's DVD and Blu-Ray players; it was closer to the nickelodeon of former years.

The Radio Pictures Corporation was organized in 1924 with a flurry of patents assigned to it.[38] Radio-pictures technology had progressed to the point where still-image broadcast devices were available for sale. The multiple prismatic rings had increased clarity, bending the light rays "so that the picture remained stationary on the screen as the film was advanced." They were unique and in many ways an improvement upon the previous metallic mechanical disks. They scanned flat, still surfaces exceedingly well. The pictures were transmitted "from a lantern slide using an ordinary nitrogen-filled lamp." The transmitters and receivers used "tuning forks," which maintained vibrations for synchronization.[39]

At this point, still images had become simple to transmit, and Jenkins' transmission was faster than the competition's, increasingly so with each demonstration. Equally important, several appliances at different price points were ready for purchase and industrial use. These systems had been tested by the radio amateurs and witnessed by navy and National Bureau of Standards officials. This new Radio Pictures Corporation was intended to promote, market, and sell devices for ARRL and industry service. The corporate mission was to "complete radio picture coverage of the whole United States; serving thousands of radio cameras in newspapers offices, photo studios, banks, municipal and government departments."[40]

Unfortunately, both companies would wither amid growing competition. The transmission of facsimiles was becoming crowded with competition, and Jenkins had moved on to work with television. With marketing companies in place, he mainly expected them to move forward on their own while he continued work on motion-picture film transmission, improving picture resolution, and efforts to put an experimental television station on the air.

Discrola Inc. and the Radio Pictures Corporation were transition points for Jenkins. Selling was not what he liked to do, unless it was related to a new project for which he needed funding. He delegated the sales to others and shifted his efforts to what he had always envisioned in broadcast television and motion pictures. He was continuing to do what he liked best—inventing things. He now thought that he could afford to make these moves, but without his direct leadership and continued enthusiasm, Discrola and the Radio Pictures Corporation would fade, only to resurface again during the Depression when Jenkins was forced to sell his companies.

Dutch Windmill Motion Pictures

In June 1925, Jenkins gave his first and most significant public demonstration of radio movies.[41] As he had always done, he solicited the cooperation of navy officials, and the navy's transmitter NOF again offered its facilities. The invited dignitaries were significant players with familiar names: Curtis D. Wilbur, the newly appointed secretary of the navy; Dr. George M. Burgess, director of the Bureau of Standards; Admiral D. W. Taylor and Captain Paul Foley from the Naval Research Laboratory; Stephen B. Davis, acting secretary of commerce; and W. D. Terrill of the Department of Commerce Radio Department. According to Jenkins, Dr. Burgess and several people from the Department of Commerce "dropped everything to come at once" when they received a phone call inviting them to witness the demonstration.[42] It was a simple one: for the first time, moving pictures of an object miles away were transmitted and received. The transmission had come from NOF, and the picture was received on a miniature screen at the Jenkins Lab. The image televised was that of a model of a "Dutch Windmill . . . erected [for the demonstration] as the blades propelled slowly by wind from an electric fan."[43] Jenkins described this as "merely a scientific test that proves we have attained our goal." Dr. Burgess's reaction was more definitive: "You've certainly got it all right, if my eyes aren't deceiving me. . . . I suppose we'll be sitting at our desks during the next war and watching the battle progress."[44] Burgess invited Jenkins to call him as further developments warranted.

The test had almost been a failure. After Burgess arrived at the lab, the equipment stopped working, and no pictures were being received. "Crestfallen, Mr. Jenkins greeted the doctor and with dragging feet led him to the rear room of the laboratory where the receiver was located." Jenkins stalled, advising Burgess that "too much should not be expected." Just before he had to apologize, the equipment began working again. The relieved inventor later wrote, "Only the pioneer will ever appreciate the relief of that moment, when anticipated chagrin was superseded by congratulations."[45] Articles describing this landmark experiment appeared in newspapers in the United States and abroad. By any measure of publicity, it was a successful demonstration. Some articles were strikingly similar, no doubt coming from press releases distributed by the NEA Service and the Associated Press. Particularly in the foreign press, the stories were shorter, each describing the model of the Dutch windmill and quoting comments from Burgess.[46] As publicity spread, the independent popular press also praised Jenkins' work.

Within the next few months, the motion experiments would progress from the model windmill to full-figure silhouettes of people in motion. In the August 1925 issue of *Popular Radio,* Charles Herndon reported, "I have just come from a shadow-show. . . . While the little crowd of neighbors and friends watched a small screen . . . there suddenly appear[ed] on it the silhouetted figure of a girl . . . the shadow-like figure moved—it danced. And those who were present seemed to realize that, simply silhouette though it was, it was really dancing at the wedding of the motion pictures with radio."[47] The excitement generated was apparent. "It works. This is the real point. . . . We are well on the way to a practical and possibly commercial adaptation of a new art."[48] This, of course, was Jenkins' intention.

The experiment did not go unnoticed by Jenkins' competitors. The radio-receiver manufacturer A. Atwater Kent agreed that the "broadcasting of motion pictures would be the next outstanding advance in the field of radio."[49] Kent was also a member of Hoover's radio conferences that were currently studying radio legislation.[50] David Sarnoff's biographer, who described Jenkins as "a squat, baggy-eyed tinker out of Dayton, Ohio, who apprenticed as a movie engineer," suggested that Jenkins had "stole[n] a march on the scientists of the great corporate laboratories by transmitting a film of a Dutch windmill in motion."[51] Jenkins was spearheading the direction of the science and attracting others with his ideas.

Not all publicity was favorable. A sports journalist for the *New York Herald Tribune* wrote a stinging commentary, humorous by today's understanding, describing Jenkins as "Another Menace to Sport." He was worried about the future of sports-event attendance and thought that what Jenkins was manufacturing would negatively impact gate receipts. "Here is Mr. C. Francis Jenkins setting in his laboratory calmly working out a scheme for elimination of customers. Perhaps, if he realized just what he was bringing about he would desist. . . . [It] seems on the way to reducing all sports to strictly amateur basis. . . . [T]he cinema-radio working at its best . . . will keep old grads away from the annual football game."[52] McGeehan proposed that the government consider the negative economic effects of television's possible stagnation of the sports industry throughout the nation, and he called on William Jennings Bryan to "stump against it."[53] An NEA Service editorial joined in, spreading the disarray. "Francis Jenkins, famous radio inventor, will go down in history as both a benefactor and a malefactor of mankind. His radio-movie invention will be the cause. . . . If it keeps us at home, we'll become a race of loafers and lazy-bones. . . . Awful, Jenkins, awful."[54] Frederic William Wile, a newspaper columnist and commentator for WRC in Washington, D.C., declared that "[w]hat will happen to Uncle Sam's business . . . when C. Francis Jenkins perfects 'radio vision,' is terrible."[55] Jenkins provided no response to his naysayers.

As the publicity grew, Jenkins continued his forward momentum technologically while also starting to produce program content. He labeled his new apparatus the "teloramaphone": "pictures at the fireside sent from distant world points will be the daily source of news; the daily instructional class; and the evening's entertainment."[56] Coverage in the popular press, positive and negative, focused generally on visions for the future. In the lab, daily demonstrations were taking place. Jenkins remained optimistic. In 1929, he was negotiating with a film-animation specialist, John R. McCrory in New York, for suitable testing film.[57] He was looking forward to programming that would "take audiences seeing across the oceans."[58] By June 1929, these film shorts were produced by Visugraphic Pictures in New York, and Jenkins had set up a studio of his own for silhouette films, where production costs could be better controlled by a laboratory-staffed facility.[59]

These experiments in the late 1920s marked a pinnacle in Jenkins' career. His transmission of still photographs featuring famous people had put him on the map, but his motion-picture experiments and the resulting press coverage were more significant, and "his was the only [wireless] game in town."[60]

While it may be true that Jenkins was alone, reception was still limited to the laboratory, those attending public demonstrations, and more specifically a growing number of ARRL ham-radio operators, who had long supported Jenkins' still-image transmissions. The hams had purchased Jenkins' equipment, and all "17,000 ... members of the American Radio Relay League [were] ... interested."[61] It was only natural that these amateur operators would play a continuing part in Jenkins' transmission of motion, as they had done in transmissions of still photography.

The ARRL in Television

The American Radio Relay League's modern ham operators provide emergency communications and public service. During natural disasters, when other means of communication often fail, ham operators have provided critical communications links. During Jenkins' days, ham-operator communications had progressed from the Morse code to voice, and for some to the visual. Jenkins helped guide their interests into still-image and then motion-image transmissions.

In the early 1920s, Jenkins believed that the ham operators were actually ahead of the commercial broadcast engineers, because they were more technically inclined. These operators had maintained proficiency in point-to-point radio communication, turning it into a worldwide popular hobby. Jenkins credits them with the "popularization of [commercial] radio broadcasting."

So, considering himself a ham operator, he began working with them. It was a significant decision, "to place [his experimentation] in the hands of the American amateur," but he did it strategically, because of what he described as the "cleverness and ingenuity . . . [of] American amateurs in particular."[62] Jenkins saw this as another path to hasten the development of television. So, as he was carrying on experiments in the transmission of still pictures with them, he was also recruiting their ideas for moving-image transmissions.

ARRL members helped Jenkins compete with the growing number of companies with a much stronger funding base and larger laboratories.[63] Jenkins was having difficulty with his prismatic rings, which had worked so well for stills, yet built-in optical errors could not be overcome at the speed needed for transmission of motion.[64] Ham frequencies provided more flexibility, because the amateurs were not under the same fixed frequency limits as the broadcast radio engineers. Thus, those needing more experimental leeway than the commercial experimental stations could find it on shortwave. Jenkins was using this opportunity in his search for solutions to the increasingly complex issues of transmitting television. He sold "tens of thousands . . . of [television kits] . . . *at cost*, for the purpose of cultivating maximum interest and ARRL cooperation in television pioneering."[65] Finally, they helped in keeping Jenkins more competitive. By the mid-1920s, RCA, Westinghouse, AT&T, and General Electric all had well-financed experimental labs. They, too, were working on television's development and had the advantage of more money and people. Jenkins could not let himself fall behind technically. He would never have caught up and could not have matched RCA's resources.[66] Writing to L. C. Herndon, the supervisor of radio in the Department of Commerce, Jenkins reflected, "I cannot afford these facilities."[67] Basically, commercial companies were dwarfing independent inventors in facilities and funding. Patent-interference lawsuits and litigation generally was a way of business that an independent inventor could ill afford.[68] Jenkins had experienced this in his Phantoscope battle with Armat, and he was not eager to get tied up again. Hence, he decided to actively network the membership of ARRL and solicit their cooperation. In this way, he acquired manpower and new ideas without appreciable expense. The amateurs were "extremely important co-workers with us in the development of this new art."[69] The pictures were most certainly novice, by modern standards, but the signal reports and suggestions that ARRL members made helped to improve picture quality and synchronization, eliminating echoing of image reception and developing consistent signal strength.

In contrast with the "discouraging experiences of [other] experimenters . . . the only system that is very practical at this time is that used by Jenkins."[70]

The cooperation from the amateurs was "absolutely essential to our experimentation."[71] G. L. Bidwell, a member of the ARRL and director of its Atlantic division, wrote with excitement in the summer of 1925, "motion pictures by radio are here!" Those pictures were just black-and-white silhouettes, but color was now the idea for the near future.[72] Jenkins' work was a thrilling challenge for the amateur wireless operators, who felt involved in television's future as Jenkins shared it with them.

By July 1928, Jenkins was working to get a broadcast station on the air and had developed a "regular program schedule of radio movies for the hams and other fans with short wave receivers." He broadcast his movies from the transmitter at the lab on Monday, Wednesday, and Friday evenings on his experimental station, 3XK, operating at 46.72 meters.[73] They were silhouettes, sometimes called "shadow-pictures or shadow-graphs." The 3XK engineers gathered at the Jenkins labs. "On one side of each studio [was] hung a white curtain"; opposite was a table full of technology.[74] They projected short films onto the screen and transmitted them to the hams, who had purchased Jenkins Radiovision receivers. The audience heard and saw the pictures: "thousands of amateurs fascinatingly watched the pantomime pictures in their receivers as dainty little Jans Marie performed tricks with her bouncing ball, Miss Constance hung up her doll wash in a drying wind, and diminutive Jacqueline did athletic dances with her clever partner Master Freemont."[75] As always, these were people Jenkins knew who had agreed to perform in front of the camera. The hams knew that theatrical films were coming in the near future. Jenkins had started it all with a cooperative and excited audience of amateurs.[76]

George H. Clark, an RCA executive, underscored the significance of the ARRL in television, asking and answering his own rhetorical questions, "Why did broadcasting begin in the United States and why did it happen in 1920?" He credited the "army of amateurs who listened . . . the vast army of amateurs who joined the Navy during the War and came back to work . . . with a new zeal." And in the United States they had the "freedom of the amateur to experiment." They moved radio and television "out of the attic and into the living room."[77]

Jenkins would continue his associations with the ARRL, looking for technical solutions as he shifted toward the larger consumer market. The multitude of QSL cards from ARRL members reflect massive communications with his experimental station. The cards and communication acknowledged reception of his signals, expressed excitement for the future, requested equipment, and exchanged engineering ideas. Jenkins would continue pressing the ARRL, industrial uses, and the transmission of still images as he improved picture quality in still-image and motion transmissions.

10

The Eyes of Radio

Long before other television pioneers would make their works public, Jenkins was already there. He successfully broadcast still pictures for a variety of military and industrial uses and transmitted motion pictures to an excited audience of ham operators. He had turned his attention to the greater market for his inventive potential—television for the home. In 1925, Jenkins would begin his first experimental television station 3XK (later W3XK-TV) in Washington, D.C. It was among the first licensed experimental television stations in the United States. Jenkins envisioned that it would follow patterns established in the growth of commercial radio. A station was the means whereby viewers could be reached.

The license efforts for station 3XK began on June 2, 1925, just before the famous motion transmission of the Dutch windmill experiment. On that day, Jenkins petitioned the Department of Commerce to "license the establishment of stations for picture transmission by radio." He requested permission to transmit visual and aural announcements as well as information and entertainment.[1] Jenkins prodded the department's commissioner of navigation, comparing his reluctance concerning experimental television's interference with radio-station operations as somewhat similar to the development of two earlier innovations—the telegraph and telephone. Jenkins casually used the name of Alexander Graham Bell in his explanation, noting that Bell had created the telephone without knowledge of telegraphy. Likewise, Jenkins was asking for understanding from "the radio Experimenter, who does not know [the elements of the visual transmission] code, for a similar opportunity to develop this ... electrical means of communications [television]."[2] Current radio-spectrum assignments were too limiting for visual experimentation because of interference with existing commercial radio stations. So Jenkins petitioned for permission to work in spectrum space above radio channels. He asked for space "above audibility . . . [as it] doubles the amount of traffic radio [the spectrum] can carry without interference." Within the week, he received permission to operate his station between 42.8 meters and 52.6 kilohertz.[3]

In June 1926, Jenkins applied for a formal license. The application described an experimental coastal station, operating from the Jenkins Laboratories.[4] The Department of Commerce granted Jenkins a six-month license with the call letters 3XK. In September, the license was extended for another three months. 3XK television was transmitting on the wavelengths of 42.02, 144.8, and 170.4 khz.[5] It was on the air irregularly in 1925 through 1927, and from 1928 to 1933 it broadcast a regular schedule of programming until it was sold to RCA.[6] The licensing of 3XK does not appear to have been important to Jenkins' public relations effort. There was comparatively little publicity surrounding the new station license—it was an extension of lab experiments. It was the first televised motion transmission, along with continuing experimentation, that would become newsworthy.

Meeting Receiver Demand

The primary audience for Jenkins' earliest experiments had been the ham operators. Jenkins sold them receiver kits for $2.50 ($32), admitting that this price was "solely for the purpose of stimulating wide interest in television."[7] The buyers assembled the receiver themselves and were able to view Jenkins' programming from 3XK. Interest spread as buyers expanded from ham to consumer households. Orders were received from the Department of Commerce, the Eastman Kodak Company, the Packard Motor Car Company, the International Projector Corporation, the Stewart-Warner Speedometer Corporation, and others in the northeastern Atlantic states, and as far west as Portland, Oregon, and south to Macon, Georgia. H. P. Hardesty, of the Packard Motor Company, wrote Jenkins that he was delighted: "I received the television outfit and the darn thing looked so simple I could not see how it would work, but it did." He had a few suggestions for Jenkins on disk noise and commented, "I am sure the large majority who are playing with television are getting a real kick out of it the same as I am, and I certainly would not like to have you discontinue sending out these broadcasts." Hardesty included postage with his letter in payment for the kit.[8]

Jenkins called his receivers "Radiovisor Kits," and they were to be attached to radio receivers. This was the beginning of aggressive marketing and sales efforts. Unfortunately, he was not ready to manufacture the quantity that

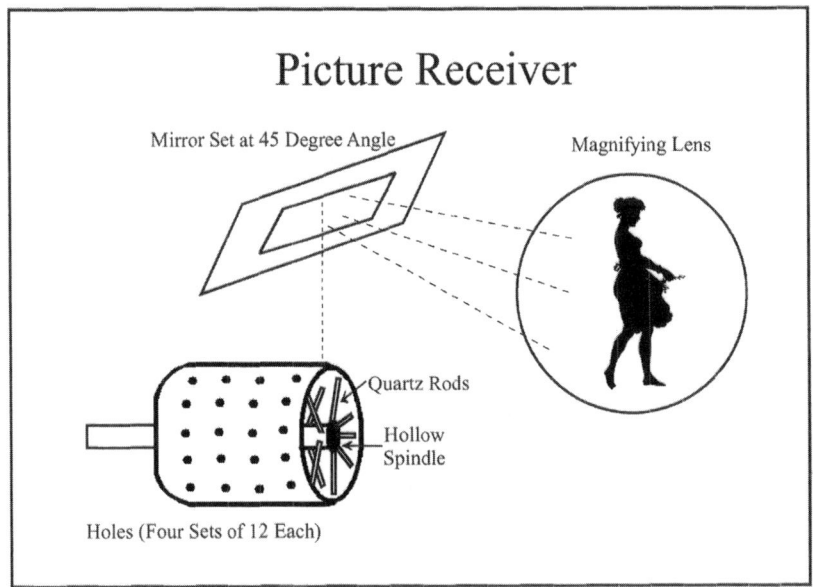

The Jenkins drum scanner was a picture receiver/transmitter that moved away from the flat-disk scanner and utilized quartz rods akin to later fiber optics. Courtesy Knight Graphics.

soon was demanded. Apologizing for his delay in sending the Radiovisor Kit to one purchaser, Jenkins noted, "We are sorry there has been any delay in shipping . . . but the demand has far over-run our expectations."[9]

Meeting manufacturing demands was not the only challenge facing Jenkins as he moved into the home-consumer market. Struggles continued in his efforts to improve receiver technology. The prismatic rings had not lived up to expectations in terms of clarity for motion pictures. The public waited impatiently, and with growing anticipation, as Jenkins promised that more sets would soon be available with improved quality. He acknowledged that picture "refinement [was] a stubborn problem, but we are making progress. . . . [T]he actual transmission of human faces almost instantaneously by radio [was] an accomplished fact."[10] But this was more continuing publicity than a marketable fact. Picture quality remained poor, wavering among experimenters between fifty and one hundred lines. Improving it was paramount. No one understood that need better than Jenkins. He was working to maintain enthusiastic momentum while stalling for time to resolve the issue.

Three problems plaguing the receivers of the time concerned the heritage of mechanical scanning disks. First, if the disks were not properly synchronized, "queer designs of black and pink dots and dashes were seen."[11] Jenkins worked consistently on synchronization to resolve the problems, and multiple patents were filed contributing to a solution.[12] Second, the disks were simply too bulky. Disks spinning at 900 rpm. were sometimes frightening, hazardous for the novice, and noisy. Disks that ranged in size from one to three feet in diameter made home-appliance designs sizably unfashionable.[13] These larger receivers were far from consumer-friendly. The third problem was that the Jenkins' prismatic rings were not producing the quality of motion-picture films. They did not sufficiently transform the film frames into television signals with the speed required in the movement of the motion-picture film or live images.

Jenkins answered these challenges by creating a comprehensive structural modification. He changed the disk from the traditional flat surface into the drum scanner.[14] Headlined by the *Christian Science Monitor* as a "radical step in television" and an "end of the disk," Jenkins was "attacking the problem in another way."[15] The drum scanner was a hollow cylinder, seven inches in diameter. New quartz rods were placed inside the drum, carrying the light to the holes on the outside cylinder "like water through a hose."[16] This invention was a classic illustration of Jenkins' forward thinking, which, according to Abramson, was "in anticipation of the modern technology of fiberoptics."[17] The drum scanner was the new element in Jenkins' Radiovisor. This design made the home receiver more compact, because the drum was smaller than the previous flat disks and complementary to the radio receiver designs already in the homes of consumers.

Jenkins announced that he was manufacturing the new Radiovisor on a commercial basis, just as the politics of licencing, competitive technology, and the industry were getting significantly more complicated.[18] The Department of Commerce, the first electronic-media regulatory agency, was about to be replaced by the Radio Act of 1927.

The 1927 Radio Act

Understanding early television during its dawn of regulation and electronic technology requires some synthesis with the life of the 1920s. These were the Roaring Twenties, and industrial growth was phenomenal—including electronic media. Commercial radio was "the youngest rider" amid the prosperity of this bandwagon.[19] Almost seven hundred commercial stations had come on the air since 1921.[20] Many inventors and companies were riding the radio bandwagon, all battling for competitive advantage. It was a hard ride. By 1928, experimental television-station-license application requests from the new Radio Commission were on the increase; Farnsworth had filed for his patent on electronic scanning; Zworykin was working on an iconoscope tube; and Baird was demonstrating mechanical color in England while advertising his appliances for sale in the United States. The gap between Jenkins and his competitors was narrowing.

In regulatory evolution, the Radio Act of 1927 grew out of the four Hoover Radio Conferences. The federal courts created the last straw when in 1926 they ruled that the Secretary of Commerce no longer had radio-station-spectrum licensing authority. Thus, some AM stations took advantage of the regulatory vacuum and were moving randomly around the spectrum, creating interference with one another and heightening the issue for radio and the newly related television experimental signals.[21] Secretary of Commerce Herbert Hoover called strongly for legislation: "We must have traffic rules, or the whole ether will be blocked by chaos."[22] He asked advice from the acting U.S. attorney general, William J. Donovan, and was told that the only alternative was to seek new legislation.[23] The result was the passage of the Radio Act of 1927 and the creation of the Federal Radio Commission.[24]

The Radio Act shifted the Department of Commerce's licensing authority to a new Federal Radio Commission (FRC). The commission was intended to last only one year, helping clear up the chaos on the spectrum within the commercial AM radio industry and then returning authority to the Department of Commerce. The complexities of the job assured that commission regulation would remain a fixture in electronic media.[25]

In 1927, the commissioners were charged with the licensing of radio stations and organizing use of the spectrum. They were largely an engineering administrative body, and to accomplish their mandate within limited time restraints, they prioritized radio, moving rapidly to clear the interference chaos created by the lack of earlier authority.

While this political debate and the FRC progressed, Jenkins already had a licensed station, 3XK, on the air, and his Radiovisor Kit was for sale. He was anxious about television and was unconcerned about radio, except where it affected television and provided television audio. Looking ahead, he wanted to develop the same positive relationship with the new FRC that he had with the Department of Commerce and the navy. He reached out to the FRC to position his work and himself to secure their favor.

Jenkins was among the first people to testify before the newly organized commission.[26] The commission's early television decisions mirrored Jenkins' beliefs and rhetoric: "It is not far distant when the eye as well as the ear will partake of the benefits of the transmissions of intelligence through the ether,"

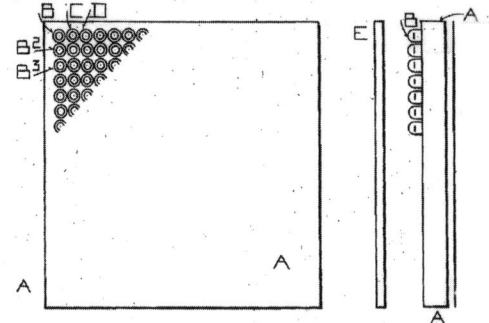

Patent diagram for a persisting luminescent screen provides front view (A) and the flat-screen edge view (E). B, C, and D are the lamps. Courtesy Virginia Roach Family Files.

the commission reported.[27] In Jenkins' testimony, he elaborated on his inventions and the preponderance of evidence that live television would soon be available in American homes. The result was that the commission established an "experimental band immediately below the present [AM] broadcasting band ... to encourage development and with a view to effective television, or transmission of pictures by radio."[28] The FRC then turned their energy to reformulating the process of radio licensing.

The Jenkins Television Laboratory

By 1928, Jenkins' 3XK experimental television licenses had been approved with each short-term renewal. It was broadcasting on 60 to 61.22 khz, with a power of five thousand watts. Two other such stations were now operating on similar frequencies. J. Smith Dodge, of Lexington, Massachusetts, was on 61.22 to 62.5 khz with five hundred watts, and he shared that frequency with 2XBU of Beacon, New York.[29] In addition, the commission had authorized construction permits for seven more television stations. All were transmitting still- and motion-image transmission with silhouette-styled films of their own. Black-and-white silhouettes and stick figures were the norm for the television experiments, as they provided the stark contrasts necessary to analyze the pictures produced. Manufactured home-television receivers were beginning to look very much like the standard radio receiver.[30] The radio audience anxiously anticipated television as a result of the prophetic articles published in the press and the fledgling business ripe with growing competition.

No longer the lone player, Jenkins was still pushing yet another competitive edge. By April 1928, that edge concerned screen size. The first television pictures were produced on small screens; ironically, they were something like the dimensions of modern cell-phone and electronic-tablet technology. Television screens in the 1920s were, at their largest, six by six inches, and Jenkins' talk of theater-sized screens at sixteen by twelve feet would have been considered a massive increase, but not one far-fetched for his photographic imagination.[31] At the 1928 spring meeting of the Society of Motion Picture Engineers, he announced that wide-screen "theater radio pictures" would soon be available.[32] Why not? He had been successful in introducing the large screen to the film world more than thirty years earlier. He hedged on his SMPE prediction, suggesting that the telecast of something like a baseball game was "unlikely to be attained short of three to five years more of research." However, he confidently declared, "radio movies or home entertainment will be available by the end of the year."[33] His approach to the larger screen was "entirely new in principle and had not yet been given the final tests."[34] It would be a screen studded with "baby photoelectric cells."[35]

The small screen size was useful in experimentation, but the desire for larger screens seemed obvious. Jenkins' desire for a large theater-sized screen dated back to his film-projection inventions of the late 1890s. So, as television and receiver technology improved, once again he turned his attention to screen size—this time for electronic-media transmission. The challenge in relation to the electronic screen size was *light*. The older flat-disk scanners were not efficient in passing light through the small holes cut within the disks, and it took a significant number of foot-candles to produce a picture comparable even to a snapshot photograph of the day under outdoor lighting. Jenkins' new plate receiver was an answer. He again predicted doing "away with disk and drum methods of scanning," as the plate receiver provided "100,000 times more [light]."[36]

The Jenkins plate receiver used what were described in the press as 2,304 "baby photo electric cells" mounted in a two-by-four-foot wood or synthetic plastic frame.[37] Each cell was said to "represent 1/2,304th part of the entire scanned area."[38] The cells lit up in accordance with the varying intensity of the signal strength received. The screen was comprised of a layer of silver and potassium in a hydrogen-filled tube within which the electrons were formed proportionate to light values applied. The results were pictures producing half-tones and increased luminance. The new principles sound like a historical ancestor of the modern liquid-crystal and motion-picture digital projection systems. In this work, the word "secrecy" appears much more often than with discussions of any prior technology. Jenkins knew that the competition was growing. A 1924 related patent for a "Light-Concentrating Device" was cited

This late 1920s/early 1930s Jenkins TV Cabinet was a part of the living-room furniture. It utilized RadioVision and the drum-scanner technology. Courtesy Virginia Roach Family.

in 1989 as a part of a storage device for the luminescent screen.[39] A group of patents was filed at this same time, all associated with the light-sensitive cell.[40] It would appear that the photo-cell optical ideas would affect more then screen size.[41] The patent for a Persisting Luminescent Screen was filed on December 28, 1927, but it was not granted until 1935, after Jenkins' death.[42] Several more patents were filed surrounding the process of the luminescent screen (see appendix A). Tests on the new photo-cell systems were conducted with little publicity. On April 9, 1928, Jenkins transmitted a Kathleen Millay poem, surprising engineers and ham operators with the broadcast in the middle of the night. This "carefully guarded test" had both Radiovision and Radiomovie orientation and was moving to replace older disks with optical technology.[43]

Ever Promoting the Future

Jenkins promoted his television as he had always done, conducting public demonstrations, touring continually, and taking advantage of all kinds of speaking engagements, particularly in New York, Philadelphia, and Washington. He spoke to local radio clubs, press groups, church gatherings, youth organizations, community and civic leagues, and at just about every opportunity. He freely gave time for radio speeches and interviews, including one over the NBC Red Network.[44] The publicity was consistent: "Jenkins

Plans Showing ... Jenkins Predicts ... Forecasts Early Television ...," the headlines read.

While by far most receiving sets were in the hands of ham operators, in the winter of 1928 Jenkins predicted that television sets for "movie transmission ... [would be] on the market by Christmas," only nine months away.[45] He forcefully declared that one hundred thousand people would be "receiving motion pictures by radio by Christmas."[46] The prediction certainly made good press.

* * *

INVENTORS' PREDICTIONS, 1928[47]

Lee de Forest, De Forest Radio Corporation: "ten more years"[48]
David Sarnoff, Radio Corporation of America: "four or five more years"
H. P. Davis, Westinghouse Electric: estimates and systems are "premature"
C. Francis Jenkins, Jenkins Laboratories: "for Christmas presents this year"

* * *

Three messages appeared consistently in Jenkins' television publicity. The first and primary message in the newspaper and popular press was the optimistic, almost prophetical, promotional rhetoric targeting a general audience. This built excitement about television's potential and the anticipation of receivers in the home. The second message aimed information at ham operators, noting the availability of equipment for sale. The third message, mostly carried in the technical and professional press, emphasized technology. The audiences for this publicity were like-minded engineers, inventors, scientists and competitors. These articles explained the equipment, the working systems, and the benefits of Jenkins' systems, and they too carried a forward-leaning tone.

Demonstration for Federal Radio Commissioners

One of the most important of Jenkins' informative and promotional presentations was before Federal Radio Commissioners on May 5, 1928. Formal invitations went out to dignitaries reading, "Mr. C. Francis Jenkins requests your presence at the birth of a new industry Radio Movies, i.e., Pantomime Pictures by Radio for Home Entertainment."[49] The invitation went to a small group of Washington, D.C., influentials, including General George O. Squiers, a former army Chief Signal Officer; members of the Federal Radio Commission, including Ira E. Robinson, chair; Sam Pickard and S. C. Hooper; Captain Guy

Radiomovie silhouettes or television shadow graphs were among the first pictures transmitted, as the stark contrast in black and white aided in experimentation. Jenkins' nephew Lewis Janney (upper right) is bouncing the ball, Jenkins is the horse (second row center), and other silhouettes are unidentified friends and family. Courtesy Lewis Janney Family.

Hill of the Army Signal Corps; and Dr. Henry D. Hubbard, secretary of the National Bureau of Standards.[50] While the demonstration used silhouettes, it also added an important new element. In addition to those gathered at the lab, receiving sets were placed in several homes in the city. Herbert Hoover was invited, but he could not attend. Other participants included the Federal Radio Commissioner Samuel Pickard, "who lived at 5514 Nebraska Avenue

NW." The homes with receiving sets were those of William P. MacCracken Jr., the assistant secretary of commerce; United Artists general counsel, William Gibbs McAdoo; and Paul Henderson, a former assistant postmaster general. "[T]he moving pictures [were broadcast] through the air to these sets" in the homes of these officials.[51]

The group in the lab gathered around a receiver, and Jenkins explained what was going to happen, the elements of the experiment, and how his equipment would work. He suggested that they should not expect too much because television technology was yet in the "cat-whisker stage."[52] After his explanation, which as always was mingled with information furthering his work, a phone call was made to start transmission, and the dignitaries were told to push a button on the cabinet. The button was on a small box looking like an ordinary radio receiver with a mirror and magnifying glass on top. Only a few people could watch at once, as the screen was still small, but the picture appeared.

The film they saw was entitled "A Day with the Children." The subjects were "two small children playing together . . . a little girl hanging up doll clothes on a wash line, and a larger girl bouncing a ball on the pavement."[53] The children acting in these early television films, as they had been in Jenkins' earlier photographic experiments, were neighborhood children of friends.[54] Lewis Janney, a nephew, identified himself as the silhouette bouncing the ball.[55] The commissioners and dignitaries were impressed. Jenkins had achieved his goal, sharing his expertise in the technology and "letting the commissioners know we have been diligent."[56]

The Radio Commission was pressed by investors and companies for television solutions and began actively seeking input on policy and development of television. They sent out questionnaires to experimental television-station applicants asking about technical specifications, plans, and priorities in development.[57] The Jenkins Labs and Radio Corporation of America (RCA) were quick to respond. Dr. Alfred N. Goldsmith, RCA's chief engineer, called on the commission to grant experimental station licenses and supported the chorus of previous Jenkins prophecies indicating that television was ready to "leave the seclusion of the research laboratory and enter into the daily affairs and uses of man."[58] Goldsmith had been in charge of RCA's engineering and television-development program since 1924. Just two months before his admonition to the FRC, he too had submitted an application for an experimental television permit.[59] In six months, Zworykin would lead RCA in taking television technology in a different direction.[60]

It was becoming a race. Jenkins still had a slight lead, with 3XK on the air, his political connections still strong, and the significant fanfare that resulted from 3XK's evolution into W3XK-TV's maiden broadcasts and the launching of regular nightly television programming.

W3XK-TV Inaugural Program

The W3XK-TV inaugural programming began July 6, 1928.[61] Broadcasts were from the Jenkins Laboratory on the frequency of 6420 kilocycles; the pictures were scanned at forty-eight lines and transmitted at a power of fifty watts.[62] The established schedule started at 8:00 P.M., and programs were broadcast nightly, except Sunday. Each program opened with an announcement regarding content. The films were a short few minutes at first, but as time progressed, they would fill the hour, and one or two longer films per night were scheduled, along with film shorts. These first silhouette pictures were sufficient "for the entertainment of friends and neighbors. . . . [They were] as entertaining as movie cartoons in the theater," and the experiment created a curiosity from "the appeal of mystery of movies by radio."[63] The stark contrasts of the black-and-white pictures were essential in testing the early systems and the circuitry.[64] The black-and-white pictures also required less bandwidth for transmission and receiving. The halftones, which provided gray scales, or gradations of black-and-white, were a more significant challenge and would be added later, as would color. So the silhouettes and stick figures used for these earliest experiments helped researchers improve picture quality, increase the number of scanning lines, better the gray-scale reproduction, and envision the coming of color.

Press reviews praised the first W3XK programs, and RadioVisor Kits were selling. Jenkins and the subject of television were again press magnets. *Radio News* endorsed the RadioVisor Kit as the "simplest fundamental kit now available . . . it really produced excellent results."[65] Different models were developed for the home and the ham operators. They sold for $7.50 ($96), and the price was rising.[66] By June 1928, another receiver sold for ten to fifteen dollars [$132 to $199]."[67] By August 1928, following W3XK's initial evening, RadioVisor Kits were selling for thirty-five dollars ($464). "Anybody who constructed the old 'cat-whisker-crystal earphone' receiver of six years ago will be able to build the new radiovisor," Jenkins claimed.[68]

Jenkins knew that few homes had sets—most of which he likely had placed there for testing—but he was pushing hard to increase those numbers. To see his television programming, one had to have a Jenkins receiver at home or use one combined with a ham-radio receiver. The same was true for RCA and any other manufacturer: viewers could only see the signals if they had the right manufacturer's receivers. There were no agreed standards. Home-receiving sets were initially scarce, but sales were growing, and the potential was immense, as radio sales had already proven. Hams remained the largest audience for the time being.

W3XK-TV's debut was received "as far west as Denver and north as Boston."[69] But getting television sets into more households was now the primary goal. Writing years later, Jenkins said that the audience had grown steadily from the initial 1925 experiments, and he predicted that "by the end of 1928, the 'lookers-in' of W3XK, were estimated at over 20,000, scattered from the Atlantic to the Pacific." As a result of signal-interference complaints from radio operators, he was forced by the FRC to temporarily suspend broadcasting in early 1929. The results "brought a veritable avalanche of inquires and urgent requests for the resumption of radio movies . . . in spite of the impediments of homemade apparatus, low power, great distances and limited subject interest in the programs themselves." Effective January 1, 1929, the FRC had prohibited television broadcasting between the hours of 6:00 and 11:00 P.M. to clear interference with radio.[70] The prohibition favored radio's prime-time schedule, but it did not last long. The order was abandoned a month later, on February 18, 1929. Despite a month's suspension of programming, the numbers had increased, as reflected in communications received at the lab.[71] Newspaper reporters joined the bandwagon, fanning excitement, especially "the critical audience of newspapermen" who had seen the first programs, some of whom speculated that even Jenkins did not fully understand the ramifications or the "innumerable aspects television and radio motion pictures promis[ed] to mankind."[72] The inaugural W3XK-TV programs were successful by any measure.

Technical Standards Advance

Jenkins played a key part in developing film standards with the SMPE. By the end of the 1920s, history repeated itself with a call for television standards. The SMPE suggested that television standards should include the number of line elements and scanning method, the number of frames, picture repetitions per second, transmitter modulation, synchronizing transmitting and receiving, and synchronizing visual and audio signals, but the SMPE was a film organization in the 1920s.[73] Standardization was necessary "before television could become a national service."[74] Without uniformity, progress would be hampered. With it, station operators, programmers, and manufacturers could move more rapidly toward what the consumer was anticipating.

On October 9, 1928, the Television Committee of the Radio Manufacturers Association (RMA) and the Radio Engineers Club met in Chicago to formulate the specifics. Jenkins was part of this group. Their recommendations were threefold: (1) The scanning standard should be the same for all stations; the recommendation was forty-eight lines. (2) The speed of the scanning disk was recommended at nine hundred revolutions per minute, producing fifteen images per second. And, (3) The "sequence" of scanning recommendation was "clockwise down," meaning scanning began top left, moving to the right, and then from top to the bottom of the frame. The committee proposed that these standards be applied to all technology, including "Jenkins' drum apparatus . . . the conventional . . . disk, and . . . a cathode ray apparatus."[75] The committee fully realized that its recommendations were only temporary. They would change, as all acknowledged that forty-eight lines produced a poor-quality picture, but it was a beginning for standardization, bringing with it benefits for industry and the consumer. Interpreting these standards in modern terms, the committee recommended a picture of forty-eight lines, with fifteen frames per second. The term "frame," originating in this meeting, "referred to a single picture area."[76] Essentially, the committee adopted the standards that were already in place at the Jenkins Labs and 3XK-TV.

Not surprisingly, some disagreed with the RMA recommendations. Protests centered around the premise that experimentation using the same standards would hinder progress rather than promote it. William W. Harper, the consulting engineer for the group, responded sarcastically to the critics, suggesting that the reverse was true without question, unless those protesting standardization wanted all to agree "that any animal having four feet was a horse."[77] Standardization was essential to progress.

In 1931, these standards were increased to "60 lines per frame . . . and 20 frames per second." The arguments for standardization remained the same. "All radiovision stations . . . must adopt the standard. . . . Otherwise each household would have to buy a receiver for each broadcast station."[78] Standardization had become a growing movement, through it would not culminate for another decade. Only in 1941 did the National Television Systems Committee recommend 525 lines per frame with scanning at a rate of thirty frames per second.[79] This standard would guide technology until the twenty-first century and digital high-definition television technology, where standards improved again.[80]

11

The Jenkins Television Corporation

The years 1926 through 1930 were perhaps the best for Jenkins and his Jenkins Laboratories. Initially, he had all of the essentials for success—except investment capital to manufacture in quantity. The audience for his television station in Washington, D.C., was growing. He had a variety of television-receiver kits for sale. He enjoyed voluminous publicity from his experiments and demonstrations. He had aggressive plans for future growth. He still dominated the industry in terms of patents and saw significant potential from future licensing royalties. His position was dominant in synchronization, particularly the transmission of pictures and sound. He controlled patented technology for projecting and transmitting film. His new photo-cell receiver showed great promise for large-screen viewing and signal processing.

These assets were expected to increase in value. Jenkins "will probably dominate . . . the television art progress," claimed Delbert Replogle, assistant to De Forest president James W. Garside.[1] The primary challenge remained that Jenkins lacked the capital to compete in full-scale manufacturing. This was soon to change.

Jenkins reported that so much press attention was given to W3XK programming that it attracted investors from New York City and Palm Beach, Florida. The result was the genesis of the Jenkins Television Corporation, organized on November 16, 1928, under the laws of the State of Delaware.[2] It was designed for manufacturing and selling equipment created by the Jenkins Laboratories, and it enjoyed substantial backing.[3] It was financed to meet the demands for receivers, a demand that Jenkins had largely created from his experiments and promotions over the years. The new firm combined Jenkins' television and de Forest's radio patents, their technology, and their salable names. Sadly, these esteemed values would be cut short by the Depression.

Significantly consequential in the organization of the new corporation, Jenkins and de Forest signed over the patents from their radio and television work, as well as those patents assigned to Jenkins Laboratories. In in exchange, Jenkins received $250,000 ($3,311,594) in cash and stock in the company.[4]

The establishment of the new corporation embodied all of Jenkins' past electronic-media assets from the sale of receivers and transmitters; wire- and facsimile-transmission apparatus; machinery for transmission of visuals and weather maps for ships at sea; royalties from licensing and "benefitted organizations; broadcasting of visual entertainment, information and instructions; sales of motion picture prism disks, ultra-speed camera and other products."[5] Although he was undoubtedly happy with the arrangement, this event marked a turning point in this independent inventor's life. No longer were his patents his own; they belonged to the corporation. He and every member of the lab signed an agreement with the new company, assigning to the company all patents, past and future. And the company provided their salary and support for their research. They further agreed not to publish or authorize publication of any information related to their patents without corporate approval.[6]

One can only speculate as to Jenkins' feelings at this time. He had been well paid for his patent rights, and now he would have a salary and the capital to launch into full-scale television manufacturing. He had been instrumental in creating the consumer demand for entertainment television for a decade, and now he was preparing to fulfill it.

The new Jenkins corporation was announced as a "$10,000,000 [$132,463,775] company to sell television." The president and vice president came from the recently reorganized De Forest Radio Corporation: James W. Garside came in as president; and A. J. Drexel Biddle, chair of the board of directors for De Forest, was the first vice president of Jenkins Television."[7] Jenkins was also vice president in charge of research as well as a member of the board. However, he remained largely with the laboratory, where he had another title, president of Jenkins Laboratories. Garside indicated that the goal of the corporation was manufacturing and selling home-movie sets, duplicating "the grandeur of the Radio Corporation of America."[8]

Who specifically approached Jenkins with this financing scheme was initially unclear. The *New York Times* generically labeled them "Wall Street Bankers" but did not identify anyone.[9] *Time* magazine later reported that "C. C. Kerr and Company, of the New York Curb Market, was offering

250,000 shares of common stock priced at $10 [$132] per share," with public shares opening as well for total capitalization of the ten million dollars.[10] Interest was aroused as far away as Los Angeles and San Francisco, where stocks had started selling.[11]

One of the financiers was Garside, a businessman from Kalamazoo, Michigan, known for rescuing businesses and restoring their health. In the summer of 1928, he had been "summoned to New York City to meet with a group of capitalists and business leaders who had just taken over the meager assets and copious liabilities of the old De Forest Radio Company." Garside described the De Forest company of the mid-1920s as at "the end of its rope, so far as finances were concerned . . . [with] funds and borrowing capacity exhausted . . . [manufacturing] equipment having been idle for a year or two . . . [and] personnel scattered to the four winds."[12] The Jersey City–based De Forest factory Garside described as "a scrap heap," but he saw it as an opportunity.[13] Garside organized the capital for resurrecting the De Forest Radio Company, largely with his midwestern associates. He was to restore the name De Forest to the industry and build from it. His team literally cleared the seventy-thousand-square-foot older De Forest factory building in Passaic, New Jersey, of debris, "even old-styled De Forest Audions, were heaped high in the plant and had to be removed."[14] He was immensely proud of the fact that "three months later . . . the new De Forest organization was going full blast, producing 8,000 good audions per day," and he wanted to increase this production to a level of nine to ten million within a year, which translated to almost twenty-seven thousand per day, thus placing the De Forest Radio Company within its "rightful place in the radio industry."[15] Garside was optimistic. The De Forest Company was reorganized, and in its first year was beginning to make a profit, which Garside would use to organize the Jenkins Corporation and jump-start it into manufacturing.[16]

Shortly thereafter, the same people who backed the reorganization of De Forest were behind creation of the new Jenkins Television Corporation. What the name De Forest meant in the radio industry, the name Jenkins denoted in television. The difference was that when Jenkins Television was created, the Jenkins Laboratory was operating with a small staff and was productive. It produced small profits and was certainly not idle, but it lacked capitalization to undertake large-scale manufacturing. This created another business opportunity for the Garside investors.

The finances of the Jenkins Television Corporation were primarily generated from the sale of public stock. The marketing campaign for their sale reveals the reasoning of Garside in merging with De Forest. While De Forest had primarily been manufacturing tubes, this new organization could launch both companies into diversified product manufacturing. Aggregating patents provided De Forest an entry into television. Jenkins was provided access to audio tubes and electronic circuitry developed by the De Forest corporation. The united operation increased the size of the engineering, production, and marketing departments. Of specific interest to Jenkins was the larger and more efficient engineering team, of which Garside said that "no other organization [could] hope to compete." The new firm also presented the opportunity to manufacture a greater array of products, including those from RadioVision, RadioMovies, Jenkins' new high-speed camera, and other equipment under experimentation. The benefits for the stockholders were that the restored security of De Forest's stock was balanced against the Jenkins company's new, more speculative rising interests in television. Jenkins' stockholders were assured an immediate and positive return.[17]

Five months following the creation of the Jenkins Television Corporation, the De Forest company reported that it was "facing its best year in the history of the company."[18] De Forest himself was not involved; it was the corporate president Garside at the helm declaring that within the next year or two "we shall have television or sight broadcasting taking its place alongside sound broadcasting."[19] The new company would expand rapidly, building another television station, going after a New York metropolitan audience, and beginning "quantity production of television receiving sets . . . [retailing] at approximately $150 [$1,987]." Most importantly, it expanded the selling of public shares.[20]

The creation of the Jenkins Television Corporation "came at a time when Wall Street was talking of nothing but the break [a dip] in the [stock] market." Thus, the launch of the new corporation did not make a stock-market financial splash, which was apparently anticipated.[21] A stir *did* occur in the press, with talk of a new corporation to rival the Radio Corporation of America. Jenkins must have been ecstatic.

He was putting his television work into this latest organization. Everything he had developed and patented was now a part of the organization, and his future was dependent upon its success. *Time* suggested that "the success of the new organization depended largely on the past and future prowess of C. Francis Jenkins."[22] But Jenkins was really no longer the sole inventor on the television playing field. And he was only one part of the new corporate administrative group. He was their idea man, the inventor of fame, he had the name in television that had become marketable—but he remained primarily involved in the lab. They were businessmen who, he hoped, would rival the major radio corporations. They would guide the sales of Jenkins' equipment already sold and the decisions of the new manufacturing corporation, directing expenditures, future sales, and the development of media resources—such as the Discrola Inc. and Radio Pictures Corporation, which already existed and had equipment for home and industrial sale. Jenkins' name now appeared in corporate publicity; it was quoted by the new corporate leadership to underscore their credibility by utilizing his notoriety and building on his prior promotional momentum. Jenkins felt that he was in proven hands.

W3XK in Wheaton, Maryland

The Jenkins Laboratories made their initial move after the reorganization, building a new station facility and studio separate from the lab. The first location of 3XK had been in the lab in downtown Washington, D.C., at 1519 Connecticut Avenue, second floor, not far from Dupont Circle and the government centers of power. The building exists today. In Jenkins' day, it was close to residential areas. As a consequence, every time 3XK went on the air, there were complaints of Jenkins' experiments interfering with home radio reception. So he began looking for another location. He found it some five miles north, in Wheaton, Maryland, and purchased the real estate from a friend, likely his brother-in-law who was in real estate development. The new W3XK would be at 10717 Georgia Avenue.[23] Construction began in February 1929, just two months following the creation of the new corporation.[24] The station was declared "the first broadcasting [facility] ever built strictly for the sending of the new television," and it was expected to be on the air by early spring, offering silhouette pictures "large enough to entertain the entire family and friends of the family." The price of these receivers was not announced.[25]

The equipment for the station was transported north from the lab and installed by members of the staff. The first programs, broadcast from the labs, brought volumes of publicity and new funding opportunities. The new location was to capitalize on those opportunities. The Wheaton station W3XK broadcast on 2850 to 2930 khz and 2000 to 2100 khz for distant and local audiences, respectively.[26] The building also still exists, a 1930s home with two tall antenna towers in the backyard, but it was then a broadcast television station, simulcasting with a new "voice station," W3XJ, broadcasting synchronous sound. Jenkins described the results, realism, and the ability to see and hear as "absolutely uncanny."

A two-hour program schedule was publicized as the first broadcast from the Wheaton station on September 10, 1931, even though the station had been on the air from the lab location for more than two years. At this event, several

W3XK in Wheaton, Md. The station, studio, and transmitter were in this house, located about five miles north of Washington, D.C. This aerial view includes the studio and transmission towers during the late 1920s. This station broadcast programs featuring live local talent. Courtesy Wayne County Historical Museum.

Ground Plan of the Station at Wheaton, MD

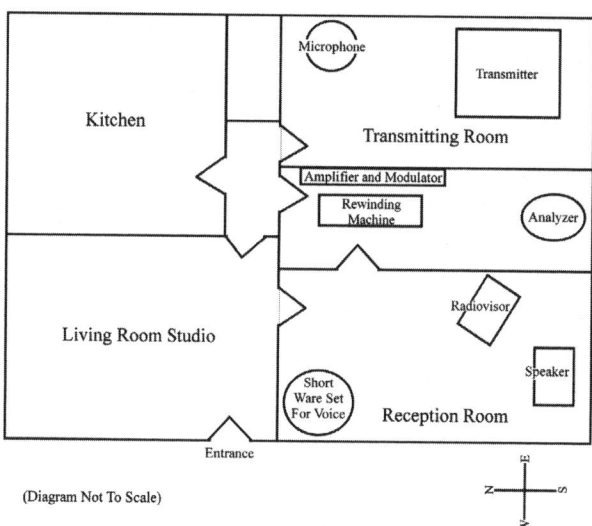

The floor plan of the station provides an idea of the process. Broadcasting began from the right-center room "analyzer." The analyzer cut the picture into component parts of light and sent it along for amplification and to the transmitter. The receiving room next door contained two receivers, which were sold by the New Jenkins Television Corporation. The receiving sets were still drum scanners, as the photo-cell receiver was still under development at this time. Courtesy Knight Graphics.

local people participated—George Clark sang; Mildred Battle whistled and sang; Ellory Stefford played the banjo; George Cook played the guitar; and eight-year-old Joe Chroney sang a song. The announcer for the program was the lab manager, Vera Hunter, who also sang the last song.[27]

Jenkins, the Jenkins Television Corporation, and Jenkins' receivers were widely touted at the trade shows for the next few years. A new receiver was launched for the Radio World's Fair in New York City, September 23–28, 1929. Jenkins had made vast progress toward the "simplification of operation." The set size came down into a small cabinet, but the screen was still only six by six inches. It utilized Jenkins' scanning drum.[28] The corporation produced almost five thousand of these sets for sale.[29]

The Jenkins station signed on and off the air, providing information regarding call letters, location, and technical information. The sign-on also requested audience reports on reception, which were used in the numerical tabulations of an estimated audience as well as any technical ideas the sender might have passed along. Each program was preceded by a short announcement. For example, the "Bouncing Ball" had become an often-used standards test. The announcement read, "You should be able to follow the movements of the girl with the ball and to see the pedestal upon which she stands. This is a silhouette."[30] As time progressed, halftones and then people began appearing on the screen. In Washington, D.C., the daily television schedule was published in the *Evening Star*. The schedule for the night of January 22, 1932, for example, read:

> **W3XK** Television, 145.1 Meters, 2,065 Kilocycles[31]
> 9:00—Television's Album of Artists.
> 9:25—Joe Koenig and Mary Carhart.
> 9:30—Sacred Program.
> 10:00—John R. Clarke, Crooner.
> 10:30 to 11:00—Studio Feature.

Jenkins continued his outreach into the community. The opera singer Dorothy Davenport of the Washingtonians sang, and thirteen-year-old Richard Davis played the violin. Singers, vocal groups, speakers, and artists of all kinds began to help in the promotions and take advantage of the opportunity to appear.[32]

One of the most interesting W3XK-TV program innovations was a contest calling for talent to perform in conjunction with the Washington Industrial Exhibition in October 1931. According to the *Washington Post*, the District of Columbia had been chosen for the exhibition because Jenkins had the ability to broadcast live from W3XK to the Washington Auditorium where the exhibition was hosted.³³ An advertisement for "amusements" at the exposition read, "Sensational Demonstration of Television, Feature of the Wash. Chamber of Commerce, Industrial Exposition, Opening Tonight, 8 P.M., the Auditorium, 19th and New York Avenue."³⁴ Jenkins even accepted the invitation to appear in person at the exhibition, provided his health, which was worsening, would allow. His lab team assembled the display of Jenkins' products for the show.

The talent contest was sponsored by the *Washington Daily News* and W3XK and simply asked for people to apply. The newspaper announcement read, "Do you think you possess television talent?" This was, of course, to promote interest, but if the readers thought they had talent, it was as simple as filling out a coupon that appeared with the article and sending it in with a photograph and "a description of ability."³⁵ Six winners were selected to perform at the Wheaton studio, and it was all staged so that they would be seen on a large screen to highlight the Industrial Exposition.³⁶ The winners performed, one each evening at 9:00. They were Arelia Colomo, headlined "Tonight's Star"; Louis Benson, "Ether Star"; Rosemarie Colomo, "Heard and Seen"; John F. Costello, "On Air Tonight"; Bertram Wissman, "Televisioner"; and Dorothy Edythe Davenport, "On Television." Interestingly, the newspaper articles identified each contestant as a winner and reported their Washington and Baltimore home addresses but did not mention their talent or describe their performance. This was likely because the excitement was focused on new technology over program content. The television listing for each night simply noted the "*News* Television Contestant" appearing at 9:00.³⁷ Reports of exhibit attendance ranged between ten to eighteen thousand.³⁸ The exhibits and the live television demonstration were clearly a success.

W3XK was improving steadily, giving the new company increasing exposure. Replogle suggested that it provided the company "a prestige in Washington that [it] could not have had in any other way." The station hosted a growing number of daily visitors from around the country, sometimes twenty-five to thirty people a day. The signal, Replogle continues, was received by amateur operators across the United States and Canada, with W3XK perceived as "the pioneer station of the United States."³⁹ Significant growth was also occurring as the Jenkins Television Corporation continued marketing Radiovisors. It had labeled itself in advertisements as "Jenkins First in Television" and crowned Jenkins as "the Father of Television."⁴⁰

Jenkins' personal leadership established the foundations and equipment for these stations. He had simulcast sight and sound from W3XK and W3XJ; he had produced film and live program content. The station had a growing audience, which gained enough attention that Jenkins started to use the station for advertising his equipment. The Federal Radio Commission accused him of broadcasting commercial messages, which were prohibited under provisions of the experimental license. He was presenting an announcement targeting ham operators, selling kits and parts for their construction. He was ordered to stop, and he did.[41]

Jenkins worked to develop theater-sized screens at the station and proclaimed them large enough to show pictures in a Washington, D.C., theater in 1931. At the same time, company executives were moving to create additional television stations for promoting a corporate presence in New York City.[42]

Stations in Jersey City and New York

The day before President-elect Herbert Hoover addressed the nation from the NBC Radio Studios, just prior to his inauguration, Jenkins' corporate president James W. Garside announced construction of a new manufacturing plant and a television station covering the New York metropolitan area. The manufacturing plant in Jersey City was to begin the production of television receivers within two weeks. These were designed to plug easily "into existing radio sets to receive the image broadcast."[43] The station, W2XCR, began broadcasting later in the spring of 1929, with the transmitter on the roof of the manufacturing plant, just across the Hudson River from New York City. Programming was intermittent, but by April, silhouette pictures would begin.[44] In the fall of 1930, it was on the air regularly with sound synchronized with WGBS New York. Simultaneously, at De Forest corporate headquarters in Passaic, a new vice president for engineering, Allen DuMont, established the station W2XCD primarily for technical experimentation for its first year.[45] Regularly scheduled programming was carried only on the stations in Wheaton and Jersey City.[46] W2XCR would move to Fifth Avenue in New York in early May 1931, the audio channel remaining on WGBS. W2XCR combined with WGBS to air "radio talkies," attracting several thousand spectators with a receiver placed in Aeolian Hall at Fifth Avenue and 54th Street.[47] It was declared "the first synchronized sound and sight broadcasting from New York."[48]

The De Forest Company vice president, W. J. Barkely, announced to his managers that in conjunction with the Jenkins Television Corporation they now had a fully practical television system. The announcement reveals something about the workings of the early Jenkins Television Corporation and the De Forest Company. While there were legal separations between the two corporations, this was primarily to facilitate the sale of stocks for each company, thus raising development funds for both. The management officers in the companies were primarily the same, and they operated as one. The De Forest Company, whose chief product had heretofore been the manufacture of audion tubes, had taken on a new, aggressive challenge—Jenkins Television was the basis for De Forest expanding into an "entirely new radio trade."[49]

On June 24, 1929, the Federal Radio Commission announced its approval of three new experimental television stations in the New York area—W2XCR-TV, the Jenkins operation, was among them. It operated on 2,100–2,200 kilocycles and with "five kilowatts power." Silhouette motion pictures were still the mainstays of programming, and the corporation was now one among several telecasting in the metropolitan area.[50] W2XCR programming could be received with a "special Jenkins Television receiver for reception of both sight and sound."[51] On May 10, movies from the Visualgraphic Pictures Corporation (which Jenkins had earlier contracted for W3XK) were producing films for broadcast, as "listeners who had radio sets equipped for television reception" tuned in with increasing enthusiasm.[52] Plans were under way for the establishment of stations in the larger urban markets across the country.[53] And for connecting these stations, Jenkins had another idea.

Aerial Television

Little is known of Jenkins' plans for W10XZ and W10XU.[54] W10XU was granted a license to operate on 2,000 to 2,100 kilocycles and with a power of ten watts. W10XZ operated at 1,608, 2,325, 3,088, 4,785, and 6,335 kilocycles at a power of six watts. Presumably, one station was visual, the other aural. They were licensed to the Jenkins Laboratories without a designation of geographical location—just labeled "airplane." This eye-in-the-sky concept put an electronic camera and a transmitter into a Stinson four-seat aircraft."[55]

The experiment was a natural extension of Jenkins' love for flying and his resulting aeronautical inventions (see chapter 6). It was first conceived for military surveillance. Jenkins was developing "an aerial radio television 'eye,' which could pry into enemy secrets hundreds of miles away . . . transmitting troop movement or fortifications to a ground receiving station . . . [and providing] a moving picture of the enemy territory, fortifications . . . and supply depots." The moving visuals recorded from the air could all be watched from military headquarters.[56] Permission to conduct these experiments apparently required a federal license and mandatory training for a permit to fly, even though Jenkins was already a licensed pilot, having flown his own planes for the last decade. Little is known about the resulting application of this military system, but within four months, Jenkins' emphasis had shifted to programming for the home viewer.

The experiments for home transmission were conducted in flights between Washington, D.C., Philadelphia, and Norfolk, Virginia. On one trip, Jenkins apparently experienced a forced landing on Long Island, and the equipment was damaged. A new plane was quickly reequipped, and experiments continued. The camera and transmitter were battery-operated and fixed to the floor of the plane. Jenkins was quiet about these experiments and would not allow the press to take pictures of the assembled equipment. They were permitted only to photograph the airplane and Jenkins himself.[57] Conducting his experiment on July 23, 1929, again before Federal Radio Commissioners, Jenkins said "that the new station would be used principally for transmitting radio movies."[58] The basic concept was now relaying and rebroadcasting signals to distant points, for defense or commercial application. At first, the actual pictures were panoramic views of Washington, D.C., but it would seem apparent that Jenkins was seeking a new way to network stations together and exchange programming information among them. W10XZ was Jenkins' flying laboratory.

Lawsuits Surround Progress

The year 1929 had started pleasantly. Tests in transmitting from the "flying laboratory" were successful. The FCC authorized the Jersey City station. W3XK was building its new studio in Wheaton, and tests involving synchronized sound and visual programming were successful. Jenkins was awarded an honorary doctorate from Earlham College, and he flew to Indiana in the laboratory plane with his wife Grace to visit family and receive the honor.[59] In contrast to this good news, Jenkins was hit with two separate lawsuits in rapid succession.

The first was filed on July 13, 1929, by Arthur D. Lord, a New York broker and attorney. Apparently Jenkins had hired Lord to promote the sale of his work. Lord asserted that he had paid Jenkins $110,000 ($1,457,101) as a part of the agreement and was asking for the recovery of $612,000 ($8,106,783) owed him "as a brokerage commission in furthering the sale." Besides Jenkins, the Jenkins Television Corporation, Discrola Inc., the Radio Pictures Corporation, the Jenkins Laboratory, Grace Jenkins, and the Park Savings Bank were all named in the proceedings. Jenkins was a member of the board of directors for the Park Savings Bank, and this was apparently where he had deposited his personal profits from the sale of his patent rights.[60] Lord asserted that his contract with Jenkins "was in effect when Jenkins sold out."[61] Lord called witnesses, including both de Forest and Garside, and pursued the action with vigor.

The second lawsuit was filed only two weeks later on July 26, 1929, by the Radio Service Corporation, seeking to recover four hundred thousand dollars $400,000 ($5,298,551) in cash and three million dollars ($39,739,133) in stock value.[62] The plaintiff requested that an injunction order be issued prohibiting "Jenkins and the other defendants from removing proceeds of the sale from [Jenkins' assets within] a safety deposit box in the Park Savings Bank."[63] Jenkins responded to the Radio Service Corporation's claims, explaining that they "did not fulfill the terms of the contract." He simply asked that the suit be dismissed.[64]

The substance of both cases was related. They claimed that the Radio Service Corporation had placed Jenkins in contact with Lord, who in turn had supposedly introduced the investment capitalists "Garside, Wiley H. Reynolds and Lee Deforrest [sic]" to Jenkins, resulting in the formation of the Jenkins Television Corporation. Thus, the claims were solely about sales commissions. The Radio Service Corporation contended that it was entitled to "a 20 per cent commission."[65] It is interesting to note that in the *Washington Post's* report on the lawsuit in 1931, the dollar figures demanded had changed dramatically, and the plaintiff had labeled the Lab as a simple holding company for Jenkins' patents, ignoring the fact that it was a working lab and had been in operation for almost a decade.

The lawsuit of Arthur D. Lord was settled out of court with "no disclosures . . . as to the amount of the settlement."[66] The suit of the Radio Services Corporation was dismissed, as the contract they had entered into with Jenkins expired in 1927.[67] As there was no continuing agreement between the Radio Services Corporation and Jenkins, he was free to negotiate with others. The opinion of Justice James Proctor, Washington, D.C., District Court, concluded that Jenkins had acted in "good faith and honorable conduct."[68]

The Stock Market and Jenkins' Corporations

The Jenkins companies were doing well in 1929, but that would change quickly when the stock market collapsed in late October. In understanding what happened to Jenkins—as a person and a corporation—it is critical to understand the financial context of the 1920s. One reason the decade was referred to as the "Roaring Twenties" was because of the spectacular changes and achievements emanating from the decade—a part of which came from Jenkins with his photographic, RadioVision, and RadioMovies operations, not to mention his own enthusiasm and prophetic publicity. It was a post–World

War I decade, and "the business of the nation was business," according to President Calvin Coolidge.[69] Businessmen were trusted, and fortunes were being made. The sale of radio sets grew from sixty million dollars in 1922 to $430 million by 1925.[70] Jenkins and all other manufacturers were sure that television would surpass these radio sales figures. The stock market too was changing—everyday people were beginning to invest. Stock market returns ranged from 26 percent in 1924 to 48 percent in 1928.[71] Times could not have seemed better. In mid-December 1928, as the Jenkins Television Corporation was created, the stock market dipped and took attention away from the corporation's stock offerings. But the market had always recovered, so industry and inventors paid it little attention.

The Jenkins Television Corporation was formed ten months before Black Tuesday, October 29, 1929, and the beginning of the Great Depression. Financial disaster would slow corporate progress in the United States and worldwide. Eventually, it would lead to the sale of the Jenkins Television Corporation. Just a few months after the stock market crash, in the summer of 1930, Jenkins proposed to the company's board of directors that, due to the financial exigency, he "would be relieved somewhat if he were to resign, as he was the highest paid officer in the company—the Board accepted the resignation." Jenkins retained responsibility for the physical properties and management of the Lab, so he could continue his work. In an urgent special-delivery letter, Delbert E. Replogle, an assistant to president Garside, spelled out corporate plans for the Jenkins Laboratories. Jenkins would take over the lease, use the "present quarters and equipment," and at "the end of your occupancy, [it would] be accounted for intact except for reasonable wear and tear and unavoidable accidents." Vera Hunter, Jenkins' office manager, was to be retained on Jenkins' payroll, and it was her responsibility to handle communications and correspondence and "maintain the sale of kits." It would appear that the corporation agreed to retain one Theodore Belotte as a transmitter operator for the Wheaton station at a salary of thirty-five dollars ($476) per week. Stuart Jenkins and Paul Thompson also remained, at $1.85 ($25) per hour, to operate the station for two-and-a-half hours each evening, "and to give time when necessary, for the maintenance of the station," when lab work allowed. Replogle, who had been in communication with Garside, noted the corporation's appreciation for Jenkins' willingness to provide special services where he or his staff could contribute. "The letter is notification that your employment . . . is to terminate August 16th, 1930."[72] Effective December 18, 1930, Allen B. DuMont would take over all operations of the Jenkins Television Corporation, assuring that "all concerned will continue

A 1930s Jenkins Laboratory staff photo of the key people in Jenkins' small lab and station organization. Front row: Sybil Almand Windridge, C. Francis Jenkins. Florence Anthony Clark. Back row: Stuart Jenks, Elwood Russey, Vera Hunter, John Ogle, Paul Thomsen. Courtesy Wayne County Historical Museum.

these arrangements with their former loyal support."[73] Jenkins also remained in charge of the Washington, D.C., station.

Corporate Fall

"Great [were] the difficulties facing radio television" at the beginning of the Depression.[74] It was only beginning to make the transition from the cat-whisker stage into commercial development. Despite the challenge of low resolution and poor picture quality, interference and low-quality receivers, the Jenkins Television stations W3XK and W2XCR were broadcasting a regular schedule to an excited and growing audience. On March 3, 1929, W3XK had broadcast President Hoover's inaugural parade. It was on a thirty-minute delay, as pictures were taken of the parade and rushed to the lab for rebroadcast, but excitement was building.[75] Jenkins Television was ahead of the race, so when David Sarnoff, RCA president, revised his prediction of television coming in the next five years to "just around the corner," and Delbert E. Replogle, Jenkins Television's vice president, "laughed at Mr. Sarnoff's statement," "it sparked a trade war," which *Time* inferred had "been going on privately for a long time."[76] Independents were on one side—including Jenkins Laboratories; Farnsworth Television in San Francisco and Philadelphia; the

Western Television Corporation in Chicago; and Shortwave and Television Laboratories in Boston. On the other side was the Radio Trust of General Electric, Westinghouse, and the Radio Corporation of America. According to *Time*, the Radio Trust group was working in secret, trying to perfect their own apparatus, while the independents were forced to seek publicity to obtain the funding for continued development. RCA was under pressure from all sides and sought to purchase competitors or discredit independent inventors and take the credit for themselves.

Jenkins' workaholic personality had resulted in orders to slow down from his doctors. He did not, and in 1931 a heart attack resulted. The timing was uncontrollable and unfortunate, but it had been progressing over the years. Working vacations did not satisfy the requirements of renewed health. At 10:00 on the morning of June 30, Jenkins suffered a serious heart attack.[77] Only a few months later, in September 1931, while on the train to a California SMPE meeting, he suffered a physical breakdown. He was helped off the train in Chicago and returned home, where he was confined to his bed. A lifetime of working long hours had led to his collapse. His workload was distributed among assistants at the lab and W3XK, and the commercial business burdens were shifted over to Jenkins Television Corporation management.[78] As a result, decisions were made without Jenkins' direct input, discussion, or influence.

Garside's management vision of rivaling RCA had revived the De Forest Radio Corporation and organized the Jenkins Television Corporation into a position of short-lived manufacturing—but it was not to be. After the stock market crash, the company stumbled. The De Forest Company had created initial cash flows through the sale of audion tubes, which were now declining. Together, the two corporations were selling stock to support expanded physical facilities for both.[79] Just thirteen months after Jenkins Television was organized, the De Forest board of directors proposed purchasing the stock of the Jenkins Television Corporation. The offering was one share of De Forest for a one and three-quarters share of Jenkins.[80] As a result, the De Forest Company increased its share to 345,680.[81] It now had complete control of the Jenkins Television Corporation. In 1932, the merging of the corporations was proposed for financial streamlining. The De Forest investment, required to launch television and support both corporate organizations, as was the original plan, was creating debt due to the deceasing demand for audion tubes and television sets. At the same time, Jenkins' participation was curtailed, diminished by health and the reorganizations. The rationale of the corporate executives was that the "indebtedness would be cancelled by absorbing the

Jenkins assets and the subsequent activities of [the] enlarged organization." Thus, they could cover the debt and continue the work. Jenkins' patents alone were "set at a value of $2,000,000. [$33,215,074]"[82] However, debtors called for the appointment of the first bankruptcy receivers for the De Forest Radio Company on June 24, 1932. The value of the De Forest assets, estimated from the patents, plant equipment, and "materials," totaled $950,000 ($15,777,160).[83] A year later, on February 1, 1933, these same bankruptcy receivers were appointed to dismantle the Jenkins Television Corporation with assets, which had dwindled to an estimated $300,000 ($5,250,011) in patent rights with indebtedness now at $388,000 ($7,790,015).[84] On March 3, 1933, the U.S. District Court for the District of Delaware authorized the sale of all assets.[85]

The Depression had forced the downfall of the Jenkins Television Corporation while aiding RCA. Previous investors were now uncertain and unresponsive. Advertising on the Jenkins experimental television stations in Washington, D.C., and New York was still prohibited. The only remaining revenue stream was from the sale of receivers, and the Depression had cut deeply into those sales. The end was self-evident. Anticipating the court's order, on March 7, 1933, RCA president David Sarnoff took the opportunity to squash the Jenkins and De Forest companies, which only a few years prior were forecast to be RCA's chief rival. He offered "25 percent of the companies' worth, including many patent rights . . . $500,000" ($8,750,020).[86] By mid-July 1933, the Jenkins Television Corporation assets were sold to the same receivers, for $200,000 ($3,500,003).[87] All of Jenkins' patent rights went with that sale. Sarnoff, who felt "no moral obligations" to anyone, had purchased the Jenkins company and De Forest operations "for no other reason than to eliminate competition . . . hundreds of patents issued to Charles Francis Jenkins would never again see the light of day."[88] This included those related to his optical-electric photo-cell transmission, reception experiments, and television for the theaters. W3XK was closed, and its two sixty-foot transmission towers were sold for scrap metal.[89] It was over. Jenkins' rights, his ideas, and his leadership had been extinguished by ill health, the economy, and a bigger corporate bat.

A few blamed the purchase of Jenkins Television for the failure of both companies. One newspaper headlined "De Forest Radio Loses $106,800 [$1,869,004] on Poor Buy," but the article referred to the purchase of radio cabinets that were unmarketable after the Depression and made no reference to the Jenkins Corporation. The headline implied an unfortunate inference at the time the companies were in receivership and being dissolved.[90] In a 1950 interview, Lee de Forest was quoted as describing the purchase of the Jenkins assets as "a wicked, utterly indefensible purchase," and at the same time he blamed earlier investors for "squandering one and one-quarter million dollars [$16,032,307]" in 1926, thus sending his company into what Garside had described as a scrap heap.[91] The blaming of Jenkins seems inaccurate. The De Forest Company of the time was already unstable.

So, what really led to the downfall? First and foremost, the Depression, which frightened off investors and decreased industry and consumer demand for products. Additionally, a series of unfortunate events, including the 1931 lawsuits from those seeking commissions, and the earlier 1928 sale of Jenkins' assets had taken an undisclosed financial settlement and an emotional toll. In 1932, a fire struck W2XCD, and the station and the adjacent laboratories were destroyed. It had started when "an electric arc burst its glass vacuum tube." Damage was estimated at sixty-five thousand dollars ($1,079,490).[92] Finally, key people were leaving the company. Allen DuMont, an engineer hired away from Westinghouse in 1928, left the corporation in 1931.[93] Jenkins had given up his salary and was concentrating on the Washington, D.C., station, the laboratory, and photo-optical and theater-sized projection. The Depression had pushed circumstances beyond management control.

Jenkins' Death

Jenkins worked too hard. Doctors had told him throughout his life that he needed to mind his health. Vacations and flying were his most common response, but too many were working vacations without physical respite. Even his love for flying was work-at-play and did not slow him down. He had never fully recovered from his first heart attack, though he continued working in the lab. Lab personnel brought a television receiver to his home so that, when his health dictated, he could direct the staff from home.[94] Grace recorded his recuperation and continued business activities in her diary. She was always optimistic that "Francis [was] getting better . . . [though he was] not able to go to the Laboratory yet"; this entry was followed less than two weeks later by one noting that he was at the lab meeting with "Warner Bros, Chicago, to see Talkie movie."[95] His health went from normal to bedridden continually over these final years of his life.

It was during this time that Jenkins wrote and published his autobiography, *The Boyhood of an Inventor*. His preface sets the stage for the significance of great inventors and himself. Written in the third person, "the boy" is Francis. "Just how the natural inventor differs mentally from the only destructively inquisitive boy may not be known [but it] may be helpful in discovering this special talent," as compared to psychology, and "this must be the excuse for the following recital of the activities of a boy who later contributes to human advancement in many fields."[96] The book was printed and distributed in 1932.

A few short months before his death, his health prohibited further participation, and Jenkins was forced to abandon his efforts. He closed the last of his namesakes—the Jenkins Laboratory.[97]

Jenkins' second heart attack occurred on March 20, 1934. It was severe. Grace sent a telegram to Francis' brothers, Atwood and William, informing them of the attack. She gave them a little hope, telling them Francis was doing better and "resting more comfortable this morning."[98] In just two weeks, the news reported his death. The telegram to Atwood, dated June 6, 1934, read only, "FRANCIS PASSED AWAY THIS MORNING AT TWO FIFTY."[99]

The doctors called it angina pectoris, a condition brought on by "untiring work over the complex problems of his laboratory that [had] overtaxed his heart and broken his health."[100] Jenkins' death occurred only a few days after his father's ninetieth birthday. The funeral included an aerial tribute with red roses tossed from the plane, services at home, and internment at the Rock Creek Cemetery.[101]

Francis willed his entire estate to his wife Grace. One press report at the time indicated that holdings were limited to his library, lab equipment, and "trinkets."[102] This description was inaccurate, likely due to the limited financial information and casual comments made prior to a financial assessment of the full estate. The bulk of Jenkins' library, "510 television books," had been already given to the Bliss School. In 1929, he had been elected to the board of directors of the Federal-American National Bank and Trust Company, "one of the few scientists in the world of finance."[103] As late as July 1932 he had purchased one hundred shares of "Telephone Stock . . . for $7,492.50 [$124,432]" and sold National City Bank stock for $4,125 ($68,506).[104] These were not the activities of a man in financial crisis. The full appraisal followed Grace's death on June 13, 1943. She had been ill for an extended period of time and was known widely for her "interest[s] in church and charitable work as was her husband."[105] She willed the estate to be divided among her own family, her nieces and nephews, and that of her husband's two brothers and the children of his brothers and sisters. The value was nearly $260,000 ($3,405,393). This included eighteen thousand dollars $18,000 ($235,758) for her home at 5502 16th Street in Washington and $214,854 ($2,985,745) in "personal property."[106] This is likely from Jenkins' sale of his patent rights, at which time he had "created a trust fund to take care of Mrs. Jenkins for the rest of her life. The rest he and she gave away outright and unconditionally to poor relations, friends, and servants whom they had had from time to time."[107] It is noteworthy that this approximates the value Jenkins received for the earlier sale of his patents and rights. He left her in good care. She distributed her inheritance from Francis, as he would have desired.

12

American Visionary

Charles Francis Jenkins was an important inventor who created breakthrough turning points in two major industries—film and television. He was always forging across traditional boundaries and looking for new ways to bring things together. He was a versatile inventor and a workaholic who seldom stopped to rest. He never gave up. In film, he sold his controversial Phantoscope projector patent, which led to large-screen movie projection, and he watched motion-picture theaters grow into a billion-dollar industry.[1] His diversified interests included the automobile, and "in 1898 he scared the wits out of [congressional] representatives' horses . . . driving a 'horseless carriage,' [one of] the Capital's first automobiles."[2] He developed instruments for aviation and almost lost his life in his flying television laboratory.[3] He kept the Jenkins Laboratory operational into the early Depression years, indeed, until his health dictated its closure.

In television, he bridged mechanical with electronic technology, later experiments related to fiberoptics, and electro-optical receivers. Press reports labeled him a "martyr to science."[4] The *Washington Herald* described him as "one of the greatest . . . and least-known inventors of our time. . . . He literally gave his life to science."[5] The *Washington Post* described him "as one of the world's foremost scientists."[6] By 1927, the *New York Times* had labeled him as "the man who was keeping America in front ranks in the development of radio vision, or 'seeing' by radio."[7] The *Christian Science Monitor* asked in 1926, "Do you realize that C. Francis Jenkins is the man who not only had the vision, but the daring to attempt to unite the two most popular subjects, radio and motion pictures?"[8] His hometown paper, the *Richmond*

Palladium and Sun-Telegram, reflected irony in the fact that his death came "exactly forty years [after] the young inventor showed motion pictures for the first time."[9] He was the only inventor who participated in the birth of both motion-picture photography and television.

Jenkins' Youth

Jenkins' agrarian upbringing created within him an independent will, an untiring work ethic, and strong character. The farm proved a haven for an innovative young mind. His aptitude for fixing and inventing things was his "gift from God," his parents told him, making sure he understood the source of his talents. His character was forged at a young age, and his Quaker heritage influenced him throughout his lifetime. The family believed in a personal God who encouraged hard work and service. Young Jenkins loved nature, the stories read daily from the Bible, and the apocryphal tales told of the Wild West he heard at his grandfather's knee, which no doubt fostered his love for travel as well as his curiosity. He was a romantic who loved people and life. As to formal education, he had only a high school diploma and one year at college, but his lifetime of innovative output eventually earned him an honorary doctorate. He was continually seeking new and practical inventions. He said, "New thing[s] always originate in a single brain, usually the brain of a poor man. . . . It is not the product of great wealth and a great laboratory. Money only develops, it never originates."[10] This would be an accurate reflection of Jenkins' work and himself. He would seesaw from rags to riches, back and forth throughout his life.

Jenkins the Man

Outside of his role as an inventor, what do we know about Jenkins the man? He was described by the *Christian Science Monitor* as a "happy warrior." Though he was "worn by almost ceaseless laboratory labor for 34 years, [he] bubble[d] over with the enthusiasm of a boy with a new electric trains."[11] He was a constant explorer and inventor. He reveled in the puzzles of how things worked, not only in the lab but at home. He saw the natural world as filled with wonderful questions—over which he speculated answers. What makes a flag flutter in the breeze? Why do a plant's leaves pop "up" from the ground? Why, when a log is thrown into a river, is it drawn to the center of the stream? How does a bird glide with the wind?[12] He was a naturally inquisitive individual.

Jenkins was also a man of deep faith who was devoted to his family. He loved children. He was a member of Quaker and Methodist congregations, and in their travels, he and his wife attended religious services when they could. Grace noted their church attendance (or nonattendance) in her diaries every week. If they missed a Sabbath service, it was usually due to illness, and this too was noted. He felt that it was his "good fortune . . . to have chosen farmer forebears, of sturdy Quaker stock," as this was the foundation of his "natural inquisitiveness [and] a mechanical intuition."[13]

Jenkins sided with the fundamentalists on the issue of man's creation. Charles Darwin's *The Origin of the Species* popularized the debates between science and religion, "modernists versus fundamentalists," before and during the Roaring Twenties.[14] Scientists expanded upon Darwin's theory of evolution, while the fundamentalists held to a strict interpretation of the Bible. Jenkins played no role in the overarching public debate, but he was not without an opinion. "We do not know how or whence life came up on this earth," he declared. Even the "scientists are not in agreement on evolutionary theory." While he would not completely discard scientific principles, he clearly saw man as a spiritual creation. "Man invents tools. Man alone uses fire. Man buries his dead. Man has commerce and trade. Man draws pictures, and can read and write. Man is a spiritual being and believes in a life after death. Man worships a power outside of himself and builds altars and temples for such worship. No animal does any of these things."[15]

Jenkins supported the prohibition of alcohol. Two years before his death, he donated "$500 [$6,304] to the Anti-Saloon League," of Washington, D.C., and other organizations seeking to suppress intoxicating beverages.[16]

Jenkins was a family man. He described the family as a characteristic that separates human life from that of the animals, as supported in Old Testament scriptures.[17] In his young-adult years, after he had left home, Jenkins returned every year to visit his parents, brothers, and sisters. As a mature adult, he lived with an extended family around him, and they worked shoulder-to-shoulder. His family and his extended families took part in his experiments. They were not only inexpensive labor, working without pay, but he enjoyed their company. When he married Grace, her family became his. His surviving letters illustrate a strong relationship with his parents, brothers (particularly Atwood), and sisters. He frequently took his father and other family members flying.[18]

Grace and Francis had no children of their own, but they were surrounded by Grace's sister's children, and they too became an important part of Jenkins' personal and experimental life. He encouraged their education and helped them financially whenever he could. He was generous with his time, money, and affections.

Flying and other travel were Jenkins' only real respite. Yet, even they were generally combined with work. The 1912 coast-to-coast automobile tour with Grace was an extended journey on which he served as one of the official photographers. He often flew himself to his out-of-town meetings and speaking engagements; as he described such a journey to his father, "at 2000 feet houses looked small . . . but the roads and waterways . . . guided our course [and] were easily enough seen."[19]

Youth-group appearances too were a part of Jenkins' public schedule. A. B. Graham of the U.S. Department of Agriculture, Extension Service, described Jenkins' effectiveness with young people. His lectures were "all such a human experience . . . that make us feel akin to each other. . . . [Y]our entire story was enlivened by happily chosen everyday examples and illustrations. The little touch of light-veined expressions that ran through it relieved it from the seriousness and tediousness of some presentations." His approach was to challenge the inventiveness of the young. "Everyone concluded, children and all, that you are a child's man—that your years are measured only by the almanac, and that your interests are in the life of today and tomorrow. That means, of course, that you are intensely interested in youth."[20] The American University Board of Trustees, of which Jenkins was an active member, summed up his enthusiasm for teaching. "His was the pure heart. He was transparent as a child . . . a devout churchman . . . intensely loyal to all the institutions of the church. . . . He kept his own heart young, and enthusiastically embraced every opportunity to promote interests of young people."[21]

Jenkins was a romantic. He loved life and enjoyed those around him. Not too many doting young men have ever "enticed away the watchman at the railhead, so that I might borrow the locomotive and take my prairie playmate for a ride."[22] His other dates were more conventional, taken in buggies, sleighs, toboggans, and on bicycles, as well as motorcycles and automobiles. The experiences in the outdoors, whether riding a train or rowing up river with Grace, were his most common activities. He did not like dancing—indeed, he was fearful of the dance floor. On the dance floor, he said once, "my hands were so big and [I had] no place to put them. . . . Of course the music had to stop when I was in the middle of the floor, a mile or so from the side lines where [a] thick rug gleefully awaited to trip me up. . . . I felt that all of Washington was looking at me."[23]

He loved the city of Washington, where he spent much of his adult life. From his initial employment with the Life Saving Service, he felt that he had moved into the center of the world's activities. He described the elegance of the evenings as "the lights of the city . . . in bloom." Touring a cousin on foot and bicycle, they were the excited and like "foolish virgins, our lamps . . . continually going out, as the owners of the wheels had neglected to give us a sufficient supply [of oil]."[24] Jenkins took his guest to the White House, the Smithsonian museums, and other sights, describing them in detail. The "grand[est] piece of work, erected by the women of America, pertain[ed to] the memory of the Father of the grandest country on earth."[25] He described Washington as a "City of Enchantment . . . a stimulus to excellence . . . [a] magnificent dream city . . . [a city of] unusual aggregation of mentality—scientific and literary and industrial. . . . Washington well deserves the pride of possession of all worthy Americans." Washington, D.C., placed Jenkins in the very center of inventive, political, and creative activities. His lab was never far from the centers of power. A phone call would bring official representatives of the navy, the Department of Commerce, and later the Federal Radio Commission. It was a city where "poets and great writers, noted scientists and renowned inventors have done their best works in an invigorating atmosphere," a gathering place of "eminent statesmen and great orators."[26]

Jenkins' adult life and experimentation reflect interesting parallels with the television inventor Philo T. Farnsworth. They were both workaholics, and both were dedicated Christians (Jenkins was a Quaker, Farnsworth a Mormon). Both were extremely strong-willed independents. RCA eventually had a hand in the end of both careers. Both made financial decisions that through no fault of their own would lead them from rags to riches and back again several times. Both worked in electronics—Farnsworth and Zworykin were becoming the leaders in electronic scanning as Jenkins moved from mechanical experiments to optical electronics. Farnsworth and Jenkins suffered financially as a result of the Depression. Both died of overwork that simply wore out their bodies.

Jenkins the Inventor

In 1894, Jenkins debuted the showing of motion-picture film on the large screen. His projector, which he perfected with Thomas Armat, transformed the film industry and the consumer's viewing experience. Jenkins' niches in photographic technology led him to create equipment for independent producers who would play major roles in the industry. He created equipment for filmmakers such as Burton Holmes, Siegmund Lubin, and Herbert J. Miles.

He manufactured and sold equipment to Carl Laemmle Sr., who founded the Independent Motion Picture Company, which would later become Universal Studios. He produced fireproof projectors for educational use. He strongly believed that film, motion-picture, and television technology had the potential to bring the world together through a common visual language.

At the same time that Jenkins was developing a reputation in motion-picture engineering, he diversified his inventive portfolio, manufacturing automobiles and creating home-heating devices, sanitary milk bottles, and aeronautical instruments. His most significant product of these many inventions was his sanitary milk bottle. These products reflected the needs of their time, and he used them to support continuing work in film and later television. Most of these inventions, he said, were "pot-boiler things I sold for from $250 to $2,500 [$3,269 to $32,685] to give me a living and pay for the greater things I had in mind."[27]

The greater things Jenkins had in mind came from the Jenkins Laboratory, established in 1921. It focused on television and would be his work center for the last decade of his life. From the lab, he transmitted wireless still pictures, and by the end of 1923 he had the world's only fully operating television system.[28] He bridged the technology of televising still photographs to motion-picture film, and his "was the only game in town."[29] By June 1925, the motion of a model windmill was seen, and motion-picture film would be next.[30] Hugo Gernsback, seeing one of Jenkins' demonstrations, declared, "television has actually arrived," and he defined it as "instantaneous sight at a distance."[31] Toward the end of the 1920s, Jenkins had his eyes on a much larger audience and another large screen. He wanted to sell television receivers to every household in America, and these would have been large, motion-picture sized television receivers.

Jenkins was a relentless worker. Even Kenneth Bilby, David Sarnoff's biographer, credited him with inspiring and "stimulat[ing] others" technologically and in establishing new television trends.[32] Jenkins is recognized as the founder of the Society of Motion Picture Engineers (today's Society of Motion Picture and Television Engineers [SMPTE]). Dr. Alfred N. Goldsmith, president of the SMPE in 1934, said that Jenkins' "vision and energy brought order into this field of motion picture engineers and enabled the society to grow . . . and his transmissions using mechanical scanning was another example of his ingenuity. . . . [He was] a distinguished and highly popular [person]."[33] At the time, Goldsmith was not only president of the SMPE but chairman of the National Broadcasting Corporation's Board of Consulting Engineers and employed at RCA as vice president and general manager. Such compliments were doubly impressive coming from a competitor.

In just about every modern mention of Jenkins' television exploits, particularly those written in retrospect, he is stereotyped with the label "mechanical television pioneer." That label reflects a limited sense of his work and his vision. Yes, Jenkins relied upon the Nipkow mechanical scanning disks.[34] It was the available technology of the time, existing from the late 1800s. However, Jenkins was never satisfied with the poor resolution or picture quality it produced. It was nowhere near comparable to his film experiences. So, over the years, he created modifications. His innovative prismatic rings were shaped like the older metastatic flat spinning disks, but they were spinning optical prisms that created variations of light to produce scanned images. They worked well for still-photography transmission but could not produce motion pictures at the speed required or at satisfactory resolution. So he developed a drum scanner, which decreased the circumference of the older disks as well as the size of transmitters and receivers. But it still produced unsatisfactory picture quality. By the late 1920s, he started experimenting with optical photo-electric cells. This was for his large-screen system, taking the six-by-six-inch disk-sized screens to a large film-sized screen for every home. He boldly declared that "the day of the disk was over" as he worked on alternative systems of scanning and signal delivery.[35] While electronic television systems developed by Farnsworth and Zworykin would eventually leave Jenkins behind, it is a mistake to label him solely as associated with older mechanical television systems. Much of his work involved electronic circuitry, the vacuum tubes necessary to create and process an electromagnetic signal, and he also developed theories for optical systems. He moved past the Nipkow disk, experimenting with various adaptations; when they led to dead ends, he progressed to experimentation with optical instruments in an effort to meld television and motion pictures.

One gets a glimpse of Jenkins' vision as he described his work as not limited by the technology of his own time but by electronic and optical communications, which would eventually connect every city and business across the nation. These networks, he declared, could "be made a new source of revenue and at a ridiculously insignificant cost." Neglecting channels that could lead into every place of business and home "is an unnecessary waste."[36] Jenkins foresaw some of the opportunities we continue to develop today.

Jenkins as a Business Entrepreneur

Jenkins was a successful entrepreneur and businessperson. It has been insinuated that he was "never a good businessman and [was] frequently pressed for cash."[37] While the latter is certainly true, the former is inaccurate. He was an entrepreneur who organized more than a dozen businesses over the course of his career. His first photographic operations included the 1890s to early 1900s Phantoscope Company, the Graphoscope Corporation, and the Marveloscope Company. These produced "peep-show" nickelodeon machines, competing with Edison, except that while Edison used storefront parlors, Jenkins targeted home sales with educational films available for rent from his catalog.

In the 1900s, other Jenkins companies produced a variety of products (see the timeline in the preface). His automobile company produced four different vehicles. The Kerosene-Gas Burner Company produced home heating implements. The Jenkins Paper Milk Bottle and the Columbia Milk Bottle Corporation manufactured and distributed not only "paper bottles" but the equipment for making a wide variety of containers. The bottle company was particularly profitable and helped sustain Jenkins' work in film and television. His RadioVision equipment for still-photographic transmission and home receivers were marketed from the Jenkins Laboratories, Discrola Inc., the Radio Pictures Corporation, and the later Jenkins Television Corporation. The laboratory would become the foundation of the Jenkins Television Corporation, which sold receivers to ham operators and was eventually the first corporation to market household receivers.

By the late 1920s, Jenkins still had the lead in the race for television, but his lead was tightening. He took a risk with the creation of the Jenkins Television Corporation, selling all of his patent rights to the company to secure the financing for manufacturing. Eventually he lost them to the Depression and RCA's competitive advantage. He kept the Jenkins Labs open and producing until a few months before his death.

Jenkins' business life was a roller coaster. He created and sold his businesses and patents, some with considerable potential. His early business partnership with Armat was a disaster. If Armat's version of the events is accurate, then Jenkins was untruthful and had committed fraud when he re-created the supposed earliest of his Phantoscope drawings. If Jenkins' version is believed, then Armat's intentions were deceptive, as he had undisclosed ideas of his own. Either way, Jenkins moved forward, watching what he believed to be his idea parleyed into significant profits, while he received the Franklin Awards for it and moved on with other business enterprises. He went bankrupt in the automotive industry, but he also profited by changing it over into related parts businesses. In the final tally, he was financially successful, selling his patents to the Jenkins Television Corporation and leaving his family a considerable estate. His career, according to the *Washington Post*, "rivaled any fiction story of a rise from humble beginnings to wealth and power."[38]

He described the stresses of invention and commercialization differently. According to Jenkins, not every good idea sees the light of success—not because it is a bad idea, but competition, fights, and the lack of money often prevent its development. Corporate directions may also be different than an individual's. Big business often bought and swallowed up smaller enterprises, preventing competition. Each of these challenges mirrors Jenkins' life. He had lacked money to develop his original film Phantoscope, and RCA would eventually purchase his television optical patents.

Jenkins as an Author

Jenkins was a prolific author. He published his own books, and he wrote for the popular press and journals of his day, documenting his work and his life. He published seven books, his first in 1897 and his last in 1931. *Picture Ribbons* and *Animated Pictures* corresponded with his first Franklin Institute Medal. The last, *The Boyhood of an Inventor* (1931), was written following his first heart attack and released during the depths of the Great Depression, just prior to RCA's acquisition of the Jenkins Television Corporation.

Jenkins also authored many popular-press articles. The first was about a "Glow Lamp Receiver" in *Cosmos*. The second, oft-cited 1913 article in *The Electrical Engineer* was his first prediction of "Transmitting Pictures by Electricity." Jenkins' speeches and papers presented before the SMPE conferences were all published in the society's journals. His work was highlighted in many trade and technical periodicals, including *Photographic Times, Electrical World, Cosmopolitan, Moving Picture News, Scientific American, Science,* and *Radio News*. His writing for the press waned in the early 1930s, when declining health took priority. His last article was for the *Yale Scientific Magazine,* "Radiovision and Television," in 1932.

It would be easy to dismiss Jenkins' writing and publications as mere self-promotion. In reality, they were directed at fund-raising. The publications focused attention on his work and made acquiring investors much easier. His publications, along with his Franklin Institute awards, provided him the notoriety needed for further invention.

Jenkins in the Future

Over the period of 1894 through 1933, Jenkins filed nearly three hundred patents, several granted after his death (see appendix A).[39] Modern inventors are still referencing his works: of Jenkins' 283 patents, twenty-six have been referenced 130 times in modern projects. Perhaps the most important is related to "electro-optical" works in modern projection and receiver technology, which were referenced twelve times (see appendix B). These engineers were researching related concepts, looking for ideas and approaches.

The most significant of Jenkins' television patents concerns his work with electro-optical technologies. Inside his drum scanner were translucent rods used in the transmission of light, and "this was clearly an anticipation of the modern technology of fibre optics."[40] Modern fiberoptics are considered "state-of-the-art technology." His photo-cell theory is also seen in the "new visual sensations that portended the special effects of today's video games and space-movie spectaculars," and Jenkins' mid-1920s Discrola was "a forerunner of the video disk."[41] The old technology of mechanical flat disks required a great amount of continuous light and rotating high-speed disks. In contrast, Jenkins' photo-cells temporarily stored the light and routed it at the desired time. The theoretical foundations of these cells relates to the modern-day Digital Light Processing (DLP) and Micro Electronic Mechanical Systems (MEM) display technology. The firms developing this technology have referenced Jenkins' patents, particularly the ones for his Electro-optical System and Method.[42]

Perhaps the most interesting relationship between Jenkins and the modern world was that his vision can be seen in large-screen motion-picture theater-projection systems. This is the system that the *Christian Science Monitor* referenced when it described Jenkins' work as "uniting motion picture and television."[43] Traditional motion-picture theaters used film to record and store the sounds and moving images and then project it using a mechanical projector. Jenkins' approach of the late 1920s is reflective of modern technology: it stored the images electronically and then projected them on the large screen. Jenkins' inventiveness was "seventy years ahead of its time."[44]

Jenkins was a genius in things mechanical and things optical. In a world of dramatic change in motion pictures and television, he was a pioneer, demonstrating energetic leadership, vision, and determination. The motto on the desk of his laboratory read, "They said it couldn't be done, but some darn fool went and did it." He had an inventive mind, and ideas flowed faster than the technology and the finances. He died on the threshold of television's electronic development.

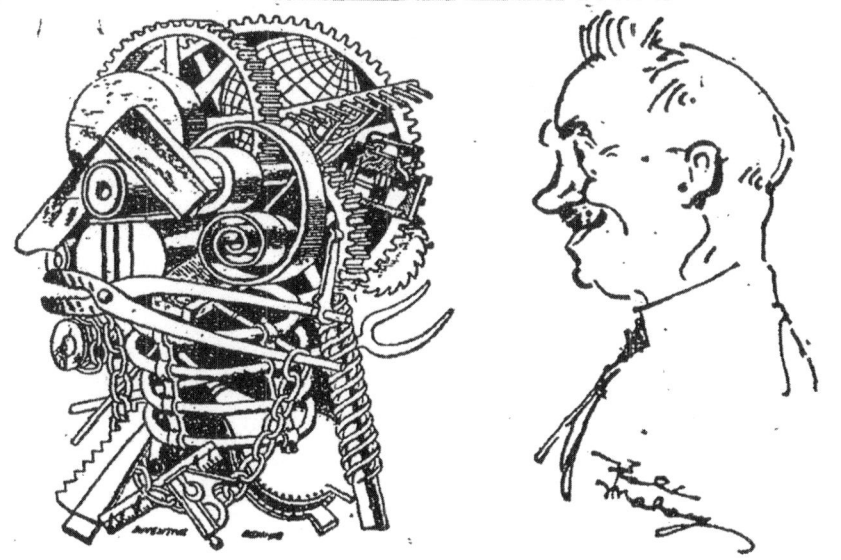

Caricature of the Inventive Genius. Courtesy *Washington Post*.

Epilogue

History is not preordained; there are alternatives that might have been. What if Jenkins had been given a stronger management role in the Jenkins Television Corporation and the overall De Forest organization? His reputation was one of inspiration and strong leadership. But he was their *idea* man, not fully integrated into the overall management-decision circles. The new corporate management wanted his vision and his highly marketable name, but they were profit-driven investors, not futuristic dreamers. The dreamer remained primarily in his lab. It would be an error to view Jenkins as a victim. In the early stages of the corporate organization, he profited handsomely with the sale of his patents to the corporation, from which he also drew a salary. He participated willingly and knowingly. He did not see the coming of the Great Depression.

What if the economic prosperity of the 1920s had continued for another decade? In 1929, Jenkins was the American industry leader. Farnsworth Television Laboratories had two key electronic patents; RCA had just hired Zworykin and was experimenting on station W2XBS; and Westinghouse had W8XAY on the air. But the new Jenkins Television Corporation had W3XK, W2XCR, and W2XCD, and more importantly, the company was already being positioned for manufacturing and distribution. The Jenkins corporate management proclaimed that in television they would achieve the "grandeur" that RCA had reached in commercial radio. The Jenkins Television Corporation had sufficient funding to move rapidly forward.

What if the Depression had not occurred? Only a few were positioned to survive its economic severity, and RCA was one of those. There is no question that Jenkins was a victim of the Depression.

What if RCA had followed Jenkins' electro-optical scanning theories instead of burying them in favor of Zworykin's electronics? Clearly, electronic scanning had surpassed all mechanical methods, including those of Jenkins. But would it have surpassed Jenkins' electro-optical system?

These questions of alterative history are speculative and rhetorical. And even though they are interesting, Jenkins' health would shortly give way. Mechanical systems were being replaced by electronic systems, and Jenkins' electro-optical theories remained largely on the drawing board with only minimal experimentation. At his death, he left behind ideas, patents, and records concerning improvement in vacuum-tube characteristics, radio-frequency measurements, scanning systems, synchronization tests, halftone amplifiers, motor research and tests, photo-cell research, color filters, and more film tests.[1] In his research, he had moved past mechanical scanning and onto electro-optical systems. His ideas remained buried from the pressures of the Depression for six decades, after which they were again referenced, the result being that modern technology looks surprisingly like a descendant of Jenkins' theories and his dreams.

Jenkins has come a long way since the western flashes of sunlight had produced the inspiration for his Phantoscope film projector. Between 1892 and 1934, he had come out of obscurity to the point where the Jenkins Lab, his experimental television station W3XK, and his patents were used in the new Jenkins Television Corporation as the foundation for one of the first television-manufacturing organizations. Jenkins envisioned a television set in every American household. He saw television and film as potentially educational and entertaining, a visual art of communication with the capability of unifying people and nations. He saw television and film as tools in the transmission of feelings and the enhancement of education. He saw cities and individuals connected by multiple systems and services, heightening communication and its speed of delivery. He saw television surpassing radio's success in terms of audience—and, of course, he grasped its profitable potential.

Between 1894 and 1913, Jenkins had challenged one of the nation's leading photographic scientists and corporate powers, Thomas A. Edison. As a result, the film industry and Edison were forced to change directions from the nickelodeon film box to the large theatrical screen of today.[2] Between 1921 and 1928, Jenkins would introduce the wireless transmission of still images. In 1925, the "motions" of a hand and a model windmill could be clearly seen in transmission experiments. By 1929, he was utilizing fundamental black-and-white silhouettes and stick-figure films he had commissioned for broadcast. He tested them on a station of his own, W3XK, attracting an audience of ARRL amateur radio operators and viewers from Washington, D.C., and across the nation. Just as Jenkins had done in film, in television he was moving the screen from six inches square to theater-size—a dramatic prediction. We smile at these pioneering predictions, because they have all come to fruition. Jenkins was a man with a vision. He was an American original, the first man working with the vision to unite television and motion pictures.[3]

Appendix A
U.S. Patents Issued to C. Francis Jenkins

Filing Date	Title	Patent No.	Issue Date
11/24/1894[1]	Phantoscope	536,569	3/26/1895
12/12/1894	Kinetographic camera	560,800	5/26/1896
8/28/1895	Phantoscope (joint patent)	586,953	7/20/1897
12/19/1896	Photographic-printing apparatus	649,730	5/15/1900
6/19/1897	Device for obtaining stereoscopic effects in exhibiting pictures	606,993	7/5/1898
3/7/1898	Stereoscopic Mutoscope	671,111	4/2/1901
5/28/1900	Vapor-burner	731,370	6/16/03
8/27/01	Steam-generator flue	926,700	6/29/09
7/11/02	Hydrocarbon-burner	711,925	10/21/02
11/18/03	Device for exhibiting pictures (Grace L. Jenkins)	765,621	7/19/04
3/7/04	Moving-picture apparatus (Grace L. Jenkins)	765,580	7/19/04
6/30/04	Moving-picture apparatus	779,364	1/3/05
8/18/04	Ball-bearing	937,811	10/26/09
2/23/05	Bag-closure	805,205	11/21/05
5/11/05	Heating apparatus	817,173	4/10/06
7/25/05	Paper receptacle	838,416	12/11/06
9/26/05	Bottle-filling machine	828,117	8/7/06
10/19/05	Automatic sign-changing letter (joint patent)	808,884	1/2/06
10/19/05	Automobile steering device	818,967	4/24/06
12/1/05	Motion-picture machine (joint patent)	819,514	5/1/06
10/22/06	Motion-picture apparatus	865,593	9/10/07
1/2/07	Steam-generator flue	926,701	6/29/09
4/6/07	Making paper-tube caps	860,385	7/16/07
6/27/07	Method of making paper bottles (Grace L. Jenkins)	927,801	7/13/09
10/19/07	Die for making closures for paper bottles	935,791	10/5/09

APPENDIX A

Filing Date	Title	Patent No.	Issue Date
10/22/07	Method of making paper-bottle closures	924,555	6/8/09
2/7/08	Telautograph	909,421	1/12/09
3/24/08	Motion-picture camera (joint patent)	934,894	9/21/09
4/8/08	Receptacle for materials to be applied with a brush	891,262	6/23/08
5/25/08	Making spirally wound tubes	916,162	3/23/09
5/25/08	Apparatus for making spirally wound tubes	921,842	5/18/09
6/18/08	Liquid-holding paper vessel	920,150	5/4/09
6/19/08	Cap for bottles	919,872	4/27/09
6/19/08	Apparatus for making and inserting bottle-caps	950,022	2/22/10
6/24/08	Apparatus for making spirally wound tubes	941,255	11/23/09
6/25/08	Making boxes or bottles	991,509	5/9/11
6/25/08	Treating paper bottles	935,848	10/5/09
7/10/08	Machine for forming and cutting off tubes	957,966	5/17/10
7/23/08	Machine for making tubes	964,162	7/12/10
8/21/08	Electrical contact device	942,647	12/7/09
9/17/08	Paper bottle	949,036	2/15/10
11/10/08	Machine for making paper bottles	933,460	9/7/09
12/5/08	Paper vessel	935,029	9/28/09
12/9/08	Paper package	925,913	6/22/09
12/14/08	Closure for paper vessels	993,170	5/23/11
12/26/08	Machine for inserting closures in receptacles	985,900	3/7/11
1/26/09	Folding box	941,256	11/23/09
1/26/09	Folding box	941,257	11/23/09
2/23/09	Device for holding bottle-bodies for the insertion of closures	958,252	5/17/10
3/22/09	Paper receptacle	954,104	4/5/10
3/22/09	Paper receptacle	954,105	4/5/10
3/22/09	Apparatus for making paper bottles	960,226	5/31/10
4/8/09	Paper receptacle	941,992	11/30/09
4/8/09	Paper receptacle	944,613	12/28/09
4/13/09	Coating paper receptacles	985,901	3/7/11
4/13/09	Machine for inserting closures in receptacles	954,106	4/5/10
4/14/09	Key and lock	947,913	2/1/10
4/29/09	Apparatus for forming and inserting receptacle-closures	954,107	4/5/10
5/4/09	Closure for paper packages	943,307	12/14/09
5/4/09	Knockdown box	944,614	12/28/09
5/4/09	Closure for paper vessels	944,615	12/28/09
5/4/09	Package for frangible articles	944,616	12/28/09
5/18/09	Apparatus for securing closures in paper tubes	954,108	4/5/10
5/18/09	Apparatus for making paper tubes	1,047,946	12/24/12
5/25/09	Bottle-closing device	954,109	4/5/10
5/28/09	Collapsible knockdown box	947,179	1/18/10

U.S. PATENTS ISSUED TO C. FRANCIS JENKINS

Filing Date	Title	Patent No.	Issue Date
6/2/09	Spirally wound collapsible receptacle	952,258	3/15/10
6/12/09	Package for chemicals	949,708	2/15/10
7/16/09	Knockdown box	981,277	1/10/11
7/19/09	Glue-applying apparatus for box-machines	985,902	3/7/11
11/24/09	Explosion-motor	972,379	10/11/10
12/4/09	Explosion-engine	972,380	10/11/10
2/1/10	Closure for receptacles	970,926	9/20/10
2/17/10	Method of making receptacle-closures	988,716	4/4/11
2/19/10	Mailing-folder	974,276	11/1/10
3/4/10	Closure-making machine	1,017,549	2/13/12
3/4/10	Adhesive-applying device	1,064,738	6/17/13
4/7/10	Receptacle-labeling apparatus	983,060	1/31/11
4/20/10	Aeroplane or flying machine	1,092,365	4/7/14
4/25/10	Warning signal	983,236	1/31/11
4/30/10	Paper-box machinery	972,767	10/11/10
5/4/10	Bottle-capping machine	979,766	12/27/10
5/25/10	Box-capping machine	1,067,431	7/15/13
6/18/10	Starting device for internal-combustion engines	1,024,077	4/23/12[2]
6/25/10	Blast-furnace	1,010,265	11/28/11
7/18/10	Flying machine	1,081,504	12/16/13
7/25/10	Starting device for gas engines	1,003,750	9/19/11
9/2/10	Flying machine	1,067,432	7/15/13
9/15/10	Paper-tube machinery	984,002	2/14/11
10/15/10	Tube-winding device	982,430	1/24/11
11/11/10	Gas-engine starter	1,047,527	12/17/12
11/18/10	Two-cycle gas engine	997,195	7/4/11
12/13/10	Aeroplane engine	1,089,645	3/10/14
1/10/11	Flying machine	1,085,263	1/27/14
1/31/11	Railway tie	1,003,751	9/19/11
2/27/11	Gas-motor starter	1,003,752	9/19/11
4/22/11	Moving-picture apparatus	1,083,016	12/30/13
5/10/11	Motion-picture apparatus	1,010,370	11/28/11
5/10/11	Motion-picture apparatus	1,047,528	12/17/12
5/22/11	Film-winding device	1,093,933	4/21/14
6/14/11	Acetylene-generating and -storing apparatus	1,010,266	11/28/11
7/3/11	Internal-combustion engine	1,098,805	6/2/14
7/20/11	Motion-picture apparatus	1,163,757	12/14/15
7/27/11	Tapering metal bodies	1,017,671	2/20/12
7/28/11	Wheel-rim	1,032,286	7/9/12
9/18/11	Tire-repair device	1,024,078	4/23/12
9/21/11	Motion-picture apparatus	1,017,672	2/20/12
12/2/11	Valve	1,047,529	12/17/12
12/2/11	Apparatus for providing boxes with closures	1,047,530	12/17/12
1/5/12	Valve	1,047,531	12/17/12
1/8/12	Motion-picture apparatus	1,091,343	12/24/14
6/27/12	Picture-film rack	1,045,410	11/26/12
6/27/12	Vehicle-tire	1,062,011	5/20/13

APPENDIX A

Filing Date	Title	Patent No.	Issue Date
6/27/12	Motion-picture camera	1,089,646	3/10/14
3/20/13	Picture-projecting device	1,090,622	3/17/14
3/20/13	Picture-projecting apparatus	1,153,163	9/7/15
3/20/13	Motion-picture apparatus	1,153,164	9/7/15
3/20/13	Motion-picture machine	1,225,636	5/8/17
6/27/13	Vehicle tire	1,197,030	9/5/16
12/12/13	View-changing device	1,153,110	9/7/15
12/16/14	Coating with fusible material	1,139,291	5/11/15
12/16/14	Motion-picture apparatus	1,152,515	9/7/15
12/16/14	Device for underwater exploration	1,156,782	10/12/15
12/16/14	Gravity-railway device	1,216,694	2/20/17
12/16/14	Repair device for pneumatic tires	1,290,566	1/7/19
3/22/15	Device for aerial warfare	1,173,522	2/29/16
12/24/15	Tire-repair device	1,174,254	3/7/16
4/6/16	Motion-picture apparatus	1,229,275	6/12/17
4/29/16	Motion-picture device	1,411,359	4/4/22
10/17/16	Film-handling device	1,234,545	7/24/17
10/17/16	Lens holder	1,258,621	3/5/18
10/17/16	Picture-projecting machine	1,302,800	5/6/19
10/17/16	Motion-picture apparatus	1,302,801	5/6/19
10/17/16	Picture-projecting apparatus	1,327,280	1/6/20
10/17/16	Motion-picture machine	1,342,681	6/8/20
4/20/17	Selective suppression of radiant rays	1,302,802	5/6/19
10/22/17	Film-reel	1,343,628	6/15/20
11/15/17	Fire extinguisher for motion-picture apparatus	1,348,565	8/3/20
4/16/18	Producing light beams	1,390,445	9/13/21
6/22/18	Picture-film reel	1,302,803	5/6/19
6/22/18	Electric meter	1,364,377	1/4/21
7/17/18	Motion-picture machine	1,281,970	10/15/18
7/17/18	Motion-picture machine	1,305,804	6/3/19
7/18/18	Motion-picture shutter	1,308,494	7/1/19
7/18/18	Mounting for picture-projecting lights	1,484,648	2/26/24
8/3/18	Motion-picture-projecting apparatus	1,311,073	7/22/19
8/3/18	Flexible reinforced transparent sheet	1,327,281	1/6/20
8/3/18	Limiting combustion of picture-films	1,348,177	8/3/20
10/1/18	Device for accumulating wave-power	1,294,808	2/18/19
10/5/18	Motion-picture framing	1,348,566	8/3/20
4/7/19	Moving-picture device	1,411,359	4/4/22
5/19/19	Motion-picture apparatus	1,378,462	5/17/21
5/19/19	Cooling device for motion-picture machines	1,408,203	2/28/22
7/9/19	Film-reel	1,322,114	11/18/19
7/9/19	Aeroplane	1,383,465	7/5/21
7/14/19	Trademark for motion-picture-projecting machines and parts thereof (joint patent)	145,372	8/2/21
7/14/19	Trademark for motion-picture-projecting machines and the parts thereof (joint patent)	147,488	10/18/21
10/23/19	Motion-picture machine	1,385,325	7/19/21
6/5/20	Lawn-mower	1,401,156	12/27/21

U.S. PATENTS ISSUED TO C. FRANCIS JENKINS

Filing Date	Title	Patent No.	Issue Date
6/5/20	Motion-picture apparatus (joint patent)	1,411,668	4/4/22
2/5/21	Motion-picture mirror of cylindrical type	1,413,333	4/18/22
2/5/21	Sleevelike refracting prism	1,440,466	1/2/23
2/5/21	Objective lens	1,544,155	6/30/25
2/5/21	High-speed motion-picture machine	1,618,090	2/15/27
4/14/21	Motion-picture carrier	1,409,004	3/7/22
4/14/21	High-speed camera	1,481,288	1/22/24
7/21/21	Design for a gaiter	60,683	3/21/22
7/21/21	Design for a gaiter	60,684	3/21/22
3/13/22	Transmitting pictures by wireless	1,544,156	6/30/25
8/30/22	Photographing oscillating sparks	1,521,188	12/30/24
8/30/22	Film reception of broadcasted pictures	1,521,189	12/30/24[3]
8/30/22	Pneumatically controlled light valve	1,525,548	2/10/25
8/30/22	Radio-picture-frequency chopper	1,525,549	2/10/25
9/11/22	Drum lens carrier	1,521,190	12/30/24
9/11/22	Prism-lens disk	1,521,191	12/30/24
9/11/22	Electroscope picture reception	1,521,192	12/30/24
9/11/22	Radio receiving devices	1,544,157	6/30/25
9/11/22	Wireless broadcasting of pictures	1,544,158	6/30/25
10/20/22	Two-way oscillating mirror	1,537,087	5/12/25
10/31/22	Flexing mirror	1,525,550	2/10/25
11/04/22	Machine for making optical prisms	1,573,609	2/16/26
12/18/22	Magnetically suspended armature	1,525,551	2/10/25
2/19/23	Device for detecting synchronism	1,537,088	5/12/25
4/21/23	Square spotlight source	1,525,552	2/10/25
4/21/23	Distant motor control	1,525,553	2/10/25[4]
9/29/23	Depth meter	1,691,719	11/13/28
2/16/24	Web-picture-message transmission	1,533,422	4/14/25
4/18/24	Altimeter	1,756,462	4/29/30
4/23/24	Radio-vision mechanism	1,530,463	3/17/25[5]
7/21/24	Rotary relay	1,662,677	3/13/28
10/22/24	Offset filament lamp	1,572,607	2/9/26
10/22/24	Light-concentrating device	1,663,308	3/20/28
10/22/24	Grid-coupled cell circuit	1,667,383	4/24/28
10/22/24	Grid-leak cell circuit	1,667,384	4/24/28
10/22/24	Resistor cell circuit	1,693,509	11/27/28
10/22/24	Oscillator cell circuit	1,704,360	3/5/29
11/19/24	Duplex radio-machine	1,694,065	12/4/28
1/2/25	Film-piercing spark unit	1,639,775	8/23/27
1/2/25	Spiral-mounted lens disk	1,679,086	7/31/28
3/2/25	Twin-blade transmitter	1,644,382	10/04/27
3/14/25	Radio-vision illumination	1,659,736	2/21/28
3/14/25	Radio-vision studio equipment	1,684,736	9/18/28
3/14/25	Radio-vision analysis	1,747,173	2/18/30
3/21/25	Twin-light-cell transmitter	1,642,733	9/20/27[6]
4/1/25	Multiple light-cell transmitter	1,641,633	9/6/27
5/5/25	Double-image radio picture	1,559,437	10/27/25
5/5/25	Web message radio	1,635,324	7/12/27
5/5/25	Radio tablet method	1,642,110	9/13/27[7]
6/5/25	Billboard radio-picture receiver	1,643,660	9/27/27
6/5/25	Light-converging-lens system	1,695,980	12/18/28

190 · APPENDIX A

Filing Date	Title	Patent No.	Issue Date
6/20/25	Plural lens-disk analyzer	1,644,383	10/4/27
7/1/25	Prism-lens unit	1,677,590	7/17/28
10/30/25	Synchronous-motor coupling	1,756,689	4/29/30[8]
1/21/26	Chronoteine or high-speed camera	1,854,742	4/19/32
1/21/26	High-speed camera	RE-17,119	10/30/28
4/22/26	Apparatus for converting electrical impulses into graphic representations	1,650,361	11/22/27
4/22/26	Electric transmission of graphic representation	1,659,200	2/14/28
4/22/26	Method of and apparatus for converting light impulses into graphic representations	1,683,136	9/4/28
6/2/26	Method of and apparatus for converting light impulses into enlarged graphic representations	1,683,137	9/4/28
8/19/26	Picture transmission	1,693,508	11/27/28
9/1/26	Multiplex radio communication	1,914,570	6/20/33
12/29/26	Prism-lens disk	RE-16,789	11/22/27
12/29/26	Electroscope picture reception	RE-16,888	2/21/28
1/10/27	Flexing mirror	RE-16,767	10/11/27
1/10/27	Pneumatically controlled light valve	RE-16,818	12/13/27
3/15/27	Radio-vision mechanism	RE-16,882	2/14/28
4/13/27	Two-way oscillating mirror	RE-16,790	11/22/27
5/21/27	Airplane stop	1,634,904	7/5/27
5/21/27	Synchronism in radio movies	1,660,711	2/28/28
6/8/27	Controlled-aperture scanning disk	1,748,383	2/25/30
6/22/27	Method of and apparatus for transmitting motion pictures	1,777,409	10/7/30
6/22/27	Gaseous light-valve	1,894,042	1/10/33
6/25/27	Radio-movie receiver	1,697,527	1/1/29
6/25/27	Light-valve transmitter	1,740,352	12/17/29
7/12/27	Airplane-launching gear	1,706,065	3/19/29
7/21/27	Synchronizing system	1,766,644	6/24/30
7/23/27	Reversing-propeller throttle control	1,694,220	12/4/28
8/29/27	Two-signal receiving device	2,017,902	10/22/35
10/31/27	Spot illumination of lens cells	1,739,312	12/10/29
10/31/27	Weather-map pen box	1,740,353	12/17/29
11/5/27	Ground-speed meter	1,711,318	4/30/29
11/29/27	Collapsible shutter for projectors	1,837,776	12/22/31
12/6/27	Pen-box ink-feed	1,835,054	12/8/31
12/28/27	Persisting luminescent screen	2,021,010	11/12/35
1/3/28	Arc lamp lens disk transmitter	1,763,357	6/10/30
1/3/28	Resistance cell-circuit	1,879,687	9/27/32
3/5/28	Code transmitter receiver	1,763,358	6/10/30
3/27/28	Multiple spot lamp	1,879,688	9/27/32
6/13/28	Helical drum scanner	1,730,976	10/8/29
6/13/28	Incandescent anode lamp	1,931,658	10/24/33
7/16/28	Cell-persistence transmitter	1,756,291	4/29/30
9/6/28	Split switching gear	1,740,354	12/17/29
9/6/28	Duplex scanning disk	1,785,262	12/16/30

U.S. PATENTS ISSUED TO C. FRANCIS JENKINS

Filing Date	Title	Patent No.	Issue Date
9/6/28	Friction drive	1,907,116	5/2/33
11/5/28	Contact scanning disk	1,740,654	12/24/29
12/8/28	Synchronism in radio movies	RE-17,221	2/19/29
2/1/29	Aeroplane engine exhaust	1,858,048	5/10/32
2/19/29	Airplane-propeller gear	1,798,740	3/31/31
2/23/29	Method of loading airplanes	1,770,700	7/15/30
4/15/29	Mapping camera	1,939,172	12/12/33
5/2/29	Scanning device	1,828,867	10/27/31
8/23/29	Airplane radio equipment	1,893,287	1/3/33
11/29/29	Method of and apparatus for converting light impulses into enlarged graphic representations	RE-17,784	8/26/30
12/17/29	Electro-optical system and method of scanning	1,859,828	5/24/32
1/9/30	Electric transmission of visual representation	1,897,481	2/14/33
1/14/30	Scanning apparatus and method	1,844,508	2/9/32
1/24/30	Electro-optical device	1,964,062	6/26/34
1/30/30	Grid-leak cell circuit	RE-17,766	8/12/30
2/1/30	Television scanning device	1,984,682	12/18/34
3/18/30	Synchronizing system	RE-19,171	5/15/34[9]
4/22/30	Elecro-optical system and method of control	1,984,683	12/18/34
5/7/30	Resistor cell circuit	RE-18,756	3/7/33
5/22/30	Synchronizing system	1,976,784	10/16/34
5/22/30	Relay for synchronizing systems	2,002,664	5/28/35
6/10/30	Cabin-airplane ventilation	1,840,393	1/12/32
6/24/30	Spiral-mounted lens disk	RE-18,452	5/3/32[10]
7/30/30	Picture transmission	RE-19,561	5/7/35
1/29/31	Radio-movie lantern slide	1,899,334	2/28/33
11/19/31	Multiplex radio communication	1,957,537	5/8/34
3/14/32	Two-signal broadcast	1,976,785	10/16/34[11]
3/22/32	Synchronizing system	RE-18,783	3/28/33[12]
4/16/32	Cell-persistence transmitter	RE-21,417	4/2/40
7/7/32	Armature winding	1,916,374	7/4/33
7/14/33	Diathermy contact pad	1,948,716	2/27/34

Appendix B
Selected Jenkins Patents Referenced in Modern Patent Applications

Design for a Gaiter (Patent 60,683) was referenced in 1993.

Motion-Picture Apparatus (Patent 865,593) was referenced in 1991 for a new card display device.

Gravity-Railway Device (Patent 1,216,694) was referenced ten times between 1990 and 2007.

Motion-Picture Apparatus (Patent 1,229,275) was referenced three times between 1990 and 1998.

Lens Holder (Patent 1,258,621) was referenced twice since 1997.

Flexible Reinforced Transparent Sheet (Patent 1,327,281) was cited in 1978 for fabric-reinforced sealing sheets.

Tapering Metal Bodies (Patent 1,017,671) was cited in 1999 for electrochemical machining.

Device for Holding Bottle-Bodies for Insertion of Closures (Patent 958,252) was cited three times between 1992 and 1994.

Machine for Inserting Closures in Receptacles (Patents 985,900 and 954,913) was cited twelve times between 1987 and 2001. A related patent (957,966) was likewise cited twelve times.

A Tube-Winding Device (Patent 982,430) was cited three times between 1984 and 2000.

Aeroplane Engine (Patent 1,089,645) was referenced in 1978 related to work on the two-stroke combustion engine.

Flying Machine (Patent 1,085,263) was referenced in 1983 for arc-wind aircraft.

Lawn-Mower (Patent 1,401,156) was referenced twelve times between 1996 and 2008.

Flexing Mirror (Patent 1,525,550) was referenced forty-six times between 1976 and 2007. It dealt with varying light intensity.

Magnetically Suspended Armature (Patent 1,525,551) was referenced in 1985; it dealt with impressing a photographic place of light with varying intensities.

Pneumatically Controlled Light Valve (RE-16,818) was referenced in 1995.

Pneumatically Controlled Light Valve (Patent 1,525,548) was referenced in 1989.

Web-Picture-Message Transmission (Patent 1,533,422) was referenced in 2006 dealing with photo, map, and message transmission.

Light-Concentrating Device (Patent 1,663,308) was referenced in 1989 as part of work on a storage device for the luminescent screen.

Synchronous Motor Coupling (Patent 1,756,689) was referenced three times between 1989 and 1997 in relation to synchronizing motors and generators.

Code Transmitter Receiver (Patent 1,763,358) was referenced in 1979 relating to a facsimile device.

Friction Drive (Patent 1,907,116) was referenced in 1983 in relation to a reversible rim-drive mechanism.

Multiplex Radio Communication (Patent 1,957,537) was referenced in 1978 in stereo sound development.

Electro-optical Device (Patent 1,964,062) was referenced twice in 1988–89. It deals with theater projection systems.

Electro-optical System and Method of Control (Patent 1,984,683) was referenced ten times. It relates to optical-receiver technology.

Notes

Prologue

1. "Phantoscope," *Richmond Telegram,* October 30, 1895, Jenkins and Motion Picture Scrapbook, 137, Franklin Institute Science Museum, Philadelphia (hereafter Franklin Institute, Jenkins Papers).

2. Charles Musser, *The Emergence of Cinema: The American Screen to 1907* (Berkeley: University of California Press, 1994), 78–79; Terry Ramsaye, *A Million and One Nights: A History of the Motion Picture,* 3d ed. (New York: Simon and Schuster, 1964), 116–18, 194–95.

3. "C. F. Jenkins, 67, Film and Radio Inventor, Dies," *New York Herald Tribune,* Jenkins Scrapbooks 1931, Wayne County Historical Museum.

Chapter 1. Jenkins' Heritage and Youth

1. Russell Conwell, "Acres of Diamonds," in *American Forum: Speeches on Historic Issues, 1788–1900,* ed. Ernest W. Wrage and Barnet Baskerville (Seattle: University of Washington Press, 1960), 263–75.

2. "Acres of Diamonds" was first delivered around 1870, and Conwell traveled the lecture circuit delivering it. See Robert Shackleton, *Russell Conwell: Acres of Diamonds, His Life and Achievements* (New York: Harper and Bros., 1915), 3–59.

3. See John M. Barry, *The Great Influenza: The Epic Story of the Greatest Plague in History* (New York: Viking Penguin, 2004).

4. Gleason L. Archer, *The History of Radio to 1926* (New York: American Historical Society, 1938), 136–37. See also John Michael Kittross, in *Historical Dictionary of American Radio,* ed. Donald G. Godfrey and Frederic A. Leigh (Westport, Conn.: Greenwood Press, 1998), 423–25.

5. "Names 'Remakers of Civilization,'" *New York Times,* September 13, 1930, 22.

6. "Wizards Gather," *Washington Herald,* September 1, 1924, qtd. in Charles Francis Jenkins, *Radio Pictures* (Washington, D.C.: Radio Pictures Corporation, 1922), 122. See Henry Sova, *Communications Serials: An International Guide to Periodicals in Communication and the Performing Arts* (Virginia Beach: Socacom, 1992), 587.

7. Jay Brien Chapman, *Story World* magazine, September 1, 1924, qtd. in Jenkins, *Radio Pictures,* 122. *Story World* magazine was published mid-1919 through August 1925. It was drawn into the *Photodramatist* in August 1923. See Henry Sova, *Communications Serials: An International Guide to Periodicals in Communication and the Performing Arts* (Virginia Beach: Socacom, 1992), 586. According to the Library of Congress online catalog, issues of *Story World* are in the Motion Picture/TV Reading Room reference collection.

8. Orrin E. Dunlap Jr., "Charles Francis Jenkins: Put Pictures on the Air," in *Radio's 100 Men of Science: Biographical Narratives of Pathfinders in Electronics and Television* (New York: Harper, 1944), 141–43.

9. William Atherton Du Puy, "A Professional Inventor," *Scientific American* 106 (March 30, 1912): 292–93. See also "Admiral Bullard and Dr. Lee de Forest Are the Deans—Forty Is the Average Age of Engineers and Manufacturers," *New York Times,* July 24, 1927, 15.

10. George H. Clark, "C. Francis Jenkins—Television Inventor," *Radio-Craft* (January 1948): 32.

11. "Predicts Vision by Use of Radio" *Boston Post,* February 11, 1924, Jenkins Scrapbooks, 1920–24, Wayne County Historical Museum.

12. Charles Francis Jenkins, *The Boyhood of an Inventor* (Washington, D.C.: National Capital Press, 1931), 31–32, 48. See also David Arthur Hollenback, "Contributions of Charles Francis Jenkins to the Early Development of Television in the United States" (Ph.D. dissertation, University of Michigan, 1983), 40–43.

13. "Radio Post," *Christian Science Monitor,* July 7, 1926, 6.

14. E. M. Haas, "Jenkins Store Celebrates Seventy-fifth Anniversary," *Richmond Palladium and Sun Telegraph,* May 16, 1928, newspaper clipping from the *Richmond Palladium-Item* news services files (hereafter *Palladium-Item* News Files).

15. History of Wayne County, 658, Virginia Roach Family Papers, Washington, D.C. Roach was a great-niece of Grace Love. Papers and writing used with permission.

16. "Charles Francis Jenkins Started Career on Farm," *Richmond Item,* June 6, 1934, *Palladium-Item* News Files.

17. Jenkins, *Boyhood of an Inventor,* 1–24.

18. *New Garden Friends Church,* historical pamphlet (New Garden, Ind.: New Garden Friends Church, n.d.), 1–4.

19. Elizabeth Maxfield-Miller, "Jenkins Genealogy: Jenkins Family," February 1984, p. 5, Philip Jenkins Family Files, Richmond, Ind.

20. Eloise Beach, "Jenkins Family Possessed Strong Inventive Streak," unidentified newspaper clipping from the *Palladium-Item* News Files. At this writing, no registered patent was located for this device under this name.

21. John R. Webb, "Amasa J. Jenkins: A Tribute," unidentified newspaper clipping, Virginia Roach Family Papers. See also Elizabeth Maxfield-Miller, "Jenkins Genealogy: Jenkins Family," February 1984, p. 4, from the Philip Jenkins Family Files, Richmond, Ind.

22. See also Acts 9:36–41.

23. Mary Ann Thomas Jenkins, Friends Archive, File FPG-VI, Lilly Library, Earlham College.

24. Luke Thomas, *History of Wayne County, Indiana*, vol. 2 (Chicago: Inter-State Publishing Co., 1884), 657–74; *Abstracts of the Records of the Society of Friends in Indiana* (Richmond, Ind.: Wayne County Genealogical Society), 179, 220, 241. There is a discrepancy in recorded wedding dates. The *History of Wayne County* clearly indicates the date of June 16, 1866, and May 21, 1866 (p. 658). However, her elegy indicates May 24, 1866, as the wedding day. See Mary Ann Thomas Jenkins, Friends Archive, File FPG-VI, Lilly Library, Earlham College.

25. "Line of Descent of Charles Francis Jenkins of Richmond, IN, and Washington, D.C." Virginia Roach Family Papers.

26. Thomas, *History of Wayne County, Indiana*, vol. 2, 670. The Thomas family were immigrant farmers of Welsh descent.

27. "C. Francis Jenkins Tells of First Projector," *Moving Picture World*, July 15, 1916, 418.

28. Jenkins, *Boyhood of an Inventor*, 2. See also John Brady, "Inventor of Radio Movies Also Invented Motion Pictures," *Boston Post*, January 28, 1923, A8.

29. Esther Cooper, "Francis Jenkins as a Boy on the Farm," *American Friend*, June 14, 1934, 214. See also Esther Cooper, "Charles Francis Jenkins Started Career on Farm," *Richmond Item*, June 6, 1934, newspaper clipping from *Palladium-Item* News Files; Mary Ellen Donat, "Messin' Paid Off for Inventor," *Richmond Palladium-Item*, March 5, 1990, A8.

30. Jenkins, *Boyhood of an Inventor*, 48.

31. Hollenback, "Contributions of Charles Francis Jenkins," 41–42.

32. Ibid., 42. The date of Jenkins' departure differs in the records. If he left in 1883, following Alvin's death (see n. 33 below), Francis would have been sixteen years of age, which is when he says he departed (Jenkins, *Boyhood of an Inventor*, 49–50). This would have given him seven years in the West before his arrival in Washington, D.C, in 1890. However, Hollenback reports that he left in 1887. The latter appears more likely, because it allows for Jenkins to finish high school and spend a little time in college. This would have given him three years in the western states, and this parallels his experiences as reported in *Boyhood of an Inventor* (48–67).

33. Family interviews infer a number of different causes for his Alvin's death, likely a farm accident in which he mistook the poisons used on the farm as a fresh drink. Whatever the cause, the tragedy was devastating. No records of the death were located.

34. Earlham College records indicate that Jenkins was the head bookkeeper for the Silver Mining Company of Lake Valley, New Mexico. From a handwritten note

in the correspondence of Thomas D. Hamm, Archivist and Associate Professor of History, Lilly Library, Earlham College, to Virginia Roach, postmarked June 12, 1992, Virginia Roach Family Papers. While there is no record of Jenkins being in the Lake Valley area, the mines of the valley were the Sierra Grande, the Sierra Bella, and the Lake Valley Mining Company. The Silver City New Mexico Museum records indicate that there were many bookkeepers working for area mines, but at the time in question there was no Silver City Mining Company. Author's correspondence with Silver City Museum, August 15, 2009.

35. "Inventor of Radio Photographs Got Big Idea While Piloting Plane," *Philadelphia Evening Bulletin,* March 5, 1923, 2.

36. Brady, "Inventor of Radio Movies Also Invented First Motion Pictures," A8. Brady indicates that the event occurred in Texas. This is likely an error. See n. 42 relative to railroads, spur lines, and the area of Jenkins' travel.

37. It is difficult to trace Jenkins' travels from his autobiography. However, the railroads crossing the country were well developed by this period. The Chicago and Northwester, the Chicago Burlington and Quincy, and the Chicago Rock Island and Pacific all might have taken Jenkins to Omaha. Here, he could have boarded the Union Pacific. Trips into Colorado would be on the Union Pacific subsidiary Union Pacific Denver and Gulf. In Ogden, Utah, he could have changed to the Central Pacific for his trip through the Sierra Nevada Mountains to Sacramento. The Southern Pacific subsidiary Oregon and California went south and north into Oregon. The Southern Pacific ran to Los Angeles, with a line continuing east across Arizona and New Mexico, and spurs would have taken him into northern Arizona and New Mexico. Lake Valley and Cook's Peak, New Mexico, which he referenced, were served by the Atchison Topeka and Santa Fe lines. Author's correspondence with Kyle Wyatt, Curator of History and Technology, California State Railroad Museum, August 17, 2009.

38. John Eustis, "Armat and Jenkins, Part I: The Dispute, 1895–1896," 5, Local History Project, Columbia Historical Society.

39. Charles Francis Jenkins to his cousin, July 20, 1891, Ann McKee Coffin Family Files. No reports could be located from the Life Saving Service.

40. Jenkins, *Boyhood of an Inventor,* 69. See also Charles Francis Jenkins to his cousin, July 15, 1891, Ann McKee Coffin Family Files. The letter describes the sights of the city from a recent tour he'd given to his cousin.

41. Jenkins, *Boyhood of an Inventor,* 70.

42. Muriel Caswall, "What It Means to Be the Wife of a Great Inventor," *Boston Post,* January 13, 1924, 1.

43. Grace and Francis' meeting was recorded in later court testimony. Grace L. Jenkins Testimony, *Motion Picture Patents Company v. Independent Moving Picture Company of America,* U.S. Circuit Court for Southern District of New York, Equity No. 5-167, p. 243 (hereafter Equity No. 5-167). The case was a part of the motion-picture "trust war." The Motion Picture Patents company was working to monopolize the technology and had tied up competitors with injunctions and interference cases. See Terry Ramsaye,

A Million and One Nights: A History of the Motion Picture, 3d ed. (New York: Simon and Schuster, 1964), 485–98, 536. Jenkins' family and associates were called to testify in this case. See also Maurice Bardeche and Robert Brasillach, *The History of Motion Pictures,* trans. Iris Barry (New York: W. W. Norton, 1938), 59, 62–77.

44. Jenkins, *Boyhood of an Inventor,* 97.
45. Untitled manuscript of Virginia Roach, p. 1, Virginia Roach Family Papers.
46. Ibid. See also Jenkins, *Boyhood of an Inventor,* 97.
47. Caswall, "What It Means to Be the Wife of a Great Inventor," 1.
48. Annotated draft of an article by C. L. Porter for the *Journal of the Society of Motion Picture Engineers* (1934), 126–30, George H. Clark Radioana Collection, Papers, Series 142, Box 143, Smithsonian National Museum of American History (hereafter the Clark Collection). The references to this collection were noted over multiple years of research. The collection has since been reorganized, with papers, boxes, and files shifted without retaining any forwarding information. The citations here remain as formatted at the time of my visits in 2003 and 2009.
49. George Clark notes, inserted in summary of C. L. Porter article for the *Journal of The Society of Motion Picture Engineers* (1934), Series 142, Box 143, Clark Collection.
50. According to Virginia Roach, Grace was engaged to a young man who was not faithful. When the other woman became pregnant, Grace broke off the relationship and moved to Washington, D.C. Virginia Roach, oral history interview, August 2003.
51. This would appear to be Lewis Eugene Breuninger (1859–1942), a successful developer in Montgomery County, Maryland. Untitled family manuscript, p. 3, Virginia Roach Family Papers.
52. Untitled family manuscript, p. 1, Virginia Roach Family Papers.
53. Different sources have the grandfather as owner and the brother-in-law as proprietor of the Log Cabin Café. The Virginia Roach manuscript indicates that a brother-in-law set Grace up as the proprietor of the café.
54. Grace L. Jenkins, testimony before the U.S. District Court, March 2, 1911, Equity No. 5-167. See also Jenkins and Motion Pictures Scrapbook, p. 28, Franklin Institute, Jenkins Papers.
55. Caswall, "What It Means to Be the Wife of a Great Inventor," 1.
56. Grace Love, personal diaries, January 22, 1899, Virginia Roach Family Papers. Grace's diaries are scattered through the family; those utilized here are in the Virginia Roach Family Papers.
57. Jenkins, *Boyhood of an Inventor,* 97.
58. Caswall, "What It Means to Be the Wife of a Great Inventor," 1.
59. Virginia Roach, untitled family manuscript, p. 2, Virginia Roach Family Papers.
60. Personal diary of Grace Love, February 14, 1899. A copy of the 1899 diary is within the Virginia Roach Papers.
61. Caswall, "What It Means to Be the Wife of a Great Inventor," 1.
62. See "Certificate of Marriage," Washington, D.C., January 30, 1902, Virginia Roach Family Papers.

63. Virginia Roach, untitled family manuscript, p. 2, Virginia Roach Family Papers.

64. This and subsequent financial comparisons are taken from S. Morgan Friedman, "The Inflation Calculator," at http://www.westegg.com/inflation. Dates used in the calculations were the original noted by Jenkins and adjusted in 2013 to the latest figures.

65. Jenkins, *Boyhood of an Inventor*, 97.

66. George Herbert Clark, "Note on Jenkins Television: Washington" (1947), p. 17, Virginia Roach Family Papers.

67. Ibid. See also Virginia Roach, oral history interview, August 2003.

Chapter 2. Early Film Experiments

1. Vernon L. Parrington, *Main Currents in American Thought*, vol. 3 (New York: Harcourt Brace, 1930), xxiii–xxix, 7–47.

2. William Atherton DuPuy, "What Inventors Are Doing," *Scientific American*, March 30, 1912, 293. See also Charles Musser, *The Emergence of Cinema: The American Screen to 1907*, vol. 1 (New York: Charles Scribner's Sons, 1990), 53.

3. See Etienne-Jules Marey, *Movement* (London: William Heinemann, 1895). Marey (1830–1904) was a French surgeon and professor of natural history. His interest in photography stemmed from his studies in physiological movement.

4. Maurice Bardeche and Robert Brasillach, *The History of Motion Pictures*, trans. Iris Barry (New York: W. W. Norton and Co., 1938), 163–75. Ironically, this patent was filed a week after Jenkins and Armat had filed their joint patent, August 28, 1895.

5. See ibid., 4.

6. For Edison's biography, see R. W. Clark, *Edison: The Man Who Made the Future* (New York: Putnam's Sons, 1977); and Paul Israel, *Edison: A Life of Invention* (New York: John Wiley and Sons, 1998).

7. Bardeche and Brasillach, *History of Motion Pictures*, 59.

8. Henry V. Hopwood, *Living Pictures: Their History, Photo-Production, and Practical Working* (New York: Arno Press, 1970), 74.

9. Albert Abramson, *The History of Television, 1880 to 1941* (Jefferson, N.C.: McFarland. 1987), 11–22. See Musser, *Emergence of Cinema*, 72. For a description of Edison's studio, see Terry Ramsaye, *A Million and One Nights: A History of the Motion Picture*, 3d ed. (New York: Simon and Schuster, 1964), 81–83.

10. Ramsaye, *Million and One Nights*, 137.

11. G. Eastman, "Photographic Film," Patent 306,594, filed March 7, 1884, granted October 14, 1884.

12. See Elizabeth Brayer, *George Eastman: A Biography* (Baltimore: Johns Hopkins University Press, 1996).

13. Albert Abramson, *Electronic Motion Pictures: A History of the Television Camera* (1955; reprint, New York: Arno Press, 1974), 23. See also Ramsaye, *Million and One Nights*, 166.

14. "C. Francis Jenkins Tells of the First Projector," *Moving Picture World*, July 15, 1916, 418–19. See also Charles F. Jenkins Testimony, Equity No. 5-167, p. 130; John Eustis, "Armat and Jenkins, Part I: The Dispute, 1895–1896," p. 5, Thomas Armat Papers, University Library Special Collections Division, Georgetown University.

15. Jenkins to H. R. Leyl, August 27, 1897, Science and Arts Committee, File, 1946, Franklin Institute. Heyl was Chairman of the Science and Arts Committee.

16. Orrin E. Dunlap Jr., *Radio's 100 Men of Science: Biographical Narratives of Pathfinders in Electronics and Television* (1944; reprint, Santa Clara, Calif.: Books for Libraries Press, 1970), 142.

17. C. F. Jenkins Testimony, p. 130.

18. C. Francis Jenkins Scrapbook, p. 76, Jenkins Papers, Franklin Institute.

19. Patent 560,800. See appendix A for patent titles and application and issue dates.

20. Charles Francis Jenkins, *Animated Pictures* (1898; reprint, New York: Arno Press, 1970), 27–29.

21. Ibid., 29–34.

22. C. Francis Jenkins Scrapbook, p. 73, Jenkins Papers, Franklin Institute. See also E. M. Haas, "Jenkins Store Celebrates Seventy-fifth Anniversary," *Richmond Palladium and Sun-Telegraph*, May 16, 1928.

23. Charles Francis Jenkins, *The Boyhood of an Inventor* (Washington, D.C.: National Capital Press, 1931), 71–72. Jenkins notes that these experiments were recorded in the *Richmond Telegraph*, June 6, 1894, and the *Photographic Times*, July 6, 1894. See also Grace L. Jenkins' testimony before the U.S. Circuit Court, S.D., of New York, Equity No. 5-167; Jenkins and Motion Pictures Scrapbook, p. 27, Franklin Institute; Homer Croy, *How Motion Pictures Are Made* (1918; reprint, New York: Arno Press, 1978).

24. Successive *Richmond Telegram* articles place June 6 as the date. See "A Genius," *Richmond Telegram*, June 6, 1894; "Charles Jenkins Has Arrived Home from Washington, D.C.," newspaper clipping, Virginia Roach Family Papers; "Movie Film," n.d. (post-1934); "First Motion Pictures Shown Flashed on Local Store Screen, 1894," September 1951; "Scientific Mag of 1896 Told of Jenkins Work with Early Movies," December 22, 1952; "Local Camera Invention Drew TV Doom Warnings," March 28, 1973; Tom Cool, "Richmond Audience in 1894 Began Movie Going," March 17, 1974; and Eloise Beach, "Jenkins Family Possessed Strong Inventive Streak," n.d., all in *Palladium-Item* News Files, file 50,034. Charles F. Jenkins Testimony, Equity No. 5-167, p. 142, would tend to confirm the June date, as Jenkins reports that he went immediately back to work after the Atlanta Exhibition. The first mention of the demonstration in the *Richmond Telegram* was October 30, 1895. There was no mention of the demonstration in an article announcing his return home from the bike trip, June 6, 1894, or following the Atlanta Exhibition, September 14, 1895. For a chronology of this debate, see David Arthur Hollenback, "Contributions of Charles Francis Jenkins to the Early Development of Television in the United States" (Ph.D. dissertation, University of Michigan, 1983), 46–57.

25. The League of American Wheelmen was formed in 1880 and is today the League of American Bicyclists.

26. The C and O Canal today is a National Historic Park. It starts in Washington and follows the Potomac River to Cumberland, Md. Constructed in 1828, the canal route has long been a favorite of cyclists.

27. Jenkins, *Boyhood of an Inventor,* 65–76.

28. Ibid., 77.

29. "A Genius," *Richmond Telegram,* June 6, 1894. See the Jenkins Phantoscope Scrapbook, Jenkins Papers, Franklin Institute.

30. "A Genius," *Richmond Telegram,* June 6, 1894, Virginia Roach Family Papers.

31. "Charles Jenkins Has Arrived Home from Washington, D.C.," *Richmond Telegram,* September 14, 1894, Jenkins and Motion Pictures Scrapbook., Jenkins Papers, Franklin Institute (also in Virginia Roach Family Papers).

32. "Phantoscope," *Richmond Telegram,* October 30, 1895, Jenkins and Motion Pictures Scrapbook, p. 137, Franklin Institute.

33. A historic plaque on the building reads: "To the Memory of C. Francis Jenkins, Inventor of the original Motion Picture Machine, who projected the first moving picture ever shown in Richmond in this building." See Ramsaye, *Million and One Nights,* 201–7.

34. The only place the title of the film is mentioned comes from Albert W. Young of Richmond, Indiana. See "First Showing," *Linns Stamp News,* June 5, 1995, 5. There were two "Annabelle Butterfly Dance" films produced by Edison in 1894. It is likely that one of these films was used in this demonstration. Musser, *Emergence of Cinema,* 78–79; Ramsaye, *Million and One Nights,* 116–18, 194–95.

35. "C. F. Jenkins, 67, Film and Radio Inventor, Dies," *New York Herald Tribune,* n.d, Jenkins Scrapbooks, 1931, Wayne County Historical Museum. This report has the date of the showing as June 6.

36. Jenkins, *Boyhood of an Inventor,* 77. Today the proving ground is the Indian Head Naval Surface Weapons Center, Indian Head, Md.

37. Ibid., 70.

38. This exhibit was also witnessed by Thomas Armat. See Jenkins to the Franklin Institute Committee on Science and Arts, January 5, 1897, and Jenkins to the Franklin Institute Committee on Science and Arts, August 27, 1897, File No. 1946, Jenkins Papers, Franklin Institute.

39. C. Francis Jenkins to Wallace Greene, November 8, 1894, Jenkins Papers, Franklin Institute. See also Ramsaye, *Million and One Nights,* 140.

40. Rebecca L. Love testimony, Equity No. 5-167, pp. 264–65.

41. Grace L. Jenkins testimony, Equity No. 5-167, p. 245.

42. Rebecca L. Love and Charles H. McLellan testimony, Equity No. 5-176, p. 271.

43. Jenkins, *Animated Pictures,* photographs, 104–95.

44. Mary E. Morrison Testimony, Equity No. 5-167, p. 259.

45. Ibid., p. 254.

46. Certificate of Death, District of Columbia, Transcript No. 9015, No. of Record, 100215.

47. Jenkins, *Animated Pictures*, 25–42.

48. Dates here are the filing dates.

49. Thomas Armat to Charles D. Wilcott, Secretary, Smithsonian Institution, May 10, 1922, 9–12, Armat Collection, George Eastman House, Rochester, N.Y. See also Eustis, "Armat and Jenkins: Part I," 18; and Hopwood, *Living Pictures*, 86–87, 108.

50. Wallace Greene, Testimony before the U.S. District Court, S.D., N.Y., Equity, No. 5-167, in Jenkins and Motion Pictures Scrapbook, p. 27, Jenkins Papers, Franklin Institute. See also Kelkres, 49.

51. Jenkins, *Boyhood of an Inventor*, 79.

52. Jenkins, *Animated Pictures*, 37–39. See also Hopwood, *Living Pictures*, 223. Jenkins says in both *Animated Pictures* and the *Boyhood of an Inventor* (80) that the cost was five cents.

53. Jenkins, *Boyhood of an Inventor*, 80.

Chapter 3. A Lifetime of Struggle

1. Charles Musser, *The Emergence of Cinema: The American Screen to 1907* (Berkeley: University of California Press, 1994), 109.

2. Terry Ramsaye, *A Million and One Nights: A History of the Motion Picture*, 3d ed. (New York: Simon and Schuster, 1964), 308–22.

3. Christopher H. Sterling and John Michael Kittross, *Stay Tuned: A History of American Broadcasting*, 3d ed. (Mahwah, N.J.: Lawrence Erlbaum, 2002), 252–53; Donald G. Godfrey, *Philo T. Farnsworth: The Father of Television* (Salt Lake City: University of Utah Press, 2001), 71–76.

4. "Armat Moving-Picture Company," pp. 1–14, Thomas Armat Papers, Box 1, Folder 7, Georgetown University Library Special Collection (hereafter Armat Papers).

5. A bit about Armat's early years is found in court records. See *Woodville Latham vs. Thomas Armat*, Court of Appeals, Washington, D.C., October 1900, Patent Appeal Docket No. 153, Armat Collection, George Eastman House, Rochester, N.Y.

6. Musser, *Emergence of Cinema*, 68.

7. Ramsaye, *Million and One Nights*, 54. See also Thomas Armat, "My Part in the Development of the Motion Picture Projector," *Journal of the Society of Motion Picture Engineers* 24 (March 1935): 243.

8. Norman Charles Raff and Frank R. Gammon were partners in the Kinetoscope Company, Washington, D.C., which marketed the Edison Kinetoscope and films for the machine.

9. Armat, "My Part in the Development of the Motion Picture Projector," 17.

10. The Bliss Electrical School, sometimes referred to as the Bliss School of Electricity, was a regularly accredited charter school specializing in the study of electricity in Washington, D.C. It was "for the young man of science bent with a fondness

for things mechanical." Founded in 1893 by Louis D. Bliss, who had worked for the Edison Company and General Electric, the school enrolled primarily government workers. Louis D. Bliss, "Concerning the School," *The Coherer* 1.1 (December 1902): 1. *The Coherer* was a bimonthly publication of the Scientific Society of the Bliss Electrical School.

11. Ramsaye, *Million and One Nights*, 79–81, 138–39.

12. Charles Francis Jenkins, *Boyhood of an Inventor* (Washington, D.C.: National Capital Press, 1931), 77.

13. Musser, *Emergence of Cinema*, 100.

14. Handwritten statement of Hugh Stuart, March 2, 1889, Exhibit A, "Contributions of C. Francis Jenkins to the Invention and Development of the Motion Picture," "Jenkins and Motion Pictures" Scrapbook, Jenkins Papers, Franklin Institute.

15. J. D. Boyce, sworn affidavit, Exhibit C, "Contributions of C. Francis Jenkins to the Invention and Development of the Motion Picture," Jenkins and Motion Pictures Scrapbook, Jenkins Papers, Franklin Institute. See also J. D. Boyle, sworn affidavit, Accession File, Photographic History Collection, National Museum of American History (hereafter Accession File).

16. John J. Hayden, sworn affidavit, Exhibit B, "Contributions of C. Francis Jenkins to the Invention and Development of the Motion Picture," Jenkins and Motion Pictures Scrapbook, Jenkins Papers, Franklin Institute. The drawings Hayden describes appear on three pages of the exhibit and are labeled on p. 5; each illustrates the "movement intermittent film-feed" machine.

17. Philo L. Bush to C. Francis Jenkins, March 13, 1923, Exhibit C, "Contributions of C. Francis Jenkins to the Invention and Development of the Motion Picture," p. 9, Jenkins and Motion Pictures Scrapbook, Jenkins Papers, Franklin Institute.

18. C. Francis Jenkins to Alexander Graham Bell, September 1, 1892, Box 130, File General Correspondence, Jenkins, Charles Francis, Alexander Graham Bell Family Papers, Library of Congress, Washington, D.C.

19. C. Francis Jenkins to Alexander Graham Bell, January 5, 1897; Alexander Graham Bell to C. Francis Jenkins, January 3, 1898, Box 130, File General Correspondence, Jenkins, Charles Francis, Alexander Graham Bell Family Papers, Library of Congress, Washington, D.C.

20. "The Marvel of the Age," Columbia Phonograph Company, undated advertising pamphlet produced by the Columbia Phonograph Company and C. Francis Jenkins to promote sales of the Phantoscope. Jenkins and Motion Pictures Scrapbook, pp. 69 and 79, Jenkins Papers, Franklin Institute.

21. Alexander Graham Bell to C. Francis Jenkins, April 24, 1896, reproduced in an advertising pamphlet, Jenkins and Motion Pictures Scrapbook, p. 79, Jenkins Papers, Franklin Institute. At this writing, there is no evidence that Jenkins demonstrated for Bell in 1892, but he certainly implied that he had a working model and a willingness to demonstrate it for the famous inventor. As time progressed, Bell would eventually provide modest assistance.

22. Arthur J. McElhone, sworn affidavit, March 3, 1898, Exhibit F, "Contributions of C. Francis Jenkins to the Invention and Development of the Motion Picture," Jenkins and Motion Pictures Scrapbook, Jenkins Papers, Franklin Institute.
23. Ibid., p. 12.
24. Milton Wright, "Successful Inventors—VIII: They Seek More Than Money, Says One of Them," *Scientific American* 137 (August 1927): 140.
25. Ramsaye, *Million and One Nights*, 208.
26. Contract between C. Francis Jenkins and James P. Freeman, January 17, 1893. Several copies of this document exist in the Jenkins Papers, Franklin Institute; the Jenkins Scrapbooks, Wayne County Historical Museum; and the Accession File. The Accession File has the clearest reproduction.
27. Emil Wellauer, sworn affidavit, February 9, 1911, "Contributions of C. Francis Jenkins to the Invention and Development of the Motion Picture," Jenkins and Motion Pictures Scrapbook, pp. 19–21, Jenkins Papers, Franklin Institute. See also Accession File.
28. Ibid. See drawings and photos of the Jenkins-Freeman and Jenkins-Armat machines, pp. 19–20.
29. Edwin Lee to C. Francis Jenkins, March 12, 1923, Jenkins and Motion Pictures Scrapbook, p. 28, Jenkins Papers, Franklin Institute.
30. Wallace Greene to C. Francis Jenkins, March 4, 1898, Exhibit F, "Contributions of C. Francis Jenkins to the Invention and Development of Motion Picture," Jenkins and Motion Pictures Scrapbook, pp. 16–18, Jenkins Papers, Franklin Institute. A photo of this exhibit also appears in Musser, *Emergence of Cinema*, 100–101. See also "Some Cooking Recipes," *Washington Star*, November 15, 1894, 9; Jenkins to the Franklin Institute Committee on Science and Arts, January 5 and August 27, 1897, File 1946, Jenkins Papers, Franklin Institute.
31. Mary E. Morrison Testimony, U.S. Circuit Court for the Southern District of N.Y., Equity No. 5-167, p. 766. See also Jenkins and Motion Pictures Scrapbook, pp. 25–26, Jenkins Papers, Franklin Institute.
32. "Chronophotography," *Photographic Times* 25 (July 6, 1894): 2. The term Phantoscope, as technology improves, is confusing. Jenkins connected several new apparatus to the term. Eustis references each of the first three as Phantoscope I, II, and III. John Eustis, "Armat and Jenkins, Part I: The Dispute, 1895–1896," p. 7, Armat Papers. Technical chronology and patent referencing are used in this writing.
33. "Chronophotography," *Photographic Times* 25 (July 6, 1894): 2.
34. C. Francis Jenkins, *The Phantoscope: A Device for Photographing and Reproducing Animated Pictures*, pp. 1–2, pamphlet produced ca. September 1894, Jenkins and Motion Pictures Scrapbook, pp. 33–34, Jenkins Papers, Franklin Institute.
35. Ramsaye, *Million and One Nights*, 141.
36. C. Francis Jenkins, "Transmitting Pictures by Electricity," *Electrical Engineer* 18 (July 25, 1894): 62–63; George Shiers and May Shiers, *Early Television: A Bibliographic Guide to 1940* (New York: Garland Publishers, 1997): 32. See also Ramsaye,

Million and One Nights, 143. A copy of this article appears in the Jenkins Scrapbook, 1920–24, Wayne County Historical Museum.

37. Mary E. Morrison Testimony, Equity No. 5-167, p. 264. There is no collaborative evidence of these showings at the school. Morrison's testimony appears in the court records and is said to be quoting from her diary records. If Armat saw these exhibitions, then he knew of Jenkins' former work; however, the records are incomplete.

38. The Phantoscope was a crude movie projector consisting of an electric lamp, a lens, a drum for feeding the film, and gears for moving the film. Abramson spells its name as Phantascope. Albert Abramson, *The History of Television, 1880 to 1941* (Jefferson, N.C.: McFarland, 1987), 20. See also C. Francis Jenkins, "Transmitting Pictures by Electricity," *Electrical Engineer* 18 (July 25, 1894); 62–63; Patent 536,569; Permanent Administration Files 1877 and 1975, Jenkins Patent listings, Box 236, File 10, Smithsonian Institutional Archives (hereafter Smithsonian Administration Files). It is noteworthy that Armat, writing the Smithsonian twenty-eight years later, would call this patent "a complete failure," but Jenkins was indeed using and conducting demonstrations with it.

39. Armat, "My Part in the Development of the Motion Picture Projector," 17.

40. L. D. Bliss to Thomas Armat, May 23, 1895, Box 2, Folder 2, Armat Papers. Bliss congratulates Armat on his drawing related to the "underground system of electrical propulsion, of which you are the inventor."

41. "The Jenkins Phantoscope," p. 32, "Contributions of C. Francis Jenkins to the Invention and Development of the Motion Picture," Jenkins and Motion Pictures Scrapbook, n.p., Jenkins Papers, Benjamin Franklin. John Warwick Daniel, a Democrat, served in the U.S. Senate from 1887 to 1910. See Richard Doss, "John Warwick Daniel: A Study in the Virginia Democracy" (Ph.D. dissertation, University of Virginia, 1955).

42. Committee on Science and Arts, Report 1897, p. 2, Jenkins Papers, Franklin Institute.

43. See Jenkins to the Franklin Institute, August 27, 1897, File 1946, Jenkins Papers, Franklin Institute; and Ramsaye, *Million and One Nights*, 142–43, 193.

44. Ramsaye incorrectly reports that the contract was signed in 1894. Musser infers that this was an attempt by Ramsaye to credit Armat as the sole inventor. See Musser, *Emergence of Cinema*, 100 n. 15; Ramsaye, *Million and One Nights*, 142. The actual contract is available in the Jenkins Papers, Franklin Institute, and the Virginia Roach Family Papers.

45. At the time, Edison's firm was making and/or licensing most of the films. Projectors were made to accommodate his thirty-five-millimeter perforations. Musser, *Emergence of Cinema*, 159.

46. "Copy of Armat-Jenkins Agreement," Jenkins Papers, Franklin Institute.

47. Thomas Armat to the Franklin Institute, January 12, 1898, p. 2, Virginia Roach Family Papers.

48. Committee on Science and the Arts, Report No. 1999, August 18, 1898, p. 2, Jenkins Papers, Franklin Institute.

49. Ramsaye, *Million and One Nights*, 140–42.

50. There is some confusion as to the amounts borrowed. Were there two investments of $1,500 and $2,000, or one investment that increased from $1,500 to $2,000? The information is based on later testimony by Armat, *Animated Projection Company v. American Mutoscope Company,* Exhibit: Thomas Armat testimony, 1:550. See also Musser, *Emergence of Cinema,* 103.

51. C. Francis Jenkins, "Radio Finds Its Eyes," *Saturday Evening Post,* July 27, 1929, 129.

52. *Motion Picture Patents Company v. Independent Moving Picture Company of America,* U.S. Circuit Court for Southern District of N.Y., Equity No. 5-167, pp. 12, 16, 141, 145.

53. Jenkins would later invent a rotary film perforator that would produce exact perforations at a rate of fifteen thousand per hour. Henry V. Hopwood, *Living Pictures: Their History, Photo-Production, and Practical Working* (1899; reprint, New York: Arno Press, 1970), 193.

54. "Mr. Edison Outdone," *Baltimore Sun,* October 3, 1895, 2.

55. Ibid. The "Washington stenographer" was Jenkins; stenography was still his full-time job at the Life Saving Service.

56. "The Marvelous Electric Phantoscope," *Atlanta Journal,* October 21, 1895, Jenkins and Motion Pictures Scrapbook, p. 30, Jenkins Papers, Franklin Institute.

57. Musser, *Emergence of Cinema,* 104.

58. Ramsaye, *Million and One Nights,* 195.

59. Jenkins, *Boyhood of an Inventor,* 77–78.

60. "Washington Picture Men at Dinner," *Moving Picture World* 18.5 (November 1, 1913): 477–79. The fire broke out in the Exhibition Plantation building, spreading to the Hagensback's animal exposition and the Phantoscope building. Between the small crowds and the fire, their investment was "wholly lost, [and] Armat pulled up stakes and came home." See C. Francis Jenkins to Edward F. Murphy, November 8, 1895, Virginia Roach Family Papers.

61. Charles F. Jenkins Testimony, Equity No. 5-167, pp. 141–42. See also Jenkins, *Boyhood of an Inventor,* 79–80. This declaration would seem to support the June date for the Richmond demonstration, because only three projectors were brought to Atlanta; if two were destroyed in the fire, then Jenkins and Armat couldn't both have taken the third to Richmond or New York.

62. C. Francis Jenkins to Edward F. Murphy, Columbia Phonograph Company, August 8 and 30, 1895, and September 7, 1895, in *Motion Picture Patents Co. v. Independent Moving Picture Company,* 1:553–59. See also Gene G. Kelkres, "A Forgotten First: The Armat-Jenkins Partnership and the Atlanta Projection," *Quarterly Review of Film Studies* 9 (Winter 1984): 45–58.

63. C. Francis Jenkins to Edward F. Murphy, November 8, 1895, Virginia Roach Family Papers.

64. Ibid.

65. Edward D. Easton deposition, December 24, 1902, *Armat Moving Picture Company v. Edison Manufacturing Company,* No. 8303, Circuit Court, Southern District

of N.Y., filed November 28, 1902. See also T. Cushing Daniel testimony, December 26, 1902, *Cushing Daniel et al. v. C. Francis Jenkins*, No. 17416, Supreme Court for the District of Columbia, filed May 27, 1896; Musser, *Emergence of Cinema*, 159–60.

66. No specific resignation documents or dates could be located. It was circa 1894–95, following the Atlanta Exhibition.

67. C. Francis Jenkins to Charles M. Jenkins, July 30, 1891, Ann McKee Coffin Family Files.

68. C. Francis Jenkins to H. R. Heyl, chairman, Sciences and Arts Committee, Franklin Institute, August 27, 1897, pp. 3–4, Jenkins and Motion Pictures Scrapbook, Jenkins Papers, Franklin Institute. Names of the attorneys were not disclosed in this correspondence.

69. Charles Francis Jenkins and Thomas Armat, Patent 586,953, filed August 28, 1895, granted July 20, 1897, lines 1–10. Jenkins filed as the sole inventor in application No. 570,010, November 25, 1895, resulting in Interference case No. 18,032. The application was thus abandoned.

70. *Motion Picture Patents Company v. the Independent Moving Picture Company of America,* August 1911, Equity No. 5–167, U.S. Circuit Court, Southern District of N.Y., Armat Papers.

71. Committee on Science and the Arts, Report No. 1999, p. 3, Jenkins Papers, Franklin Institute.

72. Ibid.; Ramsaye, *Million and One Nights*, 199–200.

73. Armat, "My Part in the Development of the Motion Picture Projector," 241–56.

74. Musser, *Emergence of Cinema*, 110.

75. Ramsaye, *Million and One Nights*, 224–25.

76. John Eustis, "Thomas Armat, Charles Francis Jenkins, and Motion Pictures in Washington, 1894–1910," p. 11, Armat Papers. Armat continued making alternations and constructed a new model. He filed for a new patent on his machine (673,992) on February 19, 1896.

77. Musser, *Emergence of Cinema*, 111–12 n.8.

78. David Arthur Hollenback, "Contributions of Charles Francis Jenkins to the Early Development of Television in the United States" (Ph.D. dissertation, University of Michigan, 1983), 63. See also Jenkins, *Boyhood of an Inventor*, 78–79.

79. Thomas Armat to Charles D. Walcott, Secretary, Smithsonian Institute, May 10, 1922, p. 18, Box 236, File 10, Smithsonian Administration Files. See also *Motion Picture Patents Company v. the Independent Moving Picture Company of America.*

80. Agreement between Armat and Jenkins, signed, n.d., Box 2, Folder 2, Armat Papers.

81. *Journal of the Franklin Institute* 141.1 (January 1896): 80. See also Musser, *Emergence of Cinema*, 104–5; and Ramsaye, *Million and One Nights*, 198–99.

82. Amended Report on the merits of Jenkins' Phantoscope, May 7, 1897, p. 1, File 1946, Franklin Institute.

83. The Benjamin Franklin awards program began in 1848, with a gift of one thousand dollars ($27,162) from the philanthropist Elliott Cresson. Today the Benjamin Franklin Institute endowment is circa eleven million dollars. See Franklin Institute Awards, "About the Awards: History of and Facts," accessed November 24, 2008, http://www.fi.edu/franklinawards/about.html. For contemporary dollar equivalents, see "The Inflation Calculator," accessed November 14, 2008, http://www.westegg.com/inflation.

84. Henry R. Heyl, the committee chair, was an inventor who held patents relating to the stapler and paper machines; E. A. Partridge was a professor of chemistry at the University of Pennsylvania; and John Carbutt was a maker of photographic materials. At this writing, little was found about the other two members of the committee, George A. Hoadley (a professor) and Edward F. Moody.

85. Report of the Chairman of the Committee on Science and the Arts, Hall of the Franklin Institute, Philadelphia, May 31, 1897, Application No. 1946, signed by James Christie, Chair, December 1, 1897.

86. Amended Report, p. 1.

87. Amended Report, p. 6.

88. Amended Report, p. 7. See also "The Phantoscope," *Journal of the Franklin Institute* 145.1 (January 1898): 78–79.

89. R. B. Owens, Secretary, Report of the Subcommittee on the Protest of Thomas Armat against the Award of the Elliott Medal to C. Francis Jenkins for his Phantoscope, p. 2; Jenkins and Motion Pictures Scrapbook; Franklin Institute Papers, p. 35.

90. Ibid.

91. Ibid.

92. Ibid.

93. "The Phantoscope," *Journal of the Franklin Institute* 145.1 (January 1898): 78–79. The award is dated 1898, but the Franklin Institute Web site lists the date as 1897. See Franklin Institute Awards, Franklin Laureate Database, accessed October 24, 2011, http://www.fi.edu/winners/detail.faw?winner_id=3182.

94. See Application for Investigation, Committee on Science and the Arts, p. 2, Franklin Institute Papers. See also Jenkins, *Boyhood of an Inventor,* 73. According to Robert Fox, this award was "one of the most important honors which American science has to offer." The awards began in 1833 to honor "men and women who by their inventions, have contributed in some outstanding way to the 'comfort, welfare and happiness' of mankind." The John Scott Medal comes with a significant monetary honorarium. See Robert Fox, "The John Scott Medal," *Proceedings of the American Philosophical Society* 112.6 (December 9, 1968): 416–30.

95. C. Francis Jenkins to R. B. Owens, Franklin Institute, November 29, 1913, John Scott Medal, Jenkins and Motion Pictures Scrapbook, Jenkins Papers, Franklin Institute.

96. C. Francis Jenkins to R. B. Owens, Franklin Institute, November 29, 1913, John Scott Medal, Jenkins and Motion Pictures Scrapbook, Jenkins Papers, Franklin Institute.

97. Ramsaye, *Million and One Nights,* 202.

98. Ibid. The exact dates of the Smithsonian Institution's motion-picture exhibit are unclear.

99. The Armat Moving-Picture Company lists the joint Jenkins-Armat patent as the first of six patents held by the company. In the papers, which appear to be written to persuade stockholder investments, Armat included newspaper clippings and "estimated the earning power of one branch of this monopoly." Armat Motion Picture Company, Box 1, Folder 7, Armat Papers.

100. Musser, *Emergence of Cinema,* 299, 333–34.

101. Armat suggested that the machine was not operable at this time, but prior demonstrations and affidavits from witnesses illustrate otherwise. Ibid., 100. See also Eustis, "Armat and Jenkins, Part I," 7.

102. Armat, "My Part in the Development of the Motion Picture Projector," 17. See also Glenn E. Mathews, "Citation of Thomas Armat," *Proceedings of Semi-annual Banquet, SMPT* (December 1935): 468–71. In the Armat Motion-Picture Company brochure, the first patent listed is the Jenkins-Armat joint patent, with no mention of Jenkins. Box 1, Folder 6, Armat Papers.

103. Jenkins-Armat Injunction, Equity No. 17416, July 23, 1896, District Supreme Court, Washington, D.C. See also C. Francis Jenkins to W. DeC. Ravenel, Administrative Assistant to the Secretary, Smithsonian Institute, October 10, 1922, Box 236, File 5, Smithsonian Administration Files.

104. Homer Croy, "Infant Prodigy of Our Industries: The Birth and Growth of the Motion Picture," *Harper's Monthly* 135.807 (August 1917): 349–57.

105. C. Francis Jenkins to Homer Croy, August 31, 1917, Jenkins and Motion Pictures Scrapbook, p. 88, Jenkins Papers, Franklin Institute.

106. Ibid.

107. Eustis, "Armat and Jenkins, Part I"; Kelkres, "Forgotten First," 57.

108. Eustis, "Armat and Jenkins, Part I," p. 7.

109. C. Francis Jenkins Testimony, Equity No. 5–167, p. 130.

110. John J. Hayden, sworn affidavit, Exhibit B, Contributions of C. Francis Jenkins to the Invention and Development of the Motion Picture, Jenkins and Motion Pictures Scrapbook, Jenkins Papers, Franklin Institute. The drawings Hayden describes appear on three pages of the exhibit (labeled p. 5); each illustrates the "first frank movement intermittent film-feed" machine. See also "C. Francis Jenkins Tells of the First Projector," *Moving Picture World,* July 15, 1916, 418–19; C. F. Jenkins Testimony, Equity No. 5–175, p. 130.

111. C. Francis Jenkins to Alexander Graham Bell, September 1, 1892, File General Correspondence, Jenkins, Charles Francis, Box 130, Alexander Graham Bell Family Papers, Manuscripts Division, Library of Congress.

112. Contract between C. Francis Jenkins and James P. Freeman, January 17, 1893, Jenkins and Motion Pictures Scrapbook, p. 18, Jenkins Papers, Franklin Institute. See also Accession File.

113. A list of these individuals can be found in Jenkins to W. DeC. Ravenel, National Museum, January 30, 1923, Box 236, Smithsonian Administrative Files. The actual affidavits are in the Jenkins and Motion Pictures Scrapbook, Jenkins Papers, Franklin Institute.

114. C. Francis Jenkins to W. DeC. Ravenel, National Museum, January 30, 1923, Box 236, Smithsonian Administrative Files; Wellauer Affidavit, Jenkins and Motion Picture Scrapbook, p. 21, Jenkins Papers, Franklin Institute.

115. J. E. Harper, Chief, Division of Appointments, Treasury Department, to H. P. Tolman, Assistant Curator, National Museum, August 14, 1922, Box 236, File 5, Smithsonian Papers.

116. See Annette Warfel, "Jenkins Inventions on Exhibit," *Richmond Palladium-Item*, July 1, 1991, A7; "First Motion Pictures Ever Shown Flashed on Local Store Screen in 1894," *Richmond Palladium-Item*, September 1951; "Richmond Audience in 1894 Began Movie Going," *Richmond Palladium-Item*, March 17, 1974, *Palladium-Item* News Files.

117. "A Genius," *Richmond Daily Telegram*, June 6, 1894, Jenkins and Motion Pictures Scrapbook, Jenkins Papers, Franklin Institute.

118. "Charles Jenkins Has Arrived Home from Washington, D.C.," R*ichmond Daily Telegram*, September 14, 1894, Jenkins and Motion Pictures Scrapbook, Jenkins Papers, Franklin Institute. There is no reference to whether this was the June 1894 article or not.

119. "Phantoscope," *Richmond Daily Telegram*, October 30, 1895, Jenkins and Motion Pictures Scrapbook, Jenkins Papers, Franklin Institute.

120. The letter actually reports that in 1895 Jenkins took the following days of leave: five in August; 11.5 in September; thirty-one in October; nine in November; and thirty-one in December. J. E. Harper, Chief, Division of Appointments, Treasury Department, to R. P. Tolman, Assistant Curator, National Museum, August 14, 1922, Box 236, File 5, Smithsonian Papers.

121. Ramsaye, *Million and One Nights*, 117–18.

122. The "'original'... crank movement intermittent film-feed 1890" appears in the Jenkins and Motion Pictures Scrapbook, pp. 5–6, Jenkins Papers, Franklin Institute.

123. R. P. Tolman, Assistant Curator, Division of Graphic Arts, to W. DeC. Ravenel, Administrative Assistant to the Secretary, Smithsonian Institute, August 17, 1922, Box 236, File 5, Smithsonian Administration Files. See also Terry Ramsaye, "The Romantic History of the Motion Picture," *New York Herald*, May 3, 1896, 44.

124. Draft letter to W. DeC. Ravenel, appears to be from R. P. Tolman, Assistant Curator, Division of Graphic Arts, July 29, 1922, Box 236, File 5, Smithsonian Administration Files.

125. See Jenkins-Freeman contract, Jenkins and Motion Pictures Scrapbook, p. 18, Jenkins Files, Franklin Institute; see also "Jenkins Phantoscope," 32.

126. Armat was asking them to remove Jenkins' materials. R. P. Tolman, Assistant Curator, Division of Graphic Arts, to W. DeC. Ravenel, Administrative Assistant to

the Secretary, Smithsonian Institute, October 30, 1922, Box 236, File 5, Smithsonian Administration Files.

127. Henry D. Hubbard, Secretary, U.S. National Bureau of Standards, Department of Commerce, to W. DeC. Ravenel, Administrative Assistant to the Secretary, Smithsonian Institute, October 12, 1922, Box 236, File 5, Smithsonian Administration Files.

128. *Thomas Armat v. C. Francis Jenkins*, Box 236, File 5, Smithsonian Administration Files.

129. For a scathing denunciation of Jenkins, see Frank L. Dyer, "The Facts Relating to the Invention of the Projection Phantoscope and the Connection of C. Francis Jenkins Therewith," Virginia Roach Family Papers. This draft copy notes that corrections were inserted by "G. E. Mathews, May 5, 1935, in accordance with article T. Armat, J. Soc. Mot. Pict. En. Vol. 24, p. 241 March 1935." It was apparently important to the authors that their reports were complementary.

130. Ward H. Goodenough [or W. Hilleadowcroft], Assistant to Edison, to Thomas Armat, December 2, 1929, Virginia Roach Family Papers.

131. Thomas Armat to Charles D. Walcott, Smithsonian Institution, July 27, 1925, Smithsonian Administration Files.

132. Ibid., pp. 2–3.

133. Handwritten letter to Thomas Armat, unsigned and undated. The handwriting appears to be Tolman's. Box 236, File 5, Smithsonian Administration Files.

134. Ramsaye, *Million and One Nights*, 202.

135. Charles Roach, "Visual Instruction in Community Center Work," *Educational Film* 3.4 (January 1920): 8–9. Interestingly Jenkins' Graphoscope projector was being advertised in this publication. See issues 3.2 (February 1929): 28; and 3.3 (March 1920): 25.

136. Ramsaye indicated that as "the years went by, [Armat] in time, after many [other] litigations with picture makers, exhibitors and others, eventually received approximately half million dollars in royalties on his projection patents." Ramsaye, *Million and One Nights*, 202. That figure would be valued at over five million in 2007 dollars.

137. Kelkres, "Forgotten First," 57.

138. Hopwood, *Living Pictures*, 86–87, 107–8.

139. Ramsaye, *Million and One Nights*, 143–44.

140. Homer Croy, *How Motion Pictures Are Made* (1918; reprint, Arno Press, 1978). Croy devoted a chapter to "How the First Motion Picture was Projected" (25–54). He chronicles Jenkins' work and the controversy but provides no new evidence or documentation.

141. Hollenback, "Contributions of Charles Francis Jenkins to the Early Development of Television in the United States," 55–56.

142. Musser, *Emergence of Cinema*, 159.

143. Ibid., 109.

144. Richard Steven Cohn, "Who Put the Magic in Movies," *SMPTE Moving Image Journal* 115:2–3 (February/March 2006): 88.

145. Unidentified author, undated handwritten document, Box 236, File 5, Smithsonian Administration Files. The charge of Armat's unlawful action is not reflected in the document but appears to be a reference to Jenkins' charge that Armat broke into his home and stole a projector. Armat's response was to claim that he had a legal right to the machine and a judge's permission to remove it from the home. Neither document has surfaced in this research.

146. "[Text illegible] . . . of City Presented Movie Academy Award," *Fredericksburg (Va.) Freelance-Star,* May 14, 1948, 12.

Chapter 4. Jenkins' Motion Pictures

1. C. Francis Jenkins to Arthur McCurdy, Bell's private secretary, January 12, 1898, Folder: Arthur W. McCardy, 1898–1903, Alexander Graham Bell Family Papers, Library of Congress, Washington, D.C.

2. Milton Wright, "Successful Inventors—VIII," *Scientific American* 137.8 (August 1927): 140.

3. This is the first of several mechanisms Jenkins labeled as Phantoscopes.

4. Patent 536,569, lines 5–10, 40–45, 46–48. See also David W. Kraeuter, *Radio and Television Pioneers: Patent Bibliography* (Metuchen, N.J.: Scarecrow Press, 1992), 181; George Shiers and May Shiers, *Early Television: A Bibliographic Guide to 1940* (New York: Garland Press, 1997).

5. The order used for this writing comes from the patent filing dates. In addition to technological detail, a patent's filing represents the climax of an inventor's creation, and collectively they provide information on the evolution of the inventor, the instruments, the solutions to problems, and specific technological claims. Thus, the patent is a primary historical document as well as a legal and technological one.

6. Patent 560,800, lines 5–10.

7. Patent 586,953, lines 6–7. Note the plural form here, referencing the prior works of Jenkins on multiple apparatus he called Phantoscopes.

8. Charles Francis Jenkins, *Animated Pictures* (1898; reprint, New York: Arno Press, 1970), 45–106.

9. Jenkins, *Animated Pictures*, frontmatter.

10. Chronophotography means that among the ideas under examination is the study of motion.

11. Jenkins, *Animated Pictures,* 17.

12. Ibid., 18.

13. Little is known of Powell's specific contacts with Jenkins' work or any direct acquaintance between the two. In *Boyhood of an Inventor,* Jenkins erroneously refers to Major John Wesley Powell as "W. B. Powell." Jenkins describes Powell "as best remembered for his voyage of discovery down the Grand Can[y]on of the Colorado."

Charles Francis Jenkins, *The Boyhood of an Inventor* (Washington, D.C.: National Capital Press, 1931), 86. Powell directed the exploration of the Green River and Grand Canyon areas of the Rockies in 1869. Lake Powell, on the Utah-Arizona border, bears his name. See John Wesley Powell, *The Exploration of the Colorado River and Its Canyons* (1875; reprint, New York: Penguin Classics, 2003).

14. Jenkins, *Boyhood of an Inventor,* 85–86. After Pinchot's creation of the Forest Service, he would go on to become the governor of Pennsylvania and later a member of the Republican and Progressive parties, although he was not apparently labeled among the Senate's controversial "Sons of the Wild Jackass." See Gifford Pinchot, *The Fight for Conservation* (New York: Doubleday, Page, and Co., 1910); Ray Tucker and Frederick R. Barkley, *Sons of the Wild Jackass* (Seattle: University of Washington Press, 1932).

15. Charles Musser, *The Emergence of Cinema: The American Screen to 1907* (Berkeley: University of California Press, 1994), 87.

16. See Robert V. Bruce, *Bell: Alexander Bell and the Conquest of Solitude* (Ithaca, N.Y.: Cornell University Press, 1990).

17. See C. Francis Jenkins, "Photography as an Aid in Teaching the Deaf," *Photographic Times* 33.3 (March 1901): 120; Jenkins Scrapbook, 1900–23, Wayne County Historical Museum.

18. Jenkins, "Photography as an Aid in Teaching the Deaf," 121; Jenkins, *Boyhood of an Inventor,* 82.

19. Jenkins, "Photography as an Aid in Teaching the Deaf," 121.

20. C. Francis Jenkins to Arthur McCurdy, January 12 and 18, 1898, General Correspondence, Folder: Arthur W. McCurdy, 1898–1903, Alexander Graham Bell Family Papers, Library of Congress, Washington, D.C.

21. C. Francis Jenkins, "The Development of Chronophotography," *Photography Times* 28 (October 1896): 449–54; C. Francis Jenkins, "The Picture Ribbons Used in Chronophotgraphy," *Photographic Times* 29 (June 1897): 259.

22. Arthur G. Bretz, "Chronophotography to Furnish Material for the Laboratory and Clinic," *Medical Record,* October 28, 1905, 706. See an obituary of Arthur G. Bretz, *Journal of American Medical Association* 90 (1928): 1645–46.

23. The Hopi photographs remain today with the Jenkins family. Just how Jenkins acquired permission to film this ceremony is puzzling, as the Hopi are an isolated and private people. He likely had some contact with the tribe during his youthful work in Arizona and New Mexico mines.

24. Jenkins mentions two Arizona Sante Fe railroad stops as his destinations: Tusayan, Arizona, and Canyon Diablo (*Boyhood of an Inventor* 86–87). Canyon Diablo is more likely correct, as it places him closer to the Hopi Nation in northeastern Arizona. Tusayan Pueblo is within the Grand Canyon National Park and was the home of the Pueblo Indians; Jenkins was filming the Hopis. The town of Canyon Diablo is today a ghost town northeast of Flagstaff and the site of the Canyon Diablo meteorite,

which struck the earth centuries earlier. See Richard O. Clemmer, *Roads in the Sky: The Hopi Indians in a Century of Change* (Boulder: Westview Books, 1995).

25. There are thirty-seven still pictures remaining from this project that have been passed through the family. They currently are in the possession of Jenkins' great nephew, Louis Janney, in Phoenix, Arizona (hereafter Janney Family Files). The film could not be located.

26. C. Francis Jenkins to Grace Love, n.d., qtd. in Jenkins, *Boyhood of an Inventor*, 87. This letter is unusually descriptive and personal in comparison to Jenkins' technical and business writings.

27. Ibid.

28. Ibid., 88.

29. Ibid.

30. The Hopi Snake Dance does not entail the worship of snakes; it is a traditional ceremony worshiping the Hopi ancestors and a prayer for rain. See D. H. Lawrence, *The Hopi Snake Dance* (Flagstaff, Ariz.: Peccary Press, 1980).

31. C. Francis Jenkins to Grace Love, n.d., qtd. in Jenkins, *Boyhood of an Inventor*, 92–93.

32. Ibid., 95–96.

33. Ibid., 96. To see what this valuable blanket might have looked like, see Steve Getzwiller, *The Fine Art of Navajo Weaving* (Tucson: Ray Manly Publications, 1984), 44–47.

34. C. Francis Jenkins to Grace Love, n.d., qtd. in Jenkins, *Boyhood of an Inventor*, 96. At this writing, there is no evidence that this motion-picture film was exhibited. Still photographs appear in promotional materials and experiments. It seems odd that, in an early period of travelogue-film development and an East Coast appetite for western film, this was not publically exhibited.

35. Musser, *Emergence of Cinema*, 159–62; Ramsaye, *Million and One Nights*, 271–74.

36. Certificate of Incorporation, Jenkins Phantoscope Company, Recorded September 26, 1904, Office of Public Records, D.C. Archives. See also the Federal Charter Company, Washington, D.C., pamphlet, handwritten date July 7, 1903 (likely the date it was acquired by Jenkins), Jenkins Scrapbook, 1900–25, Wayne County Historical Museum.

37. Certificate of Incorporation, Jenkins Phantoscope Company, Recorded September 26, 1904, Office of Public Records, D.C. Archives.

38. The American Mutoscopy Company of 1897 was later renamed the American Mutoscope and Biography Company. See Musser, *Emergence of Cinema*, 145–47. See also Paul C. Spehr, "Unaltered to Date: Developing 35mm Film," in *Moving Images: From Edison to the Webcam*, ed. John Fullerton and Astrid Soderbergh Widding (Sydney: John Libbey and Co., 2000), 3–28.

39. Herman Casler filed for a patent for his peephole device on November 21, 1894. See Herman Casler, Mutoscope, Patent 549,309, granted November 5, 1895.

40. Typewritten draft of a promotion narrative, Jenkins Phantoscope Company, Jenkins Scrapbook, 1900–25, Wayne County Historical Museum.

41. Subscription to Capital Stock, Jenkins Phantoscope Company, Jenkins Scrapbook, 1900–25, Wayne County Historical Museum.

42. Jenkins Phantoscope Company, *Washington Evening Star*, September 30, 1904, Jenkins Scrapbook, 1900–25, Wayne County Historical Museum.

43. Compare "Life Motion Pictures" and "Life Motion Pictures for Home," Jenkins Scrapbook, 1900–25, Wayne County Historical Museum.

44. Phantoscope: Life Motion Pictures for the Home, Jenkins Scrapbook, 1900–25, Wayne County Historical Museum.

45. "New X-Mas Present: The Phantoscope," *Ladies' Home Journal* (December 1904), and "Wanted . . .," *Washington Evening Star*, December 4, 1904, clippings in Jenkins Scrapbook, 1900–25, Wayne County Historical Museum.

46. Patent 765,580. The explanation for assigning the patent to Grace likely came from his commitment to his new bride; they would have been married just over two years at this filing.

47. Patent 765,580, lines 1–15.

48. Unpublished manuscript, p. 3 (most likely written by Virginia Roach in her organization and research of Jenkins' biography), Virginia Roach Family Papers.

49. Musser, *Emergence of Cinema*, 145–50.

50. Patent 779,364, lines 8–12 and 70–75. See also Patent 819,514, lines 10–20.

51. Little is known about Cahill beyond his activities with Jenkins. His name does not appear in film histories.

52. "Certificate of Incorporation Riled," *Washington Evening Star*, November 4, 1905, Jenkins Scrapbook, 1900–25, Wayne County Historical Museum. Little is known about Dieudonne or Kramme, other than the fact that they were among the investors.

53. See samples of the advertisements in Jenkins Scrapbook, 1900–25, Wayne County Historical Museum.

54. The Marvelscope Company, corporate letterhead, Jenkins Scrapbook, 1900–25, Wayne County Historical Museum.

55. Patent 819,514.

56. Patent 865,593. This patent was referenced by Patent number 5,020,899, issued June 4, 1991, for a hand grip on a new card-display device.

57. Patent 649,730, lines 10–15. See also Jenkins, *Animated Pictures*, 58–61; and C. Francis Jenkins, *Picture Ribbons* (Washington, D.C.: Press of H. L. McQueen, 1897), 40–41.

58. Patent 606,993, lines 6–7. This is the first of Jenkins' patents listing him as a resident of Washington, D.C. Earlier filings and a few yet to come listed him as from Richmond, Indiana, even thought he was living in Washington. See Musser, *Emergence of Cinema*, 15. Musser describes an early (1894) stereopticon phantoscope

projecting machine using a rotating mirror. This was a failure and no doubt led to the patent under discussion. See also Musser, *Emergence of Cinema*, 100.

59. Brian Coe, *The History of Movie Photography* (Westfield, N.J.: Eastview Editions, 1981), 78–81.

60. Patent 671,111, lines 15–45. See also Musser, *Emergence of Cinema*, 145–50, 263–71; and Spehr, "Unaltered to Date," 3–28.

61. *Richmond Daily Palladium*, August 4, 1904, Jenkins Scrapbook, 1900–25, Wayne County Historical Museum.

62. Musser, *Emergence of Cinema*, 443–44, 447.

63. C. Francis Jenkins, "For the Standard Machine with Inflammable Safety Films," *Educational Film Magazine* 3.5 (January 1920): 13–14.

64. For a photo of the projector see, Jenkins, *Boyhood of an Inventor*, 106.

65. Patent 1,225,636, lines 10–35; Patent 1,090,622, lines 10–21; Patent 1,152,515, lines 10–15; Patent 1,302,802, lines 35–40.

66. Jenkins, *Boyhood of an Inventor*, 105.

67. Ibid. The ten patents listed four projectors assigned to the Graphoscope Company. See Patents 1,225,636 and 1,229,275, lines 10–15 and 35–40. This last patent has been referenced three times between 1990 and 1998. Apparently, "standard film" was much more of a fire hazard than slow-burning film. "This apparatus allowed for use of either standard film or a special machine, in which standard film cannot be used." Five patents were filed on October 17, 1916 (see Patent 1,302,801, lines 10–15). This device assured that the shutter of the camera was brought up to speed immediately and fell when it was closed. Film Handling Device, Patent 1,234,545; Patent 1,258,621 (this patent has been referenced twice since 1897); Patent 1,327,280; Patent 1,342,681; Patent 1,311,073; Patent 1,378,462, lines 25–35 (this patent simplified the shutter and "eliminated wear"); Patent 1,281,970, lines 10–30. With projectors making their way into education, where they were used "by children or other unskilled persons," this motion-picture machine "deflect[ed] the light beam out of the projector casing." Jenkins also has a Canadian Patent (CA 198,566) that is an exact parallel to U.S. Patent 1,281,970. Why he filed this patent in both countries is unknown.

68. Patent 1,281,970, lines 10–30.

69. C. Francis Jenkins, *Expression in Photography*, pamphlet reprinted from the *Photographic Times*, 3, Jenkins Scrapbook, 1900–25, Wayne County Historical Museum. See also C. Francis Jenkins, "Expression in Photography" *Photographic Times*, 31.12 (December 1899): 545–52.

70. Jenkins, *Expression in Photography*, 4.

71. "The Annual Picture Show: Exhibition of the Capital Camera Club Opens Monday," *Washington Evening Times*, April 19, 1901, Jenkins Scrapbook, 1900–25, Wayne County Historical Museum. See also Capital Camera Club Exhibition, *8th Annual Exhibition, Capital Camera Club, May 8th to 13th, 1899"* (Washington, D.C.: Capital Camera Club, 1899). A copy is in the Smithsonian Institution Library.

72. C. Francis Jenkins, *Motion Pictures in Teaching*, (Washington, D.C.: Graphoscope Company, 1916), 1, twenty-one-page promotional pamphlet for the Graphoscope machine, Jenkins and Motion Pictures Scrapbook, Jenkins Papers, Franklin Institute.

73. Ibid., 3–5. It can be argued that the trends of visual literacy are sill evolving to establish the syntax of the visual. Chief among these contemporary publications would be Herbert Zettl, *Sight, Sound, Motion: Applied Media Aesthetics*, 12th ed. (Belmont, Calif.: Wadsworth, 2011). Zettl uses a Gestalt psychological approach to understand the language of film.

74. Jenkins, *Motion Pictures in Teaching*.

75. Musser, *Emergence of Cinema*, 2, 8–9, 109, 225–40.

76. Jenkins, *Boyhood of an Inventor*, 101; Ramsaye, *Million and One Nights*, 496–98.

77. Musser, *Million and One Nights*, 240, 306, 314, 330, 450–51.

78. Ibid., 287–306.

79. Jenkins, *Boyhood of an Inventor*, 101–2.

80. Musser, *Emergence of Cinema*, 298–99.

81. Ramsaye, *Million and One Nights*, 385.

82. Jenkins, *Boyhood of an Inventor*, 64.

83. Holmes was a native of Chicago who was said to have coined the word "travelogues." He has a star on Hollywood Boulevard marking his contributions to film. See Genoa Caldwell, ed., *Burton Holmes: The Man Who Photographed the World: Travelogues, 1892–1938* (New York: Harry N. Adams Inc., 1977). See also Musser, *Emergence of Cinema*, 9, 39, 222–23. Interestingly, according to Ramsaye, Holmes' films were licensed to Armat for distribution. Ramsaye, *Million and One Nights*, 235.

84. Eileen Bowser, *The Transformation of Cinema, 1907–1915* (New York: Charles Scribner's Sons, 1990), 44; Musser, *Emergence of Cinema*, 39.

85. E. Burton Holmes, foreword to *Burton Holmes Travelogues* (New York: McClure Company, 1901). See also C. Francis Jenkins and Oscar B. Depue, *Handbook for Motion Picture and Stereopticon Operators* (Washington, D.C.: Knega Company, 1908).

86. Musser, *Emergence of Cinema*, 181.

87. Patent 934,894, lines 10–21, 41–45.

88. Patent 1,153,110.

89. Musser, *Emergence of Cinema*, 330–34, 361. See also Bowser, *Transformation of Cinema*, 138.

90. Jenkins, *Boyhood of an Inventor*, 84; Ramsaye, *Million and One Nights*, 272, 381.

91. In *Boyhood of an Inventor*, Jenkins notes that Lubin had become a multimillionaire (84). While this is true, it is only a part of the story. Lubin's rise was meteoric, but his downfall was a tragedy created by a studio fire and business decisions that did not yield long-term success. See Joseph P. Eckhardt, *The King of the Movies: Film Pioneer, Siegmund Lubin* (Madison, N.J.: Fairleigh Dickinson University Press, 1997).

92. Jenkins, *Boyhood of an Inventor*, 83–84; Ramsaye, *Million and One Nights*, 401 and 427; Musser, *Emergence of Cinema*, 340; Bowser, *Transformation of Cinema*, 199.

93. Bowser, *Transformation of Cinema*, 74–80.

94. Musser, *Emergence of Cinema*, 9, 435–37.

95. "Memo of Agreement between Carl Laemmle and C. F. Jenkins," April 25, 1911, Jenkins and Motion Pictures Scrapbook, Jenkins Papers, Franklin Institute. What company Jenkins was to receive 25 percent in stock is unknown. It is not clear what happened to this agreement. It does not appear to have been a part of the Universal organization.

96. Grace Jenkins Diary, April 29, 1911, Virginia Roach Family Papers. See also John Drinkwater, *The Life and Adventures of Carle Laemmle* (1932; reprint, New York: Arno Press, 1978); Musser, *Emergence of Cinema*, 422.

97. Patent 1,045,410, lines 10–15, 70–75. See also Patent 1,091,343, lines 10–15, and Patent 1,089,646, lines 10–15, 100–110.

98. Patent 1,083,016, lines 10–15, 20–33.

99. Patent 1,010,370, lines 10–15, 30–35.

100. Patent 1,047,528, lines 10–15, 40–45. See also Patent 1,093,933, lines 5–15, 40–45; Patent 1,163,757, lines 15–20, 35–40; Patent 1,017,672, lines 10–20; Patent 1,153,164, lines 10–15, 30–35; Patent 1,153,163, lines 10–15, 30–35. This picture-projecting apparatus allows for "instant adjustments" of the pictures during projection.

101. Patent 1,411,359, lines 85–85.

102. Patent 1,413,333, lines 10–20.

103. Patent Re. 17,119, lines 10–15. This patent relates to 560,800.

104. "Super-Speed Camera," *Textile World*, August 27, 1927, Jenkins Scrapbook, 1927–28, Wayne County Historical Museum. See also "3,200 Pictures a Second," *Science Newsletter*, January 30, 1928, Photography section; "Film Travels at Three Miles a Minute in New Camera," *Daily Science News Bulletin*, No. 356A, January 16, 1928; "Film Appliance Expected to Aid Motor Engines," *Seattle Times*, February 12, 1928, Jenkins Scrapbook, 1927–28, Wayne County Historical Museum.

105. Research Session, *Meeting Bulletin: Society of Automotive Engineers, Inc.* 7.11 (January 17, 1928), Jenkins Scrapbook, 1927–28, Wayne County Historical Museum.

106. C. Francis Jenkins, "The Chronoteine Camera," *Meeting Bulletin: Society of Automotive Engineers* 7.11 (January 17, 1928): 1–2; "Speed Reduced to Snail's Pace by New Camera," *Christian Science Monitor*, January 28, 1928, Jenkins Scrapbook, 1927–28, Wayne County Historical Museum. Press attention was significant and repeatedly extolled the camera's speed and benefits.

107. Bobby Jones, "Slow Motion Pictures Allow Fine Study of Golfing Form, Thinks Jones," *Washington, D.C., Star*, April 26, 1931, Jenkins Scrapbook, 1931, Wayne County Historical Museum. Jenkins appears to have made several trips to the golf course, taking high-speed film of John and Florence Stewart, "Mrs. Einon and Mr. Cloud." See Diaries of Grace Love-Jenkins, November 4 and 28, 1932, Virginia Roach Family Papers.

Chapter 5. Founding the SMPTE

1. Charles Francis Jenkins, *The Boyhood of an Inventor* (Washington, D.C.: National Capital Press, 1931), 112.

2. Terry Ramsaye, *A Million and One Nights: A History of the Motion Picture*, 3d ed. (New York: Simon and Schuster, 1964), 815. See also Richard Koszarski, *An Evening's Entertainment: The Age of the Silent Feature Pictures, 1915–1928* (Berkeley: University of California Press, 1990), 198–99.

3. Eileen Bowser, *The Transformation of Cinema, 1907–1915* (New York: Charles Scribner, 1990), 29–31; Ramsay, *Million and One Nights*, 481.

4. Ramsaye, *Million and One Nights*, 815. See also Federal Motion Picture Commission, "Hearings before the Committee On Education," House of Representatives, 64th Congress HR 456, 1916.

5. C. Francis Jenkins, "The Romance of Motion Pictures," *Scientific American* Supplement 79 (May 22, 1915): 323. See also C. Francis Jenkins, "Society History," *Transactions of the Society of Motion Picture Engineers* 2.7 (November 1918): 6. *Transactions* is the forerunner of the *Journal of the Society of Motion Picture and Television Engineers* and today's *SMPTE Motion Imaging Journal*.

6. Glenn E. Matthews, "Historic Aspects of the SMPTE," *Journal of the Society of Motion Picture and Television Engineers* 75 (September 1966): 856. E. Kendall Gillett was the secretary-treasurer of a weekly publication. *Motion Picture News* 20 (November 15, 1919): 211. Nat I. Brown is unknown and was not present at the first organizational meeting in 1916.

7. C. Francis Jenkins, "Society History," *Transactions of the Society of Motion Picture Engineers* 7.2 (1918): 7.

8. Albert Abramson, *Electronic Motion Pictures: A History of the Television Camera* (1955; reprint, New York: Arno Press, 1974), 33.

9. "The Society of Motion Picture Engineers: Its Aims and Accomplishments," *Transactions of the Society of Motion Pictures Engineers* 14 (1930): iii. See also Albert Abramson, *The History of Television, 1880 to 1941* (Jefferson, N.C.: McFarland, 1987), 2, 44.

10. Certificate of Incorporation, *Transactions of the Society of Motion Picture Engineers* 1 (July 1916): 1. See also the minutes of the meeting in Matthews, "Historic Aspects of the SMPE," 856.

11. Matthews, "Historic Aspects of the SMPE," 856.

12. David Robinson, *From Peepshow to Palace: The Birth of American Film* (New York: Columbia University Press, 1995), 95; Alex Alvarez, "The Origins of the Film Exchange," *Film History: An International Journal* 17.4 (November 2006): 431–65.

13. "District Court News," *Washington Post*, August 11, 1916, 10. The *SMPTE Journal* has this date as July 24, 1916. See "Twenty-Fifth Anniversary of the Society of Motion Picture Engineers," *SMPTE Journal* 37 (July 1941): 3–5. Also note that this *Journal* issue reports membership at 1,300 members.

14. Henry D. Hubbard, "Standardization," *Transactions of the Society of Motion Picture Engineers* 1 (1916): 5–9. See also reprint in G. E. Matthews, "A Note on the Early History of the Society and Its Work on Standardization," *Transactions of the Society of Motion Picture Engineers* 72 (March 1963): 196–202.

15. Constitution and By-Laws," *Transactions of the Society of Motion Picture Engineers* 1 (July 1916): 2.

16. A later publication declared that "within a year over a hundred men joined the new venture." See "The Society of Motion Picture Engineers: Its Aims and Accomplishments," iii. This is unlikely, as records indicate a doubling in size, which would have made the group's membership around twenty-five within the first year.

17. C. Francis Jenkins, "Chairman's Address," *Transactions of the Society of Motion Picture Engineers* 2 (October 1916): 3.

18. Donald Joseph Bell, "Motion Picture Film Perforation," *Transactions of the Society of Motion Picture Engineers* 2 (October 1916): 4–6.

19. C. Francis Jenkins, "Condensers, Their Contour, Size, Location, and Support," *Transactions of the Society of Motion Picture Engineers* 2 (October 1916): 4.

20. C. Francis Jenkins, "Society History," *Transactions of the Society of Motion Picture Engineers* 7.2 (1918): 7.

21. C. Francis Jenkins, "President's Address," *Transactions of the Society of Motion Picture Engineers* 2.3 (April 1917): 3.

22. Carl Louis Gregory, "Motion Picture Cameras," *Transactions of the Society of Motion Picture Engineers* 2.3 (April 1917): 6. This was the first of several papers Gregory would present at SMPE meetings over the next few years. Carl Louis Gregory (1882–1951) worked for Technicolor in Florida, at the time of this presentation. He was a filmmaker, a popular lecturer, and a writer. See Carl Louis Gregory, ed., *Condensed Course in Motion Picture Photography* (New York: Institute of Photography, 1920); "Report of the Committee on Electrical Devices," *Transactions of the Society of Motion Picture Engineers* 2.3 (April 1917): 9.

23. Hollenback reports the membership as thirty-six. See David Arthur Hollenback, "Contributions of Charles Francis Jenkins to the Early Development of Television in the United States" (Ph.D. dissertation, University of Michigan, 1983), 96.

24. Jenkins, "President's Address," 5.

25. Ibid.

26. Ibid.

27. Ibid.

28. Report of the Committee on Optics, *Transactions of the Society of Motion Picture Engineers* 4 (July 16, 1917): 7.

29. See "Motion Picture Standards" and "Motion Picture Nomenclature," *Transactions of the Society of Motion Picture Engineers* 4 (July 16, 1917): 7–12.

30. C. Francis Jenkins, "President's Address," *Transactions of the Society of Motion Picture Engineers* 5 (October 8, 1917): 5–6.

31. Ibid., 5.

32. C. Francis Jenkins, "The Motion Picture Booth," *Transactions of the Society of Motion Picture Engineers* 5 (October 8–9, 1917): 13.

33. Gordon A. Chambers, "A Short History of Standardization in SMPTE," *SMPTE Journal* 85 (July 1976): 6.

34. C. Francis Jenkins, "President's Address," *Transactions of the Society of Motion Picture Engineers* 6 (April 1918): 5–6.

35. C. Francis Jenkins, "Condensers," *Transactions of the Society of Motion Picture Engineers* 6 (April 1918): 26.

36. C. Francis Jenkins, "President's Address," *Transactions of the Society of Motion Picture Engineers* 4 (July 16, 1917): 5.

37. C. Francis Jenkins, "President's Address," *Transactions of the Society of Motion Picture Engineers* 7 (November 18, 1918): 5–6.

38. Ibid., 8.

39. P. D. McGuire, "Impressions of Spring Meetings of Society Motion Picture Engineers," *Motion Pictures Today*, May 22, 1926, 26.

40. Herford Tynes Cowling to C. Francis Jenkins, August 21, 1931, Virginia Roach Family Papers.

41. "SMPE Meeting Most Satisfactory," *Motion Picture World: Supplement to SMPTE Bulletin* (December 1926), Jenkins Scrapbook, 1925–27, Wayne County Historical Museum.

Chapter 6. Visionary Entrepreneur

1. This epigram, attributed to Jenkins, is from the Virginia Roach Family Papers.

2. Strickland Gillilan, radio speech, April 29, 1932, transcript in the Jenkins Scrapbook, 1931, Wayne County Historical Museum. Strickland's writings are now primarily in the public domain, and his works are readily available. See Strickland Gillilan, *Sunshine and Awkwardness* (Chicago: Forbes and Co., 1920).

3. "The Jenkins Automobile Company" and "The Jenkins Automobile Company: Steam Delivery Wagons," *Motor Age*, October 25, 1900, Jenkins Scrapbook, 1900–25, Wayne County Historical Museum.

4. *Bridgeport Evening Post*, February 8, 1901, Jenkins Scrapbook, 1900–25, Wayne County Historical Museum.

5. Charles Francis Jenkins, *The Boyhood of an Inventor* (Washington, D.C.: National Capital Press, 1931), 98–100.

6. Beverly Rae Kimes, Henry Austin Clark Jr., with Ralph Dunwoodie and Keith Marvin, *Standard Catalog of American Cars, 1805–1942*, 3d ed. (Iola, Wisc.: Krause Publications, 1996), 783.

7. Steam Delivery Wagons, Jenkins Scrapbook, 1900–25, Wayne County Historical Museum.

8. Prospectus of the Jenkins Automobile Company, Jenkins Scrapbook, 1900–25, Wayne County Historical Museum.

9. "3-Ton Steam Truck," *Cycle and Auto Trade Journal* (January 1903), Jenkins Scrapbook, 1900–25, Wayne County Historical Museum.

10. "The Observation Automobile Coach," *Horseless Age* 8.25 (September 18, 1901): 527–28.

11. "You Sit in a Rotary Chair," *Motor World,* September 12, 1901; "The Observational Automobile Coach," *Automotive Review* 5.3 (September 1901); *All around Washington,* a promotional brochure created by Jenkins. *Horseless Age* indicated that this coach was manufactured by Jenkins for the Observation Automobile Company. "Observation Automobile Coach," 527; Jenkins Scrapbook, 1900–25, Wayne County Historical Museum; *Electrical World and Engineer* 37 (January 5, 1901-June 29, 1901): 953; *Automotive Review* 5.3 (September 1901): 58–59.

12. *The Jenkins Observation Automobile,* brochure, Jenkins Scrapbook, 1900–25, Wayne County Historical Museum.

13. "For Sale," *Scientific American,* January 4, 1902, Jenkins Scrapbook, 1900–25, Wayne County Historical Museum.

14. Jenkins, *Boyhood of an Inventor,* 100–101.

15. See "Jenkins—Washington, D.C.," in "Premier: 1911 The Quality Car," from the collection of the Henry Ford Museum and Greenfield Village Research Center, Virginia Roach Family Papers. These short articles describe Premier vehicles of the early 1900s, along with their respective inventors. Jenkins, *Boyhood of an Inventor,* 99–100; "The Smallest Electric Car in the World," *The Automobile,* August 1, 1901, Jenkins Scrapbook, 1900–25, Wayne County Historical Museum.

16. For a description of the miniature car, see "Miniature Electric Carriage at the Pan-American Exposition," *Scientific American,* July 13, 1901; "Miniature Automobiles, for State and Parade," *New York Clipper,* June 29, 1901; "Automobiles—Miniature and Mammoth," *Automobile Review,* July 1901; "The Littlest Auto Ever Built Will Be Seen in Buffalo This Summer," *Buffalo Evening News,* June 1, 1901, Jenkins Scrapbook, 1900–25, Wayne County Historical Museum.

17. Antonio O. Rodriguez, *Chiquita* (Doral, Fla.: Santillana USA/Alfaguara, 2008); "Jenkins—Washington D.C.," (this article incorrectly refers to Chiquita as male).

18. The Pan-American Exhibition was held in Buffalo from May 1 to November 2, 1901. See Author Goodrich, "Short Stories of Interesting Exhibits," *The World's Work: A History of Our Time* 2 (August 1901): 1054–96.

19. "Jenkins Speaks Out Like a Man," *Motor World,* April 23, 1903. See also "Jenkins's Strong Plea," *Motor World,* May 7, 1903; and C. Francis Jenkins, "The Auto Regulations," *Washington Star,* May 21, 1903, Jenkins Scrapbook, 1900–25, Wayne County Historical Museum.

20. Jenkins, *Boyhood of an Inventor,* 110–12.

21. Today the Wrights' training field is the College Park Airport.

22. "Fort Myer," unidentified newspaper clipping, Virginia Roach Family Papers.

23. Associated Press, "Wright Sees Film of Last Hop in 1908 When He Was Hurt," *Washington Evening Star,* December 16, 1928, Jenkins Scrapbook, 1927–28, Wayne County Historical Museum (also a copy in the Virginia Roach Family Papers).

24. Tom D. Crouch, *Wings: A History of Aviation from Kites to the Space Age* (New York: W. W. Norton and Co., 2004).

25. "Aeroplane or Flying Machine," Patent 1,092,365.

26. "Flying Machine," Patent 1,081,504.

27. "Flying Machine," Patent 1,067,432.

28. "Aeroplane Engine," Patent 1,089,645. This patent was referenced in later works, along with the two-stroke combustion engine, developed on June 13, 1978. See also "Internal Combustion Engine," Patent 1,098,805.

29. "Flying Machine," Patent 1,085,263. This patent was referenced in later works on arc wind aircraft on November 15, 1983.

30. Jenkins, *Boyhood of an Inventor*, 111.

31. "Device for Aerial Warfare," Patent 1,173,522.

32. It would be incorrect to assume that Jenkins was not working during this period. It is simply not reflected in patent filings because this was wartime, and secrecy was the rule as it related to defense patents, which were not granted until after the war. See Secrecy Act of October 6, 1917, ch. 95, 40 Stat. 394 (1917). What actually became of Jenkins' inventions is unknown.

33. Jenkins, *Boyhood of an Inventor*, 111–12.

34. Roosevelt was appointed assistant secretary to the navy by Woodrow Wilson. He served from 1913 to 1920.

35. Jenkins, *Boyhood of an Inventor*, 114–15.

36. Ibid., 115.

37. Ibid., 120.

38. Ibid., 119, 174.

39. C. Francis Jenkins, Pilot's License No. 6312, U.S. Department of Commerce, Aeronautics Branch, n.d., Virginia Roach Family Papers.

40. Personal Flight Log Book, August 15, 1929, Virginia Roach Family Papers.

41. Personal Flight Log Book, September 26, 1929, Virginia Roach Family Papers. There was no explanation offered for his suspicion of sabotage.

42. "C. F. Francis in Air Crash," *New York Times,* June 27, 1929, 3:1. There are discrepancies between this article and the Jenkins Flight Log Book. Jenkins has several *possibly* related notes on the adjoining page of his flight book. He dates the entry "1929, 9/26, Wash. to Wooster, 3 hr. 48 minutes," then immediately notes the New London crash. It would appear that these are two unrelated flight entries, as Wooster, Ohio, would not be on the New York–Boston route. The *Times* is assumed to be accurate, and thus the entries are interpreted as referring to different flights, one Washington to Ohio, the other the crash flight.

43. Personal Flight Log Book, Virginia Roach Family Papers.

44. John T. Brady, "Roofs of Boston's Building Can Now Be Used as Airport, Declared C. Francis Jenkins, Famous Inventor," *Boston Post,* September 18, 1927, B1. Others proposed ideas that included bowl-shaped landing fields atop the buildings, which would have required the planes to start in the center and circle around until sufficient power had been achieved for takeoff. Hugo Gernsback proposed embedding strong electromagnets on the surface of the flat roof. As the plane landed, the magnets would be engaged, slowing the plane as an iron plate fastened to the bottom of the plane would

be pulled toward the magnets. Other ideas included steel wire stretched over the rooftops to catch the plane, a method later refined for use on aircraft carriers. Powerful fans were another idea: the fans were supposed to reverse the air current, thus stopping the plane in a shorter distance. Rocket guns attached to the plane were suggested, which, fired in reverse, could slow the aircraft for landing.

45. "Aeroplane," Patent 1,383,465. In a followup patent, Jenkins would improve the gears related to propeller pitch, thus improving the safety of the landing. See "Airplane-Propeller Gear," Patent 1,798,740.

46. "Inventor Asserts Airplane Break Will Avert Crashes on Landing and During Take-Offs," *New York Times,* July 11, 1927, 6; "Washington Man Invents Airplane Braking Device," *Washington Post,* July 11, 1928, 8; "Airplane Improvements," *Washington Post,* July 12, 1927, 6; "D.C. Aerial Device Ends Landing Peril," *Washington Herald,* July 11, 1927, Jenkins Scrapbook, 1925–27, Wayne County Historical Museum.

47. "Airplane-Launching Gear," Patent 1,706,065.

48. "Better Launching and Landing," editorial, *Washington Post,* July 21, 1927.

49. "Device to Launch Plane in Second," *Philadelphia Inquirer,* July 20, 1927, in Jenkins Scrapbook, 1927–28, Wayne County Historical Museum.

50. "Invents Device to Land Planes on Skyscrapers," *Chicago Daily Tribune,* July 11, 1927. Jenkins Scrapbook, 1927–28, Wayne County Historical Museum.

51. "Invents a Runway to Launch Planes," *New York Times,* July 20, 1927, 5:3.

52. "Jenkins Give Planes New Radio Antenna," *New York Times,* January 15, 1933, 10; "Airplane Stop," Patent 1,634,904. See also related patents, "Reversing Propeller Throttle Control," Patent 1,694,220, filed July 23, 1927; "Aeronautics: Break," *Time,* July 25, 1927, 13; "Airplane Propeller Gear," Patent 1,798,740; "Aeroplane Engine Exhaust," Patent 1,858,048; "Altimeter," Patent 1,756,462; "Airplane Radio Equipment," Patent 1,893,287.

53. "Cabin Airplane Ventilation," Patent 1,840,393, filed June 10, 1930, granted January 12, 1932.

54. "Transition," *Newsweek,* June 16, 1934.

55. Emmet Dougherty, "Men Who Have Made Radio," *Radio News of Canada,* 16, Jenkins Scrapbook, 1927, Wayne County Historical Museum.

56. Cellars were small, unfinished dirt basements dug out below or just outside the house that were used for the cool storage of fresh food.

57. This was the beginning of a health epidemic that would eventually kill millions of people. See "The New Milk Bottle" and "Paper Milk Bottles" for extracts from the surgeon general's report, Jenkins Scrapbook, 1900–25, Wayne County Historical Museum.

58. "Single Service Paper Milk Bottles," *Scientific American,* June 1, 1907, 446; "Paper Milk Bottles," *Washington Star,* May 1, 1907; "Milk Inspection as Carried on at Washington," *Buffalo Times,* May 19, 1907; *Omaha Bee,* May 19, 1907; *Duluth Herald,* May 20, 1907; *Evening Wisconsin* and *Arizona Republic,* May 18, 1907, Jenkins Scrapbook, 1900–25, Wayne County Historical Museum.

59. Jenkins, *Boyhood of an Inventor*, 102–4.
60. "Facts about the Paper Milk Bottle," *Grocers' Magazine*, May 19, [n.d.], newspaper clipping from the Virginia Roach Family Papers.
61. Jenkins, *Boyhood of an Inventor*, 103.
62. "Paper Receptacle," Patent 838,416, lines 15–20.
63. Jenkins, *Boyhood of an Inventor*, 104.
64. C. Francis Jenkins to potential Investors, August 21, 1905; prospectus for the Jenkins Paper Milk-Bottle, Jenkins Scrapbook, 1900–25, Wayne County Historical Museum.
65. "Bottle-Filling Machine," Patent 828,117.
66. "Bottle-Capping Machine," Patent 979,766.
67. "Apparatus for Making Paper Tubes," Patent 1,047,946.
68. "Device for Holding Bottle Bodies for the Insertion of Closures" (Patent 958,252) was referenced three times between May 1992 and December 1904. Patent 985,900, a "Machine for Inserting Closures in Receptacles," has been referenced by twelve related modern-day patents between November 1987 and October 2001. A related patent (957,966) has been referenced a dozen times. Jenkins' Patent 982,430, a "Tube Winding Device," was referenced three times between April 1984 and March 2000.
69. See *Liquid Paper Package Company*, brochure, Jenkins Scrapbook, 1900–25, Wayne County Historical Museum.
70. See the following patents: "Receptacle for Materials to Be Applied with a Brush" (891,262), "Collapsible Knockdown Box" (947,179), "Knockdown Box" (981,277), "Folding Box" (941,256), "Mailing Folder" (974,276), "Paper Box Machinery" (972,767), "Glue Applying Apparatus for Box Machines" (985,902), and "Apparatus for Providing Boxes with Closures" (1,047,530).
71. Milton Wright, "Successful Inventors—VIII," *Scientific American* (August 1927): 141.
72. "A Few Testimonials: D. C. Holton, M.D.," Jenkins Scrapbook, 1900–25, Wayne County Historical Museum. Holton was vice president of the Brooklyn, New York, Camera Club. See "Camera Club Meeting," *Brooklyn Eagle*, March 8, 1901, 5.
73. "The Jenkins Automotive Burner," advertisement, *Automobile Review*, August 1901; "The Jenkins Automotive Burner," advertisement, *Horseless Age*, September 11, 1901, September 18, 1901, and October 11, 1901, Jenkins Scrapbook, 1900–25, Wayne County Historical Museum.
74. "Acetylene Generating and Storing Apparatus," Patent 1,010,266.
75. "Wheel Rim," Patent 1,032,286, lines 13–15.
76. "Valve," Patent 1,047,529 and 1,047,531.
77. Patent 1,327,281 (cited in 1978 for fabric-reinforced sealing sheets).
78. "Alumni Dots and Dashes: C. Francis Jenkins," *The Coherer* 1.1 (December 1902): 8.
79. "Liquid Fuel Burners" and "C. Francis Jenkins," *The Coherer* (n.d), Jenkins Scrapbook, 1900–25, Wayne County Historical Museum.

80. "Blast Furnace," Patent 1,010,265, lines 8–14.

81. "Heating Apparatus," Patent 817,173, lines 10–45.

82. "Gravity-Railway Device," Patent 1,216,694, lines 10–20. See also Patent 1,197,030, lines 10–15; William P. Taylor and James L. Howard, "Pleasure Railway," Patent 815,986, filed December 28, 1905, granted March 27, 1906, lines 10–20, 90–100.

83. Kyle K. Wyatt, Curator of History and Technology, California State Railroad Museum, to the author, October 20, 2010.

84. "The Pocket Calculator, Its Operation," "A Good Substitute for Slide Rule," Jenkins Scrapbook, 1900–25, Wayne County Historical Museum. See also "The Pocket Calculator" and "Pocket Calculator," clippings, Virginia Roach Family Papers.

85. "Christmas Tree Holder," Jenkins Scrapbook, 1900–25, Wayne County Historical Museum.

86. Edward Clifton Thomas and Charles Francis Jenkins, "Automatic Sign-Changing Letter," Patent 808,884, filed October 19, 1905, granted January 2, 1906. This patent was referenced in 1982.

87. "Talking Signs," Jenkins Scrapbook, 1900–25, Wayne County Historical Museum.

88. *Jenkins-Thomas Changing Signs,* pamphlet, Jenkins Scrapbook, 1900–25, Wayne County Historical Museum.

89. "Key and Lock," Patent 947,913, lines 10–25.

90. "Tapering Metal Bodies," Patent 1,017,671 (last cited in a 1999 patent for electrochemical machining).

91. "Electric Meter," Patent 1,364,377.

92. See, for example, Adam Gowans Whyte, *Forty Years of Electrical Progress* (London: Ernest Benn, 1930). Whyte does not mention Jenkins' gaiters.

93. "Design for a Gaiter," Patent 60,683.

94. "Electricity in the Air Harnessed for Operating Small Motor," *Washington Daily Star,* January 22, 1928. See also photograph in *Ely (Nev.) Times,* February 22, 1928, Jenkins Scrapbooks, 1927–28, Wayne County Historical Museum.

95. S. R. Winters, "A Motor Operated by Radio or Static from the Air," *Radio News* 9.11 (May 1928): 1240.

96. Paul R. Heitmeyer, "Motor Invented That Gets Power Solely from the Air," *Portland Oregonian,* February 26, 1928, Jenkins Scrapbooks, 1927–28, Wayne County Historical Museum.

97. "Success of New Motor Doubted," *Washington Evening Star,* February 29, 1928, Jenkins Scrapbooks, 1927–28, Wayne County Historical Museum.

98. Modern diathermy went beyond therapy and was used to destroy abnormal cells in the body; it was replaced by cauterization and the laser.

99. "Diathermy Contract Pad," Patent 1,948,716.

100. Jenkins, *Boyhood of an Inventor,* 108.

101. Dates of the specific tour are confused in the records. This is perpetuated by the fact that automotive touring began with the invention of the automobile at the

turn of the century. Jenkins' trip was publicized as the "first" transcontinental tour when there were, in fact, others. See the MacDonald and Campbell Perpetual Endurance Trophy of the Quaker City Motor Club, and a typewritten card, "Automobile Trip 1910," both in Virginia Roach Family Papers. Both report 1910 as the date of the trip. The *New York Times* reported the date as 1911, describing several city arrivals during the trip. Grace Jenkins' diaries also confirm the trip as taking place in 1911. George A. Weidely, "Transcontinental Trip for 1911 Finds Favor among Automobile Manufacturers," *New York Times*, June 20, 1910, 4:1.

102. "Auto Show Scheme Ocean-to-Ocean," *New York Times*, January 7, 1912, 2:1.

103. "Ocean to Ocean Tour," *New York Times*, August 27, 1911, C9. According to the *Times*, the second photographer was J. C. Bell from William Rau Photographic, Philadelphia. According to "Ocean to Ocean Assignment," *Camera Craft: A Photographic Monthly* 18.8 (1911): 397, the second photographer was D. Sargent Bell, a scenic photographer from the William H. Rau Company. Many of the images taken during this tour appear in *Boyhood of an Inventor*. See also Curt McConnell, *The Record Setting Trips by Auto from Coast to Coast, 1909–1916* (Stanford, Calif.: Stanford University Press, 2001), 73.

104. *The Automobile* reports "forty men, women and children." See the Collection of Henry Ford Museum and Greenfield Village Research Center; see also an unidentified newspaper clipping, which lists thirty-six men and women, Virginia Roach Family Papers.

105. See *The Automobile*, Henry Ford Museum and Greenfield Village Research Center (copies in the Virginia Roach Family Papers).

106. "Resume of Premier First Transcontinental Tour: World's Greatest Amateur Motoring Event," from the collection of Henry Ford Museum and Greenfield Village Research Center (copies in the Virginia Roach Family Papers).

107. Grace Jenkins Diaries, July 14, 1911, Virginia Roach Family Papers.

108. Jenkins, *Boyhood of an Inventor*, 108.

109. Grace Jenkins Diaries, July 18, 1911, Virginia Roach Family Papers.

110. Jon Guy Monihan, "The Records of a Transcontinental Trek—II," *The Automobile*, Henry Ford Museum and Greenfield Village Research Center (copies in the Virginia Roach Family Papers). Estes Park is just outside of the eastern entrance to the Rocky Mountain National Park in northern Colorado. The Forks is actually a hotel, constructed in 1874; it is in the area of Livermore, Colorado, twenty miles north of Fort Collins.

111. John Guy Monihan, "The Record of a Transcontinental Trek—III," *The Automobile*, Henry Ford Museum and Greenfield Village Research Center (copies in the Virginia Roach Family Papers).

112. Jenkins, *Boyhood of an Inventor*, 109. The Hotel Utah was completed that year. Called "the Grande Dame of Hotels," it was an elegant edifice for its time. It is today the Joseph Smith Building of the Church of Jesus Christ of Latter-day Saints' office buildings. It is the oldest commercial structure surviving in downtown Salt Lake

City with a significant history. Linda Stilletoe, *Welcoming the World: The History of Salt Lake County* (Salt Lake City: Salt Lake County, 1996): 114–17. See also Leonard J. Arrington and Heidi S. Swinton, *The Hotel: Salt Lake's Classy Lady, The Hotel Utah, 1911–1986* (Shephardville, Ky.: Publisher's Press, 1986).

113. Grace Jenkins Diaries, July 28, 1911, Virginia Roach Family Papers.

114. Grace Jenkins Diaries, August 13, 1911, Virginia Roach Family Papers. See also Jenkins, *Boyhood of an Inventor,* 109; and photo albums in the Virginia Roach Family Papers.

115. C. Francis Jenkins, "Signs on the Big Road," *Collier's Magazine,* January 10, 1914, 19.

Chapter 7. RadioVision: The Genesis and Promotion

1. The term "RadioVision" occurs in several forms throughout the records. In early papers, it appears as one compound word. Later references have it as two words: Radio Vision. Sometimes it appears in caps, and at other times not. This reflects the evolution of the technology from facsimile to television.

2. C. Francis Jenkins, *Pictures by Radio: Radio Photograms,* undated pamphlet, p. 1, Commerce Papers, Box 349, Jenkins, C. Francis, 1924–25 and undated, 1921–1928 NUCM 70–187, RLN, Herbert Hoover Presidential Library, West Branch, Iowa (hereafter Hoover Papers). The date is likely between 1924 and 1928, as Jenkins reports later experiments and overall expenditures.

3. David T. MacFarland, "Television: The Whirling Beginning," in *American Broadcasting: A Source Book on the History of Radio and Television,* ed. Lawrence W. Lichty and Malachi C. Topping (New York: Hastings House Publishers, 1975), 49.

4. Peter E. Mayeux, "James Clerk Maxwell," in *Historical Dictionary of American Radio,* ed. Donald G. Godfrey and Frederic A. Leigh (Westport, Conn.: Greenwood Press, 1998), 551–53.

5. Christopher Sterling, "Heinrich Hertz," in *Historical Dictionary of American Radio,* ed. Donald G. Godfrey and Frederic A. Leigh (Westport, Conn.: Greenwood Press, 1998), 194–95. See also Gleason L. Archer, *History of Radio to 1926* (New York: American Historical Society, Inc., 1938), 54–55; Albert Abramson, *The History of Television, 1880 to 1941* (Jefferson, N.C.: McFarland, 1987), 18; and Christopher H. Sterling and John Michael Kittross, *Stay Tuned: A History of American Broadcasting,* 3d ed. (Mahwah, N.J.: Lawrence Erlbaum, 2002), 27.

6. Albert Abramson, *Electronic Motion Pictures: A History of the Television Camera* (1955; reprint, New York: Arno Press, 1974), 18–19. See also Lynn A. Yeazel, "Color It Confusing: A History of Color Television," in *American Broadcasting: A Source Book on the History of Radio and Television,* ed. Lawrence W. Lichty and Malachi C. Topping (New York: Hastings House Publishers, 1975), 86.

7. Elliot N. Sivowitch, "A Technological Survey of Broadcasting's Pre-History, 1876–1929," *Journal of Broadcasting* 15.1 (Winter 1970–71): 4–7. See also Daniel M.

Costigan, *Electronic Delivery of Documents and Graphics* (New York: Van Nostrand Reinhold, 1978), 1–10; Clarence R. Jones, *Facsimile* (New York: Rinehart, 1949), 1–20.

8. Thomas W. Hoffer, "Nathan Stubblefield and His Wireless Telephone," *Journal of Broadcasting* 6.3 (Summer 1971): 317–29. See also Sivowitch, "Technological Survey of Broadcasting's Pre-History," 11–14; Sterling and Kittross, *Stay Tuned*, 25–26.

9. Christina S. Drale, "Lee de Forest," in *Historical Dictionary of American Radio*, ed. Donald G. Godfrey and Frederic A. Leigh (Westport, Conn.: Greenwood Press, 1998), 116–18.

10. Archer, *History of Radio to 1926*, 62–63. See also Sivowitch, "Technological Survey of Broadcasting's Pre-History," 16–18; and ElDean Bennett, "Guglielmo Marconi," in *Historical Dictionary of American Radio*, ed. Donald G. Godfrey and Frederic A. Leigh (Westport, Conn.: Greenwood Press, 1998), 249–51.

11. Sterling and Kittross, *Stay Tuned*, 28–29.

12. Peter F. Mayeux, "Reginald Aubrey Fessenden," in *Historical Dictionary of American Radio*, ed. Donald G. Godfrey and Frederic A. Leigh (Westport, Conn.: Greenwood Press, 1998), 157–58.

13. Archer, *History of Radio to 1926*, 67–69. See also Sivowitch, "Technological Survey of Broadcasting's Pre-History," 8–10.

14. For a description of Jenkins' view of "The Belin Machine," see Charles Francis Jenkins, *Radio Pictures* (Washington, D.C.: Radio Pictures Corporation, 1925), 83. Arthur Benington, "Belin, French Inventor: Has New Method of Sending Photos over Wire," *Boston Post*, January, 28, 1923, A8.

15. "Telautograph," Patent 909,421.

16. The American Marconi Wireless Telegraph Company was not returned to Marconi but to the U.S. navy, and the Radio Corporation of American was formed to control former Marconi patents and protect American interests. Thorn Mayes, "History of the American Marconi Company," *Old Timer's Bulletin* 13.1 (June 1972): 11–18. See also Thorn Mayes, "History of American Marconi Company," in *American Broadcasting: A Source Book on the History of Radio and Television*, ed. Lawrence W. Lichty and Malachi C. Topping (New York: Hastings House Publishers, 1975), 12–16.

17. C. Francis Jenkins, "Transmitting Pictures by Electricity," *Electrical Engineer*, July 25, 1894, Virginia Roach Family Papers.

18. See T. Thorne Baker, *Wireless Pictures and Television* (New York: Van Norstrand, 1927). See also Noah Arceneaux, "Radio Facsimile Newspapers of the 1930s and 40s: Electronic Publication in the Pre-Digital Age," *Journal of Broadcasting and Electronic Media* 55.3 (2011): 344–59; Jennifer S. Light, "Facsimile: A Forgotten 'New Medium' from the Twentieth Century," *New Media and Society* 8 (June 2006): 355–78.

19. C. Francis Jenkins, "Transmitting Pictures by Electricity," *Electrical Engineer*, July 25, 1894, 62–63, Virginia Roach Family Papers.

20. Albert Abramson, "Pioneers of Television—Charles Francis Jenkins," *SMPTE Journal* 95.2 (February 1986): 225–27.

21. "Telautograph," Patent 909,421, lines 9–20 and 40–45.

22. Jenkins, *Radio Pictures*, 28.

23. C. Francis Jenkins, "Motion Pictures by Wireless," *Moving Picture News* 8 (October 4, 1913): 17–18, Jenkins Scrapbook, 1920–24, Wayne County Historical Museum. Hollenback notes that this idea was not original to Jenkins, but that S. C. Gilfillan had published similar ideas for home theater. See David Arthur Hollenback, "Contributions of Charles Francis Jenkins to the Early Development of Television in the United States" (Ph.D. dissertation, University of Michigan, 1983), 76; S. C. Gilfillan, "The Future Home Theater," *The Independent* 73 (October 3, 1912): 886–91. No description or documentation could be located on the Hearst transatlantic transmission.

24. C. Francis Jenkins, "Motion Pictures by Wireless," *Moving Picture News* 8 (October 4, 1913): 17–18.

25. Jenkins, *Radio Pictures*, 11.

26. "Motion Picture Machine," Patent 1,385,325. This patent represents Jenkins' first attempts at televised motion. It is also related to his 1921 patent for "Transmitting Pictures by Wireless" (1,544,156). See also John Brady, "Radio Movies in Your Home," *Boston Post*, January 28, 1923, A3; C. Francis Jenkins, "Continuous Motion Picture Machines," *Transactions of the Society of Motion Picture Engineers* 11 (May 1920): 97; C. Francis Jenkins, "Prismatic Rings," *Transactions of the Society of Motion Picture Engineers* 14 (May 1923): 78; Hollenback, "Contributions of Charles Francis Jenkins to the Early Development of Television in the United States," 107.

27. Brady, "Radio Movies in Your Home," A3. See also Jenkins, *Radio Pictures*, 25.

28. Jenkins, *Radio Pictures*, 14–16, 26–28.

29. Ibid., 27.

30. Ibid., 40–41. See also Brady, "Radio Movies in Your Home," A3.

31. Donald G. Godfrey, *Philo T. Farnsworth: The Father of Television* (Salt Lake City: University of Utah Press, 2001), 10–16.

32. For Jenkins' view of "The Dr. Korn Machine," see Jenkins, *Radio Pictures*, 79.

33. Jonathan Coopersmith, "The Failure of Fax: When Vision Is Not Enough," *Business and Economic History* 33.3 (Fall 1994): 272–75.

34. Jenkins, *Radio Pictures*, 39 (emphasis added).

35. Ibid.

36. Ibid., 39–40.

37. Edwin Emery, *The Press and America: An Interpretative History of Journalism*, 2d ed. (Englewood Cliffs, N.J.: Prentice-Hall, 1962), 494–96.

38. Sterling and Kittross, *Stay Tuned*, 160.

39. "Photographic Radio Transmission," damaged, unidentified newspaper clipping (from the Sunday Magazine of the *Washington Evening Star*), Jenkins Scrapbook, 1920–24, Wayne County Historical Museum.

40. Charles Francis Jenkins, *Pictures by Radio: Radio Photograms*, undated pamphlet, p. 4, 1921–28 NUCM 70-187, RLN, Commerce Papers, Box 349, Jenkins, C. Francis, 1924–25, and undated, Hoover Papers.

41. Herbert Hoover to C. Francis Jenkins, December 1, 1924, Hoover Papers. Hoover also gave permission to utilize the letter in promoting Jenkins' experiments.

42. C. Francis Jenkins to Paul Henderson, October 1, 1924, in Charles Francis Jenkins, *The Boyhood of an Inventor* (Washington, D.C.: National Capital Press, 1931), 136. Col. Paul Henderson was the second assistant postmaster general at the time of the letter. He is known today as helping to establish overnight mail service. See *Congressional Directory*, 67th Cong., 47th Sess., December 1922 (Washington, D.C.: Government Printing Office, 1922), 278.

43. "Inventor Discusses Method of Making Movies by Radio," *Washington Herald*, September 5, 1922, Jenkins Scrapbook, 1920–24, Wayne County Historical Museum.

44. L. C. Porter, "Edison Mazda Lamps Used in the Transmission of Chines by Radio," *The Edison Sale Builder* (ca. 1923–24), Jenkins Scrapbook, 1920–24, Wayne County Historical Museum.

45. Jenkins, *Radio Pictures*, 6.

46. *Moving Picture News*, December 23, 1922, Jenkins Scrapbook, 1920–24, Wayne County Historical Museum.

47. Brady, "Radio Movies In Your Home," A3.

48. L. C. Porter, "Edison Mazda Lamps Used in the Transmission of Chines by Radio," *The Edison Sale Builder* (ca. 1923–24, Jenkins Scrapbook, 1920–24, Wayne County Historical Museum).

49. "Seven-League Camera," *Time*, June 2, 1924.

50. Orrin E. Dunlap Jr. "Seeing around the World by Radio: The Development of a New Combination Photo-electric Cell and Vacuum Tube Has Created an 'Eye' for Wireless," *Scientific American* 134 (March 1928): 162.

51. Jenkins, *Pictures by Radio*, 6.

52. Jenkins' predictions varied between a month or two and eighteen months. This was true of most independent inventors, who tended to be more hopeful than realistic.

53. "Predicts Vision by Use of Radio" *Boston Post*, February 11, 1924, Jenkins Scrapbook, 1920–24, Wayne County Historical Museum.

54. "Drum Lens Carrier," Patent 1,521,190 (this patent dealt with increasing the speed at which the picture was scanned); "Prism Lense Disk," Patent 1,521,191; "Electroscope Picture Reception," Patent 1,521,192; "Flexing Mirror," Patent 1,525,550 (this patent was a means of "varying the intensity of the light falling on a given area," utilizing fluctuating current; it was referenced forty-six times since 1976); "Magnetically Suspended Armature," Patent 1,525,551 (this patent was for the broadcasting of radio pictures and provided a means for "impressing on a photographic plate a light of varying intensity"; it was referenced once in 1985); "Wireless Broadcasting of Pictures," Patent 1,554,158 (this patent improved picture distortion through synchronization).

55. Jenkins, *Radio Pictures*, 118.

56. "Youth Service Jenkins," *Washington Daily News*, November 13, 1923, Jenkins Scrapbook, 1920–24, Wayne County Historical Museum.

57. Albert Abramson, *Zworykin: Pioneer of Television* (Urbana: University of Illinois Press, 1995), 41–47; Godfrey, *Philo T. Farnsworth*, 12–15.

58. Russell Burns, *John Logie Baird: Television Pioneer* (London: Institution of Electrical Engineers, 2000): xiv–xvii, 37, 88–110.

59. "Photographing Oscillating Sparks," Patent 1,521,188 (sparks were recorded on a light-sensitive surface); "Pneumatically-Controlled Light Wave," Patent RE- 16,818 (this patent controlled the "beam of light in a more elastic manner than" was heretofore seen; it was referenced in 1995 for the development of an integrated galvanometer scanning device); "Pneumatically-Controlled Light Wave," Patent 1,525,548 (this patent was referenced again in 1989); "Film Reception of Broadcasted Pictures," Patent 1,521,189 (this patent was an improvement in the broadcast reception of motion pictures; it allowed for "covering the whole area of each discrete picture sent and received, while maintaining the quality of the picture obtained").

60. "Moving Picture Shows by Radio Perfected by Local Inventor," *Washington Evening Star,* May 19, 1922, Jenkins Scrapbook, 1920–24, Wayne County Historical Museum.

61. Regis Tucci, "WEAF, New York City," in *Historical Dictionary of American Radio,* ed. Donald G. Godfrey and Frederic A. Leigh (Westport, Conn.: Greenwood Press, 1998), 410–11. See also Sterling and Kittross, *Stay Tuned,* 64.

62. "Sleevelike Refracting Prism," Patent 1,440,466.

63. "Objective Lens," Patent 1,544,155.

64. "High-Speed Motion-Picture Machine," Patent 1,618,090. These prisms were used in an earlier patent. See "Motion-Picture Machine," Patent 1,385,325.

65. NOF was a historic AM-radio station of the navy. Originally established for research and the broadcast of Greenwich time signals, it also broadcast music, educational information, health information, veteran's reports, and children's programs. See C. Austin, "The Romance of the Radio Telephone," *Radio Broadcast* 1.1 (May 1922): 16. See also S. R. Winters, "The Passing of NOF as a Broadcast Station," *Radio News* 4 (March 1923): 1623, 1742–43.

66. Jenkins, *Radio Pictures,* 118; Jenkins, *Boyhood of an Inventor,* 122–23. It would be two years before AT&T would conduct a similar public demonstration. See "Seven-League Camera," *Time,* June 2, 1924.

67. Jenkins, *Boyhood of an Inventor,* 123.

68. "Motion Pictures by Radio, Latest Achievement of Science," *Rochester (N.Y.) Post Express,* October 12, 1922, 1–2. The navy's participation in this experiment could not be documented. Perhaps this was due to navy secrecy, as the "official demonstration," on December 12, 1922, was yet to be conducted.

69. "Improvement in Radio Pictures," *Motion Picture News* (October 1922): 2.

70. The film exchange was created in the early 1900s for distributing films to the theaters and traveling exhibitors. Usage was growing rapidly, and the exchange provided motion-picture rentals to the various venues. See Charles Musser, *The Emergence of Cinema: The American Screen to 1907* (Berkeley: University of California Press, 1994), 365–68.

71. The Native American photos are unique among the group, as they are older photos from Jenkins' trips West.

72. White House to C. Francis Jenkins, December 5, 1922, in Jenkins, *Radio Pictures,* 52.

73. Charles F. Marvin was chief of the National Weather Bureau from 1913 to 1934.

74. For drawings, messages, and photos used in these experiments, see the Jenkins Scrapbook, 1920–24, Wayne County Historical Museum. Each is placed with a handwritten note of description relating to the test purpose.

75. Elihu Thomson to C. Francis Jenkins, November 28, 1922, in Jenkins, *Boyhood of an Inventor,* 134 (also reproduced in Jenkins, *Radio Pictures,* 51). See also Archer, *History of Radio to 1926,* 67.

76. George A. Hoadley to C. Francis Jenkins, March 8, 1923, in Jenkins, *Boyhood of an Inventor,* 135.

77. Herbert Hoover to C. Francis Jenkins, February 1, 1924, in Jenkins, C. Francis, Box 349, Commerce Paper Series, Commerce Period, 1921–1928 NUCM 70–187, RLIN, Hoover Papers (also reprinted in Jenkins, *Boyhood of an Inventor,* 129).

78. The photos are in Box 349, Hoover Papers.

79. Bryan was secretary of state under Woodrow Wilson and a three-time presidential nominee of the Democratic party. See Ernest J. Wrage and Barnet Baskerville, *American Forum: Speeches on Historic Issues, 1788–1900* (Seattle: University of Washington Press, 1960), 343–51; Ernest J. Wrage and Barnet Baskerville, *Contemporary Forum: American Speeches on Twentieth-Century Issues* (Seattle: University of Washington Press, 1962), 93–95, 107; Robert W. Cherny, *A Righteous Cause: The Life of William Jennings Bryan* (Boston: Little Brown and Co., 1985).

80. William Jennings Bryan to C. Francis Jenkins, July 29, 1924, in Jenkins, *Boyhood of an Inventor,* 133. Bryan was approaching the famous John Scopes trial at the time of this letter.

81. Jenkins, *Boyhood of an Inventor,* 145.

82. See John D. Alden, *American Steel Navy: A Photographic History of the U.S. Navy from Introduction of the Steel Hull in 1883 to the Cruise of the Great White Fleet* (Annapolis, Md.: Naval Institute Press, 1985).

83. Jenkins, *Boyhood of an Inventor,* 142. Jenkins lists "General John A. Pershing." This is likely John J. Pershing (1860–1948), who at the time was chief of staff of the U.S. army. Mason Mathews Patrick (1863–1942) was chief of the Army Air Service at the time. "W. A. Moffit" was likely W. A. *Moffett* (1869–1933), chief of the newly organized Bureau of Naval Aeronautics. See Gene Smith, *Until the Last Trumpet Sounds: The Life of General of the Armies John J. Pershing* (New York: Wiley, 1998).

84. No documentation could be found inferring that Jenkins was there to film this event. However, the event was filmed for military purposes.

85. Jenkins, *Boyhood of an Inventor,* 145. For Preston Roger Bassett (1892–1992), see Peter Jessup, Interview with Preston R. Bassett on Motion Pictures and Blind Flying, July 1980, Columbia University, Oral History Project. Bassett was a prolific author.

86. Preston R. Bassett, "Arc Lamp," Patent 1,632,161, filed July 24, 1923, granted June 14, 1927.

Chapter 8. Radio Pictures: Going Operational

1. C. H. Claudy, "Motion Pictures by Radio," *Scientific American* 127 (November 1922): 350.

2. Jenkins makes a passing reference to this experiment and the people who attended it in a presentation before the Institute of Radio Engineers on June 24, 1927. In those remarks he identified June 13, 1925, as the "first public demonstration of radio vision." This would appear to be a reference to a later experiment. See C. Francis Jenkins, "Radio Vision," *Proceedings of the Institute of Radio Engineers* 15.11 (1927): 958–64.

3. "Marvel of Marvel: Photographs Picked from the Air," *Washington Evening Star*, December 22, 1922, Jenkins Scrapbook, 1920–24, Wayne County Historical Museum. See also Charles Francis Jenkins, *Radio Pictures* (Washington, D.C.: Radio Pictures Corporation, 1925), 118; and Charles Francis Jenkins, *Boyhood of an Inventor* (Washington, D.C.: National Capital Press, 1931), 123. See also photos and documenting notes in Jenkins Scrapbook, 1920–24, Wayne County Historical Museum. This naval operation was likely familiar to Jenkins, as he had recently purchased a float plane, and it was during this time that the base was used for testing sea planes and aviation. Today the base has been reconfigured as a naval-support facility.

4. "Pictures Seen and Heard in Radio Demonstration," *Washington Sunday Star*, January 14, 1932, 6; "Broadcasting Photographs Is the Newest Radio Feat," newspaper clipping, Jenkins Scrapbook, 1920–24, Wayne County Historical Museum. Note that the published report follows the event by more than a month. Hays's representation at the meeting represents both government and industry. He was sent to Hollywood amid growing concerns about the effects of sex, crime, violence, and advertising in film. John M. Joy, "Film Mutilation," *Journal of the Society of Motion Picture Engineers* 26 (November 1926): 5.

5. "Cabinet Pleases Congress Leaders: Tentative Selection for Mr. Harding's Cabinet," *New York Times*, February 23, 1921, 2.

6. "Pictures Seen and Heard in Radio Demonstration," *Washington Sunday Star*, January 14, 1932, 6. See also Jenkins, *Boyhood of an Inventor*, 123; and Jenkins, *Radio Pictures*, 118.

7. "Radio May Broadcast Movies," unidentified newspaper clipping, Jenkins Scrapbook, 1920–24, Wayne County Historical Museum. See also Jenkins, *Radio Pictures*, 40, 45–51.

8. Jenkins, *Boyhood of an Inventor*, 123. What specific aid the navy provided beyond facilities and cooperation is unknown.

9. "First Radio Photos in World Received Here from Capital," *Philadelphia Evening Bulletin*, March 3, 1923, 1. The *Bulletin* published from 1847 to 1982 with more than a half-million readers in 1923. See Otto Wilson, "Transmitting Photographist by Radio," *Wireless Age* 10 (July 1923): 67–68; Jenkins, *Boyhood of an Inventor*, 123.

10. This fact may be true in terms of American experiments and electronic scanning transmission, but as noted earlier, other city-to-city link experiments had already

taken place. See "Inventor of Radio Photographs Got Big Idea while Piloting Plane," *Philadelphia Evening Bulletin*, March 5, 1923, 2. The North American Newspaper Alliance was a newspaper syndicate operating and distributing news from 1922 to 1980.

11. S. R. Winters, "The Transmission of Photographs by Radio," *Radio News* 4 (April 1923): 1772–73; Commander Stanford C. Hooper, "Photo Broadcasting Seems Early Reality," *Washington Evening Star*, March 3, 1923, Jenkins Scrapbook, 1920–24, Wayne County Historical Museum. Hopper was the head of the navy's Radio Division, Bureau of Engineering. See also Otto Wilson, "Transmission Photographs by Radio," *Wireless Age* 10 (July 1923): 67–68.

12. L. C. Porter, "President's Address," *Transactions of the Society of Motion Picture Engineers* 16 (May 1923): 21.

13. The National Press Club, founded in 1908, is a private club of journalists and communications professionals located in Washington D.C. See Gilbert Klein, *Reliable Sources: 100 Years of the National Press Club* (Nashville: Turner Publishing, 2009).

14. "Sending Photos by Radio Is Explained by Inventor," *Washington Post*, April 29, 1923, Jenkins Scrapbook, 1920–24, Wayne County Historical Museum.

15. Jenkins, *Radio Pictures*, 35.

16. John Brady, "RadioVision—New Marvel May Be Born in Hub," *Boston Post*, December 30, 1923, A7. See also "Sending Oriental Symbols through Air Made Possible," *Washington Sunday Star*, September 9, 1923, Jenkins Scrapbook, 1920–23, Wayne County Historical Museum.

17. See Jenkins, *Radio Pictures*, 37–38. The message was signed by Ambassador Zhaoji, November 15, 1923. Zhaoji was a Chinese ambassador to the United States from 1921 to 1929.

18. It would appear that the message sent is referenced in Jenkins, *Boyhood of an Inventor*, 137. See also Jenkins, *Radio Pictures*, 37–38. The signed date is November 1924. Note the different year reported in the translation.

19. "Letter in Chinese Is Sent over Radio," *Washington Post*, February 20, 1927, sec. F.

20. David Arthur Hollenback, "Contributions of Charles Francis Jenkins to the Early Development of Television in the United States" (Ph.D. dissertation, University of Michigan, 1983), 130.

21. Jenkins, *Boyhood of an Inventor*, 138.

22. "Sending of Oriental Symbols through the Air Made Possible," *Washington Sunday Star*, September 9, 1923, Jenkins Scrapbook, 1920–24, Wayne County Historical Museum.

23. "Thomson Radio Club," *Lynn (Mass.) Works News*, January 18, 1924, 19–20.

24. Ibid., 19.

25. "See Inventor of Radio Photography Here," *Omaha World-Herald*, August 25, 1923, Jenkins Scrapbook, 1920–24, Wayne County Historical Museum.

26. Jenkins, *Radio Pictures*, 3. A group photo of Jenkins' staff at this time also appears in the *Washington Star*, December 24, 1922, Features.

27. Jenkins, *Radio Pictures*, 5.

28. "See Inventor of Radio Photography Here," *Omaha World-Herald*, August 25, 1923, Jenkins Scrapbook, 1920–24, Wayne County Historical Museum.

29. Jenkins, *Boyhood of an Inventor*, 138–42.

30. "See Inventor of Radio Photography Here," *Omaha World-Herald*, August 25, 1923, Jenkins Scrapbook, 1920–24, Wayne County Historical Museum. Jenkins, *Radio Pictures*, 119, would appear to place the date as December 1924. Jenkins, *Boyhood of an Inventor*, 138–39, has the date as December 15, 1923, as do press reports.

31. Jenkins, *Boyhood of an Inventor*, 138. See H. M. Lord, "The National Budget System and the Financial Situation Facing the United States," *National Municipal Review* 12.2 (1923): 61–66.

32. "Radio Post" *Christian Science Monitor*, July 7, 1926, 6.

33. Ibid.

34. "Transmission of Motion Pictures by Radio Likely," *Ottawa Citizen*, October 3, 1923, 4.

35. Daniel Stashower, "A Dreamer Who Made Us Fall in Love with the Future," *Smithsonian* 21.5 (October 1979): 44–55.

36. Keith Massie and Stephen D. Perry, "Hugo Gernsback and Radio Magazines: An Influential Intersection in Broadcast History," *Journal of Radio Studies* 9.2 (2002): 264–81.

37. H. Gernsback, "Radio Vision," *Radio News* 4 (December 1923): 681, 823.

38. Ibid., 823.

39. Watson Davis, "The New Radio Movies" *Popular Radio* 4.6 (December 1923): 437, 443.

40. Newspaper clipping, Jenkins Scrapbook, 1920–24, Wayne County Historical Museum. Lynn is northeast of Boston, a smaller town with a history of industry. Invitations to the Institute and the Radio Club in Jenkins Scrapbook, 1920–24, Wayne County Historical Museum.

41. "Inventor of Radio Pictures at WNAC," *Boston Globe*, December 28, 1923, Jenkins Scrapbook, 1920–24, Wayne County Historical Museum.

42. Jenkins, *Radio Pictures*, 119.

43. The number of pictures taken varies in different reports. See ibid., 125; "Thomson Radio Club," *Lynn (Mass.) Works News*, January 18, 1924, 20. Jenkins made several SMPE presentations during these years, all focused on high-speed cameras. See C. Francis Jenkins, "Continuous Motion Projector for Taking of Pictures at High Speed," *Transactions of the Society of Motion Picture Engineers* 12 (May 1921): 126; C. Francis Jenkins, "100,000 Pictures per Minute," *Transactions of the Society of Motion Picture Engineers* 13 (October 1921): 69; C. Francis Jenkins, "Motion Picture Camera Making 3,200 Pictures per Second," *Transactions of the Society of Motion Picture Engineers* 17 (October 1923): 77. The *Washington Daily News* reported 3,200 pictures per minute. See "Local Man Inventor of Fastest Motion Picture Camera in World," *Washington Daily News*, November 2, 1923, Jenkins Scrapbook, 1920–24, Wayne County Historical Museum.

44. C. H. Claudy, "Two Hundred Thousand Photographs per Minute," *Scientific American* 126, April 9, 1921, 288, 297.

45. There is little information about these demonstrations other than what is reported in *Scientific American*. The article also points out the construction and film damage. See ibid.

46. Robb Weller, Weller-Grossman Productions, to the author, December 7, 2011. In possession of the author.

47. Carl Akeley to Jenkins, March 16, 1925, in Jenkins, *Radio Pictures*, 66. Carl Ethan Akeley is known for his developments in the field of taxidermy. His work is displayed across the nation, including at the American Museum of Natural History in New York. His photography has been published in numerous books particularly focused on Africa. See Carl Akeley, *In Brightest Africa* (New York: Garden City, 1923).

48. George Eastman to Jenkins, February 18, 1924, in Jenkins, *Radio Pictures*, 63.

49. C. Francis Jenkins to Herbert Hoover, September 4, 1924, Box 349, Hoover Papers.

50. Jenkins, *Radio Pictures*, 95–118. This company should not be confused with the RKO Radio Pictures Corporation, formed in 1928 by Westinghouse, General Electric, and RCA, or the Radio Pictures Corporation, formed by John V. L. Hogan in 1929. See Joseph H. Udelson, *The Great Television Race: A History of the American Television Industry, 1925–1941* (Tuscaloosa: University of Alabama Press, 1982), 63.

51. C. Francis Jenkins, "Recent Progress in the Transmission of Picture by Radio," *Transactions of the Society of Motion Picture Engineers* 17 (October 1923): 15–17, 84–85.

52. Little information was located regarding the specific activities of the Radio Pictures Corporation, likely due to the fact that it would not have lasted long with the rapid development of television approaching and Jenkins' attention turning in that direction.

53. This acknowledgment appears in Jenkins "Radio Vision," 122. See also Jenkins, *Radio Pictures*, 122. It is the same reproduction in both publications, from Jay Brien Chapman, *Story World*, September 1, 1924. Written by and for screenwriters, *Story World* magazine was published from mid-1919 through August 1925. It was drawn into the *Photodramatist* in August 1923. See Henry Sova, *Communications Serials: An International Guide to Periodicals in Communication and the Performing Arts* (Virginia Beach: Socacom, 1992), 586. The editor of *Story World*, Jay Brien Chapman, wrote fiction stories from 1923 to 1932 and published in *Argosy*. See Martin Levin, ed., *Hollywood and the Great Fan Magazines* (New York: Arbor House, 1970). According to the Library of Congress online catalog, issues of *Story World* are available in the Motion Picture/TV Reading Room's Reference Collection.

54. John Lockwood, "Reaching Out to the Red Planet," *Washington Post*, March 17, 2004, A24, Virginia Roach Family Papers. The "Red Planet" article seems a little out of place, as it is dated 2004; it is a review of the events of August 22–23, 1924. See also the photo of Dr. David Todd and Jenkins, *Washington Herald*, August 22, 1924, 1.

55. Silas Dent, "Mars Invites Mankind to Reveal His Secret," *New York Times*, August 17, 1924, 6. Distance estimates vary throughout the reports; the most direct was that

Mars came within 34,630,000 miles of Earth at this time. See "Calls Attempt Nonsense" and "Mars Nearest at 7 P.M.," *New York Times,* August 23, 1924, 9.

56. "Mars Sails by Us without a Word," *New York Times,* August 24, 1924, 30.

57. "Listening for Mars: Heard Anything?" *New York Times,* August 22, 1924, 13. See also "78 Army and Navy Radios Fail to Hear Mars Signal," unidentified newspaper clipping, Virginia Roach Family Papers.

58. "Radio Hears Things as Mars Nears Us," *New York Times,* August 23, 1925, 9. WRC-AM was a part of the NBC Red Network at this time.

59. "Code Expert Ready for Message," *New York Times,* August 23, 1925, 1. Friedman had apparently been instrumental in deciphering coded messages for the Senate oil investigations of the 1920s.

60. "Radio Hears Things as Mars Nears Us," *New York Times,* August 23, 1925, 9.

61. "Calls Attempt Nonsense," *New York Times,* August 23, 1924, 9.

62. "Dots on a Film: Has Mars Signaled? U.S. Professor's Theory," *Edinburgh Scotsman* newspaper clipping, International Press-Cutting Bureau, Box 66, Folder 497, Yale University Library Manuscripts and Archives (hereafter Yale Papers).

63. For the best description of the setup, see "The Mars(?) Record," typewritten description, Box 66, Folder 494, Yale Papers.

64. David P. Todd to Charles Francis Jenkins (confidential), December 14, 1928, p. 1, Box 66, Folder 490, Yale Papers. Different sources report different recording times. A photo caption for the *Washington Evening Star,* August 22, 1924, 13, reports "the device runs for 100 hours." "Seeks Sign from Mars in 30-Foot Radio Film" reports that twenty-nine hours were recorded; see unidentified clipping, Virginia Roach Family Papers. "Radio Fans Listen in Vain for Messages" reports that "more than 20 hours" were recorded; see unidentified clipping from Virginia Roach Family Papers. John Lockwood, "Reaching Out to the Red Planet," *Washington Post,* March 17, 2004, A24, reports that the film paper was two hundred feet long and recorded for twenty hours.

65. "The Mars (?) Record," p. 2, typewritten description, Box 66, Folder 494, Yale Papers. See photo of the apparatus in *Washington Evening Star,* August 22, 1924, 13.

66. For photos of these experiments, see Box 66, Folders 490 and 496, Yale Papers.

67. David P. Todd to Charles Francis Jenkins (confidential), December 14, 1928, p. 1, Box 66, Folder 490, Yale Papers.

68. "Seeks Sign from Mars in 30-foot Radio Film," *New York Times,* August 28, 1924, 6. See also David P. Todd to Charles Francis Jenkins (confidential), December 14, 1928, p. 3, Box 66, Folder 490, Yale Papers. For photos of the signals received, see Box 66, Folders 490 and 496, Yale Papers.

69. Christopher H. Sterling and John Michael Kittross, *Stay Tuned: A History of American Broadcasting,* 3d ed. (Mahwah, N.J.: Lawrence Erlbaum, 2002), 44, 48, and 100. See also William F. Lyon, "Short Wave," in *Historical Dictionary of American Radio,* ed. Donald G. Godfrey and Frederic A. Leigh (Westport, Conn.: Greenwood Press, 1998), 362.

70. A ham operator today is licensed in the United States by the Federal Communications Commission.

71. See Gleason L. Archer, *Big Business and Radio* (1939; reprint, New York: Arno Press, 1971), 132–65. See also L. S. Howeth, *History of Communications Electronics in the United States Navy* (Washington, D.C: Government Printing Office, 1963).

72. "Biggest Radio Show to Open Tomorrow," *New York Times*, May 21, 1922, 36.

73. See Kathleen M. O'Malley, "Ham Operators," in *Historical Dictionary of American Radio*, ed. Donald G. Godfrey and Frederic A. Leigh (Westport, Conn.: Greenwood Press, 1998), 191–92. See also Sterling and Kittross, *Stay Tuned*, 100. Today's British Broadcasting Corporation and the Voice of America use shortwaves to reach a worldwide audience.

74. "Amateurs to Meet with Army Radio Men," *New York Times*, November 3, 1928, 26.

75. "Radio Messages Sent 12,000 Miles in 5 Min." *New York Times*, November 23, 1923, 2.

76. See Clinton B. DeSoto, *Two Hundred Meters and Down: The Story of Amateur Radio* (West Hartford, Conn.: American Radio Relay League, 1963), for the story of the conflicting relationships.

77. "Amateur Organization Expanding Rapidly," *New York Times*, February 13, 1927, 18.

78. Dan C. Wilkerson, "Visible Radio Communication," *QST* (May 1925): 15.

79. G. L. Bidwell, "Television Arrives," *QST* (July 1925): 9–14. While Bidwell has the byline here, the article is distinctly Jenkins' and even includes his old nickel explanation of how scanning worked.

80. "Picture Transmission Permitted under General Amateur License," *QST* (July 1925), in Jenkins Scrapbook, 1925–27, Wayne County Historical Museum.

81. Jenkins, "Award Announcement for Radio Suggestions," *QST* (May 1925): 18.

82. "Sending Picture by Radio Provides New Thrills for Amateur Wireless Artists," *Dallas Morning News,* December 6, 1926, Jenkins Scrapbook, 1925–28, Wayne County Historical Museum. See also *QST* (May 1925): 18; and "Book Prizes Also," *QST* (July 1925), Jenkins Scrapbook, 1925–27, Wayne County Historical Museum.

83. "Jenkins' Awards," *QST* (October 1925): 21.

84. Jenkins, "Award Announcement for Radio Suggestions," *QST* (May 1925): 18.

85. "Jenkins' Awards," 21.

86. There appears to be no record as to the specifics of this idea.

87. "U.S. Inventor's Offer to Amateurs," *New South Wales Sun,* Jenkins Scrapbook, 1925–27, Wayne County Historical Museum.

88. S. R. Winters, "Amateurs Take Up Radio Vision," *Radio Age* 5 (April 1926): 17–18. See also "Cartoons Sent by Radio," *Christian Science Monitor,* May 10, 1926, Radio Section.

89. "Radio Pen Draws Pictures from the Air," *Popular Mechanics Magazine* 45.5 (May 1926): 705–6.

90. Thornton P. Dewhirst, "Practical Picture Transmission," *QST* (December 1925): 12.

91. "World Wide Radio Service Aids Weather Prediction," *Science News-Letter* 10.294 (November 27, 1926): 133–34.
92. Jenkins, *Boyhood of an Inventor*, 150–51.
93. Ibid., 150–56. The location of the Chicago receiver was not recorded by Jenkins. It was likely at the Naval Station Great Lakes or the Navy Pier, both in Chicago on the shores of Lake Michigan. The *Trenton*, named after the city of Trenton, N.J., was the second ship christened "Trenton." The ship referenced by Jenkins was launched in 1923 and was part of the Atlantic fleet. See "Trenton," *Dictionary of American Naval Fighting Ships*, accessed July 17, 2013, http://www.history.navy.mil/danfs/t8/trenton-ii.htm. The *Kittery* was a cargo vessel acquired by the navy for World War I. During Jenkins' experiments, the ship worked Caribbean routes of the Atlantic. See "USS *Kittery*," *Department of Navy—Naval History and Heritage Command*, accessed July 17, 2013, http://www.history.navy.mil/photos/sh-usn/usnsh-k/ak2.htm.
94. This hurricane killed 373 people in Miami and was responsible for $105 million ($1.347,000,000) in damage. Jay Barnes, *Florida's Hurricane History* (Chapel Hill: University of North Carolina Press, 1998): 111–26. See also Jenkins, *Boyhood of an Inventor*, 156–57; "Static Hurricane Warnings," American Association for the Advancement of Science, *Science News-Letter*, January 29, 1927, 69–70. Jenkins was an exhibitor at the association's Philadelphia Convention only a month earlier, December 27–31, 1926.
95. "Static Hurricane Warnings," 69–70. See also unidentified clipping from *Popular Radio* (May 1927), Jenkins Scrapbook, 1925–27, Wayne County Historical Museum.
96. "Attending Annual Session of American Association for the Advancement of Science," *Philadelphia Evening Bulletin*, December 28, 1926; and "Great Scientists Assemble Here for Convention," *Philadelphia Record*, December 28, 1926, Jenkins Scrapbook, 1925–27, Wayne County Historical Museum.
97. Emmet Dougherty, "Weather Maps Prepared Ashore Sent in 15 Minutes to Ships," *New York Herald*, April 2, 1927, IX:3; "Wonders!" *New York American*, December 26, 1926; "Progress," *Philadelphia Public Ledger*, December 28, 1926; "Radio Weather Maps under Study at Sea," *Washington Post*, February 13, 1927, Jenkins Scrapbook, 1925–27, Wayne County Historical Museum.
98. The Lakehurst Station is known primarily for the Hindenburg disaster on May 6, 1937. See Lawrence W. Lichty, "Hindenburg Crash Coverage," in *Historical Dictionary of American Radio*, ed. Donald G. Godfrey and Frederic A. Leigh (Westport, Conn.: Greenwood Press, 1998), 197–98. A copy of the recording is in the Milo Ryan Phono Archive, Tape 3986, cut 2, National Archives, Washington, D.C. See also Sterling and Kittross, *Stay Tuned*, 195–96.
99. "Dirigible to Receive Weather Maps by Air," *Washington Post*, April 3, 1927, M3.
100. Robert D. Heinl, "Airfields to Have Maps of Weather Received by Radio," *Washington Post*, March 27, 1927, F8.
101. "Airships to Receive Maps of Weather," *New York Times*, April 2, 1927, 20.

102. Heinl, "Airfields to Have Maps of Weather Received by Radio," F8. See also "Board Refuses to Change Radio Wave Lengths," *New York Herald,* April 6, 1927, Jenkins Scrapbook, 1925–27, Wayne County Historical Museum.

103. Jenkins, *Boyhood of an Inventor,* 157.

104. Ibid.

105. Noah Arceneaux, "Radio Facsimile Newspapers of the 1930s and 1940s: Electronic Publication in the Pre-Digital Age," *Journal of Broadcasting and Electronic Media* 55.3 (2011): 344–59.

Chapter 9. Television: Seeing by Electricity

1. The first use of the word "television" in Jenkins' work appears in C. Francis Jenkins, "Seeing across the Oceans by Radio," *Illustrated Mechanics* 10.4 (July 1925), cover, 3, 12. Here the singular term "RadioVision" is also changed to two words, "Radio Vision."

2. The etymology of the word *television* begins with *tele* [pronounced *teela*]. In the dictionaries of the time, *tele* is defined as a prefix of Greek origin meaning "far off, far away, far from."

3. C. Francis Jenkins, *Radiomovies, Radiovision, and Television* (Washington, D.C.: National Capitol Press, 1929), preface, 9–10. See also George Shiers, "Television Fifty Years Ago," *Journal of Broadcasting* 19.4 (1975): 6, 13.

4. "Televisionary," *Radio News* 8 (June 1927): 1484.

5. Albert Abramson, *Zworykin: Pioneer of Television* (Urbana: University of Illinois Press, 1995), 46.

6. Ibid., 39–61.

7. Russell Burns, *John Logie Baird: Television Pioneer* (London: Institution of Electrical Engineers, 2000), 75–76.

8. Donald G. Godfrey, *Philo T. Farnsworth: The Father of Television* (Salt Lake City: University of Utah Press, 2001), 12–14, 28–32.

9. David Arthur Hollenback, "Contributions of Charles Francis Jenkins to the Early Development of Television in the United States" (Ph.D. dissertation, University of Michigan, 1983), 153.

10. "New Radio Developments Expected to Occur in 1925," *New York Times,* December 28, 1924, 12.

11. James L. Kilgallen, "Peep Hole Television Near End," *Milwaukee News,* January 3, 1929; see also Philo T. Farnsworth Scrapbooks, 1929, in possession of the author.

12. "Admiral Bullard and Dr. Lee de Forest Are the Deans—Forty Is the Average Age of Engineering and Manufacturers," *New York Times,* July 24, 1927, 15.

13. C. Francis Jenkins, *Radiomovies, Radiovision, Television* (Washington: D.C.: National Capital Press, 1929), 9.

14. C. Francis Jenkins, "Transmitting Pictures by Electricity," *Electrical Engineer* 18 (July 25, 1894): 62–63.

15. Albert Abramson, "Pioneers of Television—Charles Francis Jenkins," *SMPTE Journal* (February 1986): 225–26.
16. "The Telectroscope," *Scientific American* 40 (May 17, 1879): 309.
17. Albert Abramson, *The History of Television, 1880 to 1941* (Jefferson, N.C.: McFarland, 1987), 7–9, 20, 23.
18. "Motion Pictures by Wireless," *Moving Picture News* 8 (September 27, 1913), n.p. The followup article, with technical details, could not be found. See also Abramson, "Pioneers of Television," 224–38.
19. Abramson, "Pioneers of Television," 227.
20. "Motion Picture Machine," Patent 1,385,325. See also Abramson, "Pioneers of Television," 226–27; "Transmitting Pictures by Wireless," Patent 1,554,156. The first patent was the prism, and the second utilized overlapping disks and a new cell for modulating light and deflecting it onto the screen.
21. S. R. Winters, "The Transmission of Photographs by Radio," *Radio News* 4 (April 1923): 1772–73.
22. See Abramson, "Pioneers of Television," 238, n.18.
23. C. Francis Jenkins, "Recent Progress in the Transmission of Motion Pictures by Radio," *Transactions of Society of Motion Picture Engineers* 17 (October 1923): 81–85. There was no reference to the point of origination of this signal; however, it was likely from the navy's station NOF, given the distance noted.
24. "Chemists," *Time* 16 (April 20, 1925): 2.
25. Jenkins, "Recent Progress in the Transmission of Motion Pictures by Radio," 83–85.
26. C. Francis Jenkins, "Radio Movies," *Transactions of the Society of Motion Picture Engineers* 21 (May 1925): 7–12. The Kinetoscope camera was patent number 560, 800.
27. "Radio Motion Pictures Shown Successfully by D.C. Inventor," *Washington Evening Star,* June 15, 1923, 1–2. See also "Long Distance Cinema," *Time* 17 (June 25, 1923): 17; "Says Device Sends Movies through Air," *Washington Post,* June 15, 1923, Jenkins Scrapbooks, 1920–24, Wayne County Historical Museum.
28. Shiers, "Television Fifty Years Ago," 6.
29. H. Gernsback, "Radio Vision," *Radio News* 5 (December 1923): 681, 823–24.
30. Watson Davis, "The New Radio Movies," *Popular Radio* 4 (December 1923): 436–43.
31. "Visual Radio Possibilities are Discussed," *Christian Science Monitor,* August 30, 1927, 13.
32. There are no reports related to the organization of the Jenkins Laboratories, the number of employees, or their specific assignments. Seasonal help was hired for the Christmas holiday seasons, and photos of lab personnel include only a few people. It was a small operation in comparison to the labs of major manufacturers.
33. C. Francis Jenkins, *The Boyhood of an Inventor* (Washington, D.C.: National Capital Press, 1931), 122.

34. C. Francis Jenkins, "Prismatic Rings," *Transactions of the Society of Motion Picture Engineers* 14 (May 1922): 65–66.

35. C. Francis Jenkins, "The Discrola," *Transactions of the Society of Motion Picture Engineers* 16 (May 1923): 234–36.

36. George Shiers, *Early Television: A Bibliographic Guide to 1940* (New York: Garland, 1997), 68. Shiers lists three patents assigned to Discrola Inc. In a general search, Patent 1,409,004 was located and noted as assigned to Discrola Inc.

37. "Sleevelike Refracting Prism," Patent 1,440,466 (this patent was not a flat dishlike disk but a cylindrical, sleevelike or collar-like prism); "Objective Lens," Patent 1,544,155 (this patent replaced the usual lens and bends the light rays along their proper path); "High-Speed Motion-Picture Machine," Patent 1,618,090 (this patent overcomes defects in previous prisms); "Motion-Picture Carrier," Patent 1,409,004 (this patent was for carrying pictures).

38. The patents assigned to Discrola or the Radio Pictures Corporation were: "Pneumatically Controlled Light Valve," Patent 1,525,548; "Radio Picture-Frequency Chopper," Patent 1,525,549; "Drum Lens Carrier," Patent 1,521,190 (this patent placed lenses on a spiral drum for broadcasting and receiving pictures); "Prism-lens Disk," Patent 1,521,191; "Electroscope Picture Reception," Patent 1,521,192; "Film Reception of Broadcasted Pictures," Patent 1,521,189.

39. Abramson, "Pioneers of Television," 230–31.

40. Jenkins, "Recent Progress in the Transmission of Picture by Radio," 15–17, 84–85.

41. A preview of this public demonstration had occurred on May 30, 1925—again, a private demonstration before the public one. "The Jenkins Television System," *Popular Wireless and Wireless Review*, May 30, 1925, 657. See Jenkins, *Radiomovies, Radiovision, Television,* 9; Jenkins, *Boyhood of an Inventor*, 145–49.

42. "Radio Shows Far Away Objects in Motion; Washington Officials See Test of Invention," *New York Times*, June 14, 1925, 1:4; Jenkins, *Boyhood of an Inventor*, 145.

43. "First Motion Pictures Transmitted by Radio Are Shown in Capital," *Washington Post*, June 14, 1925, 1–2. See also "Radio Vision Shown First Time in History by Capital Inventor," *Washington Sunday Star*, June 14, 1925, 1–2. The *Post* and Jenkins' *Boyhood of an Inventor* both report the invitation of a George M. Burgess; this would actually be George K. Burgess. The *Washington Post*'s online archive lists the headline as, "First Motion Pictures Are Received . . ."; the actual headline was "First Motion Pictures *Transmitted*. . . ." See Jenkins, *Boyhood of an Inventor*, 146–47.

44. "Radio Vision Shown First Time in History by Capital Inventor," *Washington Sunday Star*, June 14, 1925, 1–2. See also "Science: Radio Cinema," *Time*, June 29, 1925, 25.

45. Jenkins, *Boyhood of an Inventor*, 149; "Radio Vision Test Nearly a Failure," *Washington Evening Star*, June 15, 1925, Jenkins Scrapbooks, 1925–27, Wayne County Historical Museum.

46. S. R. Winters, "Broadcasting the Movies," *New York Sun*, June 20, 1925, Radio Section, 1–2; "Radio Motion Pictures Transmitted," *Christian Science Monitor*, June

17, 1925, 14. The Jenkins Scrapbooks contain pages of clippings from the national and world press. See, for example, "Apparatus Tested in Washington Shows by Radio Moving Object at Distance of Several Miles," *San Antonio Express;* "Radio Film Test Proves Successful," *Seattle Post-Intelligencer,* June 14, 1925; "Movies Transmitted by Radio Five Miles in Inventor's Test," *New York World,* June 14, 1925; "Radio Motion Pictures Are Transmitted," unidentified newspaper clipping, Jenkins Scrapbooks, 1925–27, Wayne County Historical Museum.

47. Charles Allen Herndon, "Motion Pictures by Ether Waves," *Popular Radio* 8 (August 1925): 107–13. The article includes photos of the apparatus, the prismatic disks, and the units for motion-picture transmission.

48. W. B. Arvin, "See with Your Radio," *Radio News* 6 (September 1925): 278, 384–87.

49. "Radio Pictures Due, Kent Asserts," *Washington Times,* August 6, 1925; *Washington Star,* August 8, 1925, Jenkins Scrapbook, 1925–27, Wayne County Historical Museum. A. Atwater Kent (1873–1949) was a leader in radio manufacturing. His products made by the Atwater Kent Manufacturing Company of Philadelphia were among the cadillacs of the industry. See Alan Douglas, *Radio Manufacturers of the 1920s* (Chandler, Ariz.: Sonoran Publishing, 1995): 65–89; Ralph Williams and John P. Wolkonowicz, *A. Atwater Kent: The Man, the Manufacturer, and His Radios* (Chandler, Ariz.: Sonoran Publishing, 2002).

50. Hoover's office sponsored four radio conferences during 1922–25 to study and propose radio legislation. The result was the passage of the 1927 Radio Act. See C. M. Jansky, "The Contributions of Herbert Hoover to Broadcasting," *Journal of Broadcasting* 1.3 (Summer 1957): 241–49; Daniel E. Garvey, "Secretary Hoover and the Quest for Radio Regulation," *Journalism History* 3.3 (Autumn 1976): 66–70.

51. Kenneth Bilby, *The General: David Sarnoff and the Rise of the Communications Industry* (New York: Harper and Row, 1986), 118. Bilby's comments reflect the RCA bias, and they are inaccurate. Jenkins was born in Dayton but moved as a baby to Richmond, Indiana, where he grew up. At the time of the windmill experiment, he had lived in Washington, D.C., for more than twenty years. The broadcast used a model of a windmill.

52. W.O. McGeehan, "Down the Line: Another Menace to Sport," *New York Herald Tribune,* n.d., 1925, n.p., Jenkins Scrapbook, 1925–27, Wayne County Historical Museum.

53. Ibid. McGeehan, (1879–1933) was a well-known syndicated sports reporter for the *New York Herald Tribune.*

54. NEA Service, "At Last, We Can Attend Opera in Shirt Sleeves," ca. June 30, 1925, Jenkins Scrapbook, 1925–27, Wayne County Historical Museum.

55. Frederic William Wile, "Adventures of . . .," *Washington Morning Star,* October 14, 1925, Jenkins Scrapbook, 1925–27, Wayne County Historical Museum. Wile (1873–1941) was an author, columnist, and editorial writer.

56. Jenkins, *Radiomovies, Radiovision, Television,* 34, 39. See also Charles Allen Herndon, "Motion Pictures by Ether Waves," *Popular Radio* 8 (August 1925): 112.

57. John R. McCrory to C. Francis Jenkins, January 15, 1929 and March 4, 1929, indicate that testing and film production were in progress. Box 47, File 2, George H. Clark Radioana Collection, Archive Center, National Museum of American History, Washington, D.C. (hereafter Clark Collection).

58. C. Francis Jenkins to E. A. Weishaar, United Publishing Company, April 18, 1925, in C. Francis Jenkins, "Seeing across the Oceans by Radio," *Illustrated Mechanics* 10.4 (July 1925): 3, 12.

59. "Films Are Made for Television," *New York Herald*, n.d., Accession File. No explanation was found as to the change from McCrory to Visugraphic Pictures. However, in correspondence with McCrory, there were technical problems with what was produced. See Jenkins, *Boyhood of an Inventor*, 168–69. Jenkins indicated that soon after his SMPE paper was read, a "similar studio" was set up in Hollywood. The reference here is unclear. At this time, Chicago and New York filmmakers were transitioning to Los Angeles.

60. Daniel Stashower, *The Boy Genius and the Mogul: The Untold Story of Television* (New York: Broadway Books, 2002), 58–61.

61. "Construction Is Simple," *Christian Science Monitor*, May 11, 1925, 1.

62. C. Francis Jenkins, "Television about to Take Its Place in the Home," draft article for the *Cleveland Plain Dealer*, sent from the Jenkins Television Corporation, Jersey City, N.J., ca. 1929, Box 47, File 2, Clark Collection.

63. "Transmission of Colored Pictures by Radio Now Theoretically Possible," *Technical Engineering News*, Jenkins Scrapbook, 1925–27, Wayne County Historical Museum. This publication appears to be a trade journal produced by the Massachusetts Institute of Technology.

64. Abramson, "Pioneers of Television," 226–27.

65. "Television," copy for new television booklet, likely authored by Jenkins, p. 5, Box 47, File 4, Clark Collection. See also C. Francis Jenkins to Arthur M. Pohl, Detroit, November 9, 1928, and November 21, 1928, Series 5, Box 37, File 4, Clark Collection. This is the only source indicating that equipment was sold "at cost." Newspaper accounts promoting Jenkins' activities and pictures of the equipment would certainly have left the reader with the idea that Jenkins was targeting the general public.

66. For a comparison of this anxiety, see Elma G. Farnsworth, *Distant Vision: Romance and Discovery on an Invisible Frontier* (Salt Lake City: Pemberly Kent Publishers, 1989), 199. See also Donald G. Godfrey and Alf Pratte, "Elma J. Gardner-Farnsworth: The Pioneering of Television," *Journalism History* 20.1 (Summer 1994): 75–79.

67. C. Francis Jenkins to L. C. Herndon, June 27, 1930, Box 334, Radio Vision, General Correspondence, 1929–1932, File Jenkins Television Corporation, Records of the FCC, National Archives, RG 173 (hereafter National Archives Jenkins Television Corporation Files).

68. For examples of battles between different inventors in this period, see Godfrey, *Philo T. Farnsworth*, 71–76.

69. S. R. Winters, "Shadowgraph Television," *New York Sun,* August 25, 1928, Radio Section.

70. "Seeing through Television," *New York Herald,* n.d., Jenkins Scrapbook, 1927–28, Wayne County Historical Museum.

71. C. Francis Jenkins to L. C. Herndon, June 27, 1930, p. 2. Box 334, National Archives Jenkins Television Corporation Files.

72. G. L. Bidwell, "Television Arrives," *QST* 9 (July 1925): 9–14. See also "Foresees Colored Radio Pictures," *New York Times,* June 21, 1925, 7.

73. "Engineer Seeks to Interest Amateurs in 'Movies' by Radio," *New York Herald-Tribune,* July 15, 1928. See also Jenkins, *Radiomovies, Radiovision, Television,* 10; "Let's Stay Home and See the Movies," *Christian Science Monitor,* July 7, 1928, Jenkins Scrapbook, 1927–28, Wayne County Historical Museum.

74. "Radio Amateurs of Washington Receive Movie Broadcasts by Queer Invention," *St. Louis Post-Dispatch,* Jenkins Scrapbook, 1927–28, Wayne County Historical Museum.

75. Jenkins, *Radiomovies, Radiovision, Television,* 11.

76. Ibid., 10. See also "Radio Amateurs of Washington Receive Movie Broadcasts by Queer Invention," *St. Louis Post-Dispatch,* Jenkins Scrapbook, 1927–28, Wayne County Historical Museum.

77. This untitled article appears to be written by George H. Clark, RCA Show Division Executive, n.d., Series 5, Box 379, File 3, Clark Collection. Clark was an engineer who had served in the navy and worked for the Marconi Telegraphy Company of America and then RCA. He was an avid collector of information, resulting in the George H. Clark Radioana Collection at the National Museum of American History. During television's genesis, Clark was in charge of the RCA Show Division, which exhibited technology across the country for RCA. He is an RCA loyalist and promoter, so some care must be taken with his statements.

Chapter 10. The Eyes of Radio

1. C. Francis Jenkins to an unidentified Commissioner of Navigation, June 2, 1925, Box 334, Jenkins Television Corporation File, National Archives, Washington, D.C. (hereafter Jenkins Television Corporation File).

2. C. Francis Jenkins to D. S. Carson, Commissioner of Navigation, June 9, 1925, Box 334, Jenkins Television Corporation File.

3. Three stations were then assigned in this frequency range. Two were owned by Westinghouse, in Sharon and Pittsburgh, Pa.; and the third, by RCA, was staged from New Brunswick, N.J. The Westinghouse stations were on 49 and 49.5 meters, and the RCA station was on 43 meters. C. Francis Jenkins to Acting Commissioner of Navigation A. J. Tyrer, July 1, 1925, Box 334, Jenkins Television Corporation File.

4. "Applicant's Description of Apparatus," Department of Commerce, Bureau of Navigation, Radio Service, June 3, 1926, Series 139, Box 543, File 4, Clark Collection.

5. "Station License" and letter from the Department of Commerce to the Jenkins Laboratories, September 9, 1926, Box 41, File 3, Clark Collection. See also Donald G. Fink, "Television Broadcasting Practice in America, 1927–1944," *Proceedings of the IEE* 92 (January 1945): 145.

6. "Report Station W3XK, Washington, D.C.," November 11, 1930. p. 1 (written by D. E. Replogle, who would become chief engineer, treasurer, and then vice president of the Jenkins Television Corporation), Box 385, File 4, Clark Collection.

7. "A Kit for Two and One Half Dollars," December 8, 1928, 1777, unidentified magazine page, Jenkins Scrapbook, 1927–28, Wayne County Historical Museum.

8. H. P. Hardesty, Office of the General Service Manager, Packard Motor Company, Detroit, to C. Francis Jenkins, November 22, 1928, Series 5, Box 41, File 9, Clark Collection.

9. Jenkins Laboratories, Secretary, to Robert J. Daugherty, October 16, 1928, Series 5, Box 47, File 2, Clark Collection. See also Virginia Roach Family Papers.

10. "Radio Will Bring Movies into Home Says Inventor," *Washington Morning Star,* January 8, 1928, part 1. See also "Local Inventor Sees Low-Priced Radio Transmission Vision within Next Year," *Washington Morning Star,* December 24, 1927, Jenkins Scrapbook, 1927–28, Wayne County Historical Museum.

11. "Offers Public Radio Movies," *New York Telegram,* August 25, n.d., newspaper clipping, Jenkins Scrapbook, 1927–28, Wayne County Historical Museum.

12. "Device for Detecting Synchronism," Patent 1,537,088 (synchronizing "the prismatic rings at the sending station . . . with rotation of the prismatic rings at the receiving station"); "Synchronism in Radio Movies," Patent, Re. 17,211 (synchronizing "distantly-separated motors" between transmission and reception); "Synchronizing System," Patent Re. 18,883 (synchronizing the speeds between transmission and receiving); "Synchronizing System," Patent 1,766,644 (improvement on synchronization); "Relay for Synchronizing Systems," Patent 2,002,664 (related to other synchronization improvements). A number of these patents were granted after Jenkins' death in 1934. By 1935, RCA owned these patents but did nothing with them.

13. C. Francis Jenkins, *Radiomovies, Radiovision, Television* (Washington: D.C.: National Capital Press, 1929), 61.

14. C. Francis Jenkins, "The Drum Scanner in Radiomovies Receivers," *Proceedings of the IRE* 17 (September 1929): 1576–83. See also C. Francis Jenkins, "The Jenkins Radiovisor," *Radio* 10 (December 1928): 18; C. Francis Jenkins, "Developments Which Led to Jenkins' Radio Movies," *Boston Globe,* August 26, 1928, Radio Section.

15. "Radical Step in Television is Announced," *Christian Science Monitor,* August 30, 1928, Radio Section; "Jenkins Sees End of Disk Vision Idea," *New York Herald-Tribune,* September 9, 1928, 1 and 9.

16. "Method of and Apparatus for Converting Light Impulses into Enlarged Graphic Representations," Patent 1,683,137. See also Jenkins, *Radiomovies, Radiovision, and Television,* 59–71; "Rods of Quartz for Radio Movies," *Science News-Letter,* May 19, 1928, 311.

17. Albert Abramson, "Pioneers of Television—Charles Francis Jenkins," *SMPTE Journal* (February 1986): 233–34. See also "Rod of Quartz Make Radio Movies Possible," Science Service, Press Release 371F, Sheet 3, May 5, 1928, Jenkins Scrapbook, 1927–28, Wayne County Historical Museum.

18. "Radio Movies and Television for the Home," *Radio News* 10 (August 1928): 116–18.

19. Frederick Lewis Allen, *Only Yesterday* (New York: Harper and Row, 1964), 136–37.

20. Christopher H. Sterling and John Michael Kittross, *Stay Tuned: A History of American Broadcasting*, 3d ed. (Mahwah, N.J.: Lawrence Erlbaum, 2002), 827.

21. *United States v. Zenith Radio Corporation, et al.,* 12F.2d 614 (N. D. Ill.), April 16, 1926.

22. Herbert Hoover, "Opening Address," *Third National Radio Conference, Recommendations for Radio Regulation* (Washington, D.C.: Government Printing Office, 1924), 1–2.

23. Attorney General's Opinion, Department of Justice, 35 Ops. Attorney General 126, July 8, 1926.

24. The Radio Act of 1927, Public Law 632, 69th Cong., February 23, 1927. The Radio Act was passed by Congress and signed into law by President Calvin Coolidge on February 23, 1927.

25. The FRC became permanent in 1929. In 1934, the Communications Act updated the charges of the FRC and created the Federal Communications Commission (FCC), which remains today the regulatory agency of electronic media. See Louise M. Benjamin, "Working It Out Together: Radio Policy from Hoover to the Radio Act of 1927," *Journal of Broadcasting and Electronic Media* 42.2 (Spring 1998): 221–36; Donald G. Godfrey, "Senator Dill and the 1927 Radio Act," *Journal of Broadcasting* 23.4 (Fall 1979): 477–89; Donald G. Godfrey and Val E. Limburg, "The Rogue Elephant of Radio Legislation: William E. Borah," *Journalism Quarterly* 67.1 (Spring 1990): 214–24.

26. U.S. Federal Radio Commission, *Annual Report for the Fiscal Year Ended, June 30, 1927* (Washington, D.C,: Government Printing Office, 1927), 1.

27. "Wave Band Taken to Aid Television," *New York Times,* April 6, 1927, 23.

28. Ibid., 1 and 23.

29. Little is known about these stations or their owners. J. Smith Dodge was part owner of WLEX and W1XAY in Lexington, Mass.; 2XBU was owned by Herold E. Smith of Beacon, N.Y. See Joseph H. Udelson, *The Great Television Race: A History of the American Television Industry, 1925–1941* (Tuscaloosa: University of Alabama Press, 1982), 37–40.

30. "Stations Licensed for Television," *New York Times,* July 21, 1928, 2 (18). Here we begin to notice some variances in the *New York Times* online files pagination, as compared to the hard-copy newspaper; the online pagination is in parentheses. See also "Visual Broadcasting," *Time,* August 27, 1928, Science Section.

31. Douglas Godfrey to the author, June 19, 2011. During the late 1920s and early 1930s, Douglas Godfrey was a projectionist at the Empress Theater in Magrath, Alberta, Canada. Today, he is retired senior Chevron research engineer.

32. C. Francis Jenkins, "Pantomime Pictures by Radio for Home Entertainment," *Transactions of the Society of Motion Picture Engineers* 12 (April 1928): 110–16.

33. "Television Is Evolving at Slow but Sure Pace," *New York Times*, April 22, 1928, 15 (145).

34. "Jenkins Plans Large-Size Radio Motion Pictures," *New York Herald-Tribune*, August 5, 1928, Jenkins Scrapbook, 1927–28, Wayne County Historical Museum.

35. "Many Lights Jenkins' Idea in Television," *Christian Science Monitor*, September 22, 1928, 4.

36. "Jenkins Solves Light Problem for Television," *New York Herald-Tribune*, September 23, 1928, Jenkins Scrapbook, 1927–28, Wayne County Historical Museum.

37. "Many Lights Jenkins' Idea in Television," 4. In various writings, these photo cells are referred to as light bulbs, flash-light bulbs, and photo-electric cells.

38. H. W. Somers, "Invents New System of Television," *New York Sun*, September 23, 1928, Radio Section.

39. C. Francis Jenkins, "Light-Concentrating Device," Patent 1,663,307, filed October 22, 1924, granted March 20, 1928.

40. On October 22, 1924, six additional patents were filed relating to the "light sensitive cell": "Grid-Coupled Cell Circuit," Patent 1,667,383; "Resister Cell Circuit," Patent Re. 18,756; "Grid-Leak Cell Circuit," Patent 1,667,384; "Resister Cell Circuit," Patent 1,693,509; "Grid-Leak Cell Circuit," Patent Re. 17,766; and "Offset Filament Lamp," Patent 1,572,607.

41. "Cell-Studded Televisor, Jenkins Invents New Apparatus to Radio Outdoor Scenes," NEA Service Release, October 1928, Jenkins Scrapbook, 1927–28, Wayne County Historical Museum. See also William Hartge Fifer, "The Inventions of Dr. C. Francis Jenkins," University of Maryland Beta Chapter of the Tau Beta Pi Association, the Honorary Engineering Fraternity, January 10, 1930, University Records, ARCV 72-115.68, Records of Phi Mu, UMCP HBK Maryland Room Archives, University of Maryland at College Park Libraries (hereafter University of Maryland Archives).

42. "Persisting Luminescent Screen," Patent 2,021,010.

43. "Photo-Radio Carries Poem," *New York Times*, April 9, 1928, 23. Tests in secrecy seem almost incongruent with Jenkins' past, when he sought publicity; his work with the photo-electric cells is the first mention of it. It is likely that he was becoming increasingly conscious of his competitors catching up to him with their own technological achievements. Yet, it cannot have been too much of a secret as articles about it appear in the popular press.

44. "C. Francis Jenkins, under the Auspices of the National Radio Institute, Will Talk on 'Modern Development in Television and Radio Vision,' 8 p.m., October 23, 1928, over the Red Network," Announcement, Jenkins Scrapbook, 1927–28, Wayne County Historical Museum.

45. "Jenkins Plans Vision Showing in Three Cities," *New York Herald Tribune*, ca. April 1928; "Lecture on Television," *University Hatchet* (George Washington University), April 4, 1928; "Mount Pleasant Men's Club Lists Radio Lecture," Jenkins Scrapbook, 1927–28, Wayne County Historical Museum.

46. "Movies by Radio Soon Will Be Ready for Amateurs, Says C. F. Jenkins," *Richmond Palladium*, August 24, 1928, 7.

47. Predictions from "Television Today," *Literary Digest*, October 27, 1928, 24–25.

48. This is an interesting prediction, given the fact that the De Forest Corporation was experiencing hard times itself and would purchase the Jenkins Corporation within the next year.

49. See a copy of the invitation, Jenkins Scrapbook, 1927–28, Wayne County Historical Museum; "Inventor Broadcasts Radio Motion Pictures," *Washington Post*, May 6, 1928, 7.

50. "Capital Inventor Shows Radio Movies to Experts," *Washington Evening Star*, May 5, 1928, 1. Major General George Owen Squier was a pilot as well as an inventor. He was known for his inventions related to multiplexing in telephone and telegraph communications. See George Squier, *Multiplex Telephone and Telegraphy by Means of Electrical Waves* (Washington, D.C.: Government Printing Office, 1919). Ira Ellsworth Robinson was a politician and lawyer, the first chair of the Federal Radio Commission. Sam Pickard was a native of Kansas, influential in rural broadcasting. He would leave the commission to become a vice president of the CBS Radio Network. Stanford Caldwell Hooper was a rear admiral in the navy prior to his appointment to the commission. He was an engineer for the Radio Division of the navy's Bureau of Engineering. See Susan J. Douglas, "The Navy Adopts the Radio, 1899–1919," in *Military Enterprise and Technological Change: Perspectives on the American Experience*, ed. Merritt Roe Smith (Boston: Massachusetts Institute of Technology Press, 1985), 154–74. Guy Hill was a civilian radio expert. And finally, this was the same Henry D. Hubbard who earlier had been pushing the motion-picture industry to create consistent standards.

51. "Capital Inventor Shows Radio Movies to Experts," *Washington Evening Star*, May 5, 1928, 1; "Inventor to Broadcast Movies by Radio Today," *Washington Post*, May 5, 1928, 16. There is a discrepancy between these two newspaper reports. The *Post* indicates that the films were to be broadcast within the next week, and the *Evening Star* indicates that they had already been broadcast to these homes prior to the formal demonstration. Jenkins offered to place a receiver in Hoover's home, but the offer was declined due to time constraints. See B. D. Nash, secretary to Mr. Hoover, to C. Francis Jenkins, May 5, 1928, Box 349, Hoover Papers.

52. The "cat's whisker" is a reference to early radio experimentation with crystal receivers to detect radio signals. See Sterling and Kittross, *Stay Tuned*, 36–38.

53. "Silhouettes Dance on Mirror Attached to New Radio Device," *New York Times*, May 13, 1928, 17 (141); "Capital Inventor Shows Radio Movies to Experts," *Washington Evening Star*, May 5, 1928, 1–2; "Inventor to Broadcast Movie by Radio Today," *Wash-*

ington Post, May 5, 1928, 16; "Broadcast Pictures: Jenkins Exhibited Device before Washington Experts," *New York Times,* May 6, 1928, 3; "Inventor Broadcasts Radio Motion Picture," *Washington Post,* May 6, 1928, Jenkins Scrapbook, 1927–28, Wayne County Historical Museum.

54. "Jenkins Plans Large-Size Radio Motion Pictures," *New York Herald-Tribune,* August 5, 1928, Jenkins Scrapbook, 1927–28, Wayne County Historical Museum.

55. Annotated pages from Charles Francis Jenkins, *Boyhood of an Inventor* (Washington, D.C.: National Capital Press, 1931), 162. Jenkins' nephew, Lewis LeMar Janney, has marked pages in the book noting his participation with his uncle (Janney Family Files; copies also in possession of the the author).

56. "Officials See Jenkins Test," *New York Herald-Tribune,* January 12, 1929, Jenkins Scrapbook, 1927–28, Wayne County Historical Museum. Note that this news article appeared eight months after the first demonstration to the Radio Commission and describes other demonstrations that had apparently taken place. At this writing, there was no further information relative to other demonstrations before the commissioners.

57. Federal Radio Commission, *Second Annual Report* (Washington, D.C.: Government Printing Office, 1928): 256–57.

58. Ibid., 252–53.

59. Albert Abramson, *The History of Television, 1880–1941* (Jefferson, N.C.: McFarland, 1987), 112.

60. Television technology was centered around mechanical scanning at this time. Jenkins was exploring the use of photo-electric cells, but Zworykin and Farnsworth would change that direction to electronic scanning. See Albert Abramson, *Zworykin: Pioneer of Television* (Urbana: University of Illinois Press, 1995); and Donald G. Godfrey, *Philo T. Farnsworth: The Father of Television* (Salt Lake City: University of Utah Press, 2001).

61. Jenkins, *Boyhood of an Inventor,* 158, indicates the start date was July 2. Both dates appear in the press reports. We have utilized the press dates in this writing, as they are closer to the event.

62. "Radio Movies to Begin in Washington Tonight," *Washington Post,* July 6, 1928, 18; "Broadcast of Movies by Radio Here July 2," *Washington Post,* June 23, 1928, 2: and "Radio Silhouette Movie Inaugurated in Capital," *Washington Post,* July 7, 1928, 1.

63. "Radio Movies to Begin in Washington Tonight," *Washington Post,* July 6, 1928, 18.

64. Farnsworth and Zworykin later used *Felix the Cat* and *Steamboat Willie* (the first Mickey Mouse) in their experimentation. These cartoons were professionally created and had progressed to the point of having storylines, but were still simple black-and-white with high contrast.

65. H. M. Bayer, "Equipment for Television Experimenters," *Radio News* 10 (December 1928): 582.

66. Jenkins, *Boyhood of an Inventor,* 158.

67. "Broadcast of Movies by Radio Here July 2," *Washington Post,* June 23, 1928, 2 (note that the price has increased).

68. "Promises Home Television Soon," *New York Telegram,* August 22, 1928, Jenkins Scrapbook, 1927–28, Wayne County Historical Museum.

69. "After Nine Years Planning," *National Daily Washington Times,* July 7, 1928, Jenkins Scrapbook, 1927–28, Wayne County Historical Museum.

70. Federal Radio Commission, *Third Annual Report* (Washington, D.C.: Government Printing Office, 1929), 22, 55–56.

71. Charles Francis Jenkins, *Television the Eye of Radio* (Jersey City, N.J.: Jenkins Television Corporation), 8–10. See also, "Malden Man Picks Radio Movies from the Air," *Boston Globe,* n.d., Radio Broadcast Section, Jenkins Scrapbook, 1927–28, Wayne County Historical Museum.

72. "Radio Pictures," *Washington Post,* July 8, 1928, 27.

73. Julius Weinberger, Theodore A. Smith, and George Rodwin, "The Selection of Standards for Commercial Radio Television," *Proceedings of the Institution of Radio Engineers* 17.9 (September 1929): 1584.

74. Ibid.

75. William W. Harper, "The Progress of Television: A Resume of Television Standardization Activities Affecting the Field of Popular Broadcasting Experimentation," *Radio* (November 1928), Jenkins Scrapbook, 1927–28, Wayne County Historical Museum.

76. "48 Lines Radiovision Standard," *Science News-Letter,* November 3, 1928, Jenkins Scrapbook, 1927–28, Wayne County Historical Museum.

77. Harper, "Progress of Television." Harper was a member of the Television Committee of the Radio Manufacturers' Association and chair of the Television Committee of Radio Engineers Club of Chicago.

78. Jenkins, "Television as a Field for Standardization," *Commercial Standards Monthly* (December 1931): 166.

79. Alexander B. Magoun, *Television: The Life Story of Technology* (Westport, Conn.: Greenwood, 2007), 65–68. See also Eric Freedman, "Standards," in *Encyclopedia of Television,* 2d ed., ed. Horace Newcomb, vol. 4 (New York: Fitzborn, 2004), 2185–86.

80. Albert Abramson, *The History of Television, 1942 to 2000* (Jefferson, N.C.: McFarland, 2003), 175, 200, 219, 259.

Chapter 11. The Jenkins Television Corporation

1. D. E. Replogle, "Analysis of Jenkins Patent Situation," December 29, 1930, p. 1, Series 5, Box 43, File 7, Clark Collection. This document is a report that followed the organization of Jenkins Television after the stock market crash of October 29, 1929. It would appear to be an updated assessment, given the rapidly changing economic climate of the Depression.

2. Many corporations organize under Delaware law because the state's rules are more favorable to business needs.

3. Charles Francis Jenkins, *Boyhood of an Inventor* (Washington, D.C.: National Capital Press, 1931), 169.

4. "Annual Report, 1928. Jenkins Television Corporation," July 1, 1929, Folio No. 2464-15, State of Delaware, Division of Corporations, Dover. See also "Inventor Denies Television Debts," *Washington Post*, September 11, 1929, 8; "Heart Attack Closes Career of Dr. Jenkins," *Washington Post*, June 7, 1934, 11.

5. "$10,000,000 Concern to Push Television," *New York Times*, December 5, 1928, 3. It is interesting that many of these assets seemed to fade from priority under the new Jenkins Television Corporation. The new emphasis would be getting the receivers ready for sale and expanding station operations.

6. P. H. Diehl, treasurer, Jenkins Television Corporation, to C. Francis Jenkins, August 26, 1929, Series 5, Box 42, File 8, Clark Collection. The letter apparently went to each engineer, the lab personnel, and plant workers with a contractual agreement they were to sign and return.

7. "Stock Offered—Jenkins Television Corporation," *Commercial and Financial Chronicle* 127 (December 8, 1928): 3256–57; "$10,000,000 Concern to Push Television," *New York Times*, December 5, 1928, 3; "$10,000,000 Company to Sell Television," *Washington Star*, December 4, 1928, Jenkins Scrapbook, 1927–28, Wayne County Historical Museum.

8. "Business and Finance: Televisionary Biddle," *Time*, December 1, 1928, 25.

9. "$10,000,000 Concern to Push Television," *New York Times*, December 5, 1928, 3.

10. "Business and Finance: Televisionary Biddle," *Time*, December 1, 1928, 25. Nothing is known about C. C. Kerr and Company. The New York Curb Market is the forerunner of the American Stock Exchange. See Maury Klein, *Rainbow's End: The Crash of 1929* (New York: Oxford University Press, 2001).

11. "Interest Aroused in Jenkins Television," *San Francisco Chronicle*, March 18, 1929, Philo T. Farnsworth Scrapbooks, 1929, in possession of the author.

12. "Former Kalamazoo Man Now Radio Leader: Tells How He Reorganized the De Forest Radio Company and Organized Jenkins Television Corporation," *Kalamazoo Gazette*, n.d, 5, Series 5, Box 388, File 2, Clark Collection.

13. Ibid., 1.

14. Ibid., 3–4.

15. Ibid., 6. Given the Depression, it is doubtful that these quotas were ever met. At this writing, no production figures were found. The audion tube, an instrument that facilitated voice transmission, gave De Forest his fame. See Michael H. Adams, "De Forest, Lee, 1873–1971," in *Encyclopedia of Radio*, vol. 1, ed. Christopher G. Sterling (New York: Fitzroy Dearborn, 2004), 440–43.

16. Thomas Temple Hoyne, "Garside Sees Great Future for Jenkins Television," February 1929, Philo T. Farnsworth Scrapbooks, 1929, in possession of the author.

17. Draft communications "To Our Clients," pp. 1–3, in Thomas C. Mulhull and Co., Austin C. Lescarboura and Staff, to James W. Garside, October 7, 1929, Series 5, Box 42, File 10, Clark Collection.

18. Press Release, New York News Bureau Association, Los Angeles Tickers, March 25, 1929, Series 5, Box 37, File 3, Clark Collection.

19. "Former Kalamazoo Man Now Radio Leader."

20. "Latest Television Sets to Cost $150," unidentified newspaper clipping, Jenkins Scrapbook, 1927–28, Wayne County Historical Museum.

21. "Business and Finance: Televisionary Biddle," *Time,* December 1, 1928, 25.

22. Ibid.

23. "Early TV Station Operated in County," *Montgomery County Sentinel,* March 15, 1956, B4.

24. Herbert William Cooper, "The Construction and Operation of the Jenkins Television Laboratory at Wheaton, MD," April 17, 1931, University of Maryland Archives. This research paper is significant, as its primary sources were individual interviews with Jenkins and two laboratory staff members, Paul Thomas and Theodore Delote.

25. "Jenkins to Build Television," *New York Times,* January 20, 1920, 19. The likelihood of this being the "first broadcasting station ever built" is a debate for another venue; there are several that could probably vie for the title. See John P. Arnold, "Radio Picture Transmission and Reception" *Radio* (December 1928): 17–18. At this time, Arnold, the editor of *Radio,* lists three stations in Schenectady, N.Y.; two in New York City; three in Chicago; one in Los Angeles; one in Lexington, Mass.; and one, the Jenkins station, in Washington, D.C.

26. Jenkins, *Boyhood of an Inventor,* 194.

27. "D.C. Television Radio First Time," *Washington Daily Star,* October 11, 1931, Accession File. At this writing, little is known about these performers. It is assumed that they were local talent.

28. J .M. Verralls, "America and Television Progress," *Television* 2 (December 1929): 483; "In the Radio Marketplace," *Radio Broadcast* 15 (September 1929): 274.

29. Verralls, "America and Television Progress," 483.

30. See sample of announcements in the Clark Collection, Series 5, Box 385, File 4.

31. "W3XK Television," *Washington Evening Star,* January 22, 1932, Jenkins Scrapbook, 1931, Wayne County Historical Museum.

32. See "Television," *Washington Daily News,* October 26, 1931; "Central High Student on Television Program," *Washington Star,* November 22, 1931, Jenkins Scrapbook, 1931, Wayne County Historical Museum.

33. "Television Showing Planned in Capital," *Washington Post,* October 18, 1931, 12; "Trade Exposition Has 120 Displays," *Washington Star,* October 25, 1931, Jenkins Scrapbook, 1931, Wayne County Historical Museum.

34. Advertisement in *Washington Post,* October 26, 1931, Jenkins Scrapbook, 1931, Wayne County Historical Museum.

35. "Television Contest Closes Tonight: Auditions at 7," *Washington Daily News,* October 24, 1931. See also "News Offers Opportunity to Budding Television Stars," October 22, 1931, and "Applications Being Received in News' Television Contest,"

October 23, 1931. Both of these clippings appear to be from the *Washington Daily News,* Jenkins Scrapbook, 1931, Wayne County Historical Museum.

36. "Television Contest Closes Tonight: Auditions at 7," *Washington Daily News,* October 24, 1931, Jenkins Scrapbook, 1931, Wayne County Historical Museum. The large screen was a part of the Jenkins photo-cell screen experiments, which would have made the screen two feet by two feet square at this time.

37. The *Washington Daily News* pictured the winners each evening from Monday, October 26, through Saturday, October 31, 1931. "Television Contest Closes Tonight: Auditions at 7," *Washington Daily News,* October 24, 1931, Jenkins Scrapbook, 1931, Wayne County Historical Museum.

38. "Industrial Show Attracts 10,000," *Washington Daily Star,* October 28, 1931; "Industrial Show Attendance 18,000," *Washington Daily Star,* October 19, 1931, Jenkins Scrapbook, 1931, Wayne County Historical Museum.

39. D. E. Replogle, "Report Station W3XK," November 11, 1931, p. 2, Series 5, Box 385, File 4, Clark Collection.

40. Advertisement in *Radio News* 13 (July 1930): 59. Advertisements appear throughout Jenkins Scrapbook, 1931, Wayne County Historical Museum.

41. Joseph H. Udelson, *The Great Television Race: A History of the American Television Industry, 1925–41* (Tuscaloosa: University of Alabama Press, 1982), 52–53.

42. "New Television Idea Hastens Development," unidentified clipping, Jenkins Scrapbook, 1931, Wayne County Historical Museum.

43. "Television Due Here Soon," *New York Times,* January 11, 1929, 16 (21).

44. "Television Movies to Start in Month," *New York Times,* March 11, 1929, 30 (34); "Three Stations Offer Television," unidentified newspaper clipping (Hartford, Conn., June 24, 1929), Farnsworth Scrapbooks, 1929 newspaper clippings, Philo T. and Elma G. Farnsworth Papers, Ms0648, Special Collections Department, J. Willard Marriott Library, University of Utah.

45. DuMont was an engineer who had worked at Westinghouse until 1928, when he took the position as chief engineer at De Forest to assist in its reorganization under Garside's direction. DuMont's leadership at the De Forest Company led to the production of "up to 30,000 tubes per day" and earned him a promotion to vice president in charge of engineering and manufacturing. See David Weinstein, *The Forgotten Network: DuMont and the Birth of American Television* (Philadelphia: Temple University Press, 2004), 2, 10.

46. D. E. Replogle, "Report Station W2XCD," November 11, 1931, p. 1, Series 5, Box 385, File 4, Clark Collection.

47. "Radio Talkies Put on Program Basis," *New York Times,* April 27, 1931, 4.

48. D. E. Replogle, "Report Station W2XCR," November 11, 1931, p. 2, Series 5, Box 385, File 4, Clark Collection.

49. D. E. Replogle, "Report Station W3XK," November 11, 1931, p. 2, Series 5, Box 385, File 4, Clark Collection.

50. "3 More Stations Offer Television," *New York Times,* June 24, 1929, 21 (28). Other stations on the air included W2XCL, owned by the Pilot Radio and Tube Company in Brooklyn, and W2XBE, owned by WAAM Inc. in Newark, N.J. According to the Web site Early Television Museum, there were at this time five stations in Schenectady, N.Y. owned by General Electric; one CBS station; two stations owned by Hugo Gernsback; and four owned by RCA. Accessed July 28, 2011, http://earlytelevision.org/mechanical_stations.html.

51. W. J. Barkley, vice president, to De Forest Radio Company District Managers, May 4, 1931, Series 5, Box 37, File 1, Clark Collection; D. E. Replogle, "Report Station W3XK," November 11, 1931, Series 5, Box 385, File 4, Clark Collection.

52. "Movies by Television to Be Broadcast Soon," *New York Times,* May 10, 1929, 34 (39).

53. D. E. Replogle, "Jenkins Planning Commercial Debut," *The Audion,* December 1929. Replogle was an assistant to President Garside. *The Audion* appears to be an in-house publication. Series 5, Box 388, File 2, Clark Collection.

54. Federal Radio Commission, *Third Annual Report* (Washington, D.C.: Government Printing Office, 1929), 100; see also *Radio Service Bulletin* 151 (October 21, 1929): 5.

55. C. Francis Jenkins to Paul C. Staake, De Forest Radio Company, June 12, 1929, Series 5, Box 42, File 6, Clark Collection.

56. "Jenkins to Test an Aerial Television 'Eye'; Would Give Army Pictures of Enemy from Air," *New York Times,* April 13, 1929, 19 (11); "Television's New Aerial 'Eye' for War and Peace Broadcasting," *Detroit Free Press,* July 21, 1929, 7; "Aerial Eye Puts Ground in Relief," unidentified newspaper clipping, July 23, 1929; "Capital Scenes to Be Broadcast from 'Aerial Eye,'" Accession File. See also "Hopes to Apply Television to Aerial Use as War Aid," unidentified newspaper clipping, April 21, 1929, Farnsworth Scrapbooks, 1929 newspaper clippings, Philo T. and Elma G. Farnsworth Papers, Ms0648, Special Collections Department, J. Willard Marriott Library, University of Utah.

57. C. Francis Jenkins to Paul C. Staake, De Forest Radio Company, June 12, 1929, p. 2, Series 5, Box 42, File 6, Clark Collection.

58. "Plane Will Attempt Televising Broadcast" *New York Times,* July 22, 1929, 28 (31).

59. "Two Honorary Degrees Awarded," *The Earlhamite* 50.4 (June 1929): 4. *The Earlhamite* is a quarterly publication of Earlham College. See also "Jenkins to Receive Honorary Degree," *Washington Daily Star,* June 13, 1929, Accession File.

60. See Park Savings Bank list of the Board of Directors, Jenkins Scrapbook, 1931, Wayne County Historical Museum.

61. "Television Inventor Faces Second Suit," *New York Times,* July 27, 1929, 6 (11); "To Given Evidence Here in Television Lawsuit," *New York Times,* December 7, 1929, 16 (24).

62. "Jenkins Replies in Television Suit," *New York Times,* August 18, 1929, 20. See also "Television Inventor Faces Second Suit," *New York Times,* July 27, 1929, 6 (11).

63. "Jenkins Replies in Television Suit," *New York Times*, August 18, 1929, 20. See also "Television Inventor Sued," *New York Times*, October 15, 1929, 51 (56).

64. "Jenkins Replies in Television Suit," *New York Times*, August 18, 1929, 20.

65. "Capital Inventor Sued for $550,000," *Washington Post*, November 20, 1931, 22. No information regarding Wiley H. Reynolds was located.

66. "$612,500 Television Litigation Settled," *Washington Post*, December 18, 1929, Accession File.

67. If this contract expired in 1927, it would appear to have been more likely connected to Discrola Inc. or Jenkins' Radio Pictures Corporation. Discrola was supposedly selling appliances for viewing rented movies. The Radio Pictures Corporation was supposed to be selling facsimile apparatus for industry-wide application. There is no evidence that these corporations were even moderately successful.

68. "Television Patent Action Is Dismissed," *Washington Post*, May 13, 1931, 22. See also "Television Action Ends," *Wall Street Journal*, May 13, 1931, Jenkins Scrapbook, 1931, Wayne County Historical Museum.

69. Calvin Coolidge, "Government and Business," in *Foundations of the Republic: Speeches and Addresses* (Freeport, N.Y.: Books for Libraries Press, 1968), 317–32.

70. Frederick Lewis Allen, *Only Yesterday* (1931; reprint, New York: Harper and Row, 1965), 165.

71. See "Chart History of Stock Market," Amateur Investors, accessed August 6, 2013, http://www.amateur-investors.com/Chart_History_of_Stock_Market.htm.

72. D. E. Replogle, assistant to the president, Jenkins Television Corporation, to C. Francis Jenkins, August 15, 1930, Series 4, Box 14, File C. Francis Jenkins, Clark Collection.

73. De Forest Radio Company, Inter-Office letter from Allen B. DuMont, signed C. G. Mun, December 18, 1930, Series 5, Box 37, File 8, Clark Collection.

74. "Television," *Time*, May 18, 1931, Cinema Television Section.

75. "Parade Pictures to Be Broadcast," March 3, 1929, Philo T. Farnsworth Scrapbooks, 1929, in possession of the author.

76. "Television," *Time*, May 18, 1931, Cinema Television Section.

77. Telegram from Grace Jenkins to Jenkins Brothers, June 30, 1921, Jenkins Scrapbook, 1931, Wayne County Historical Museum.

78. "Dr. C. F. Jenkins Recovering Here," *Washington Daily Star*, September 14, 1931, Accession File.

79. "Rights by De Forest Radio Company," *New York Times*, May 11, 1929, 26. Additional stock was sold to raise ond million dollars. "De Forest Radio Expands Capital," *New York Times*, September 24, 1929, 47 (authority to raise three million dollars from selling shares was authorized).

80. "De Forest Radio Makes Offer," *New York Times*, September 28, 1929, 27 (31).

81. "More De Forest Radio Stock Out," *New York Times*, October 22, 1929, (45).

82. "Television Merger Is Contemplated," *Washington Post*, February 1, 1932, (4).

83. "Receivers Named in De Forest Radio," *Washington Post*, June 24, 1932, (15).

84. "Receiver Gets Television," *Washington Times,* February 1, 1933. See also "Receivers Named for Jenkins Co.," Jenkins Scrapbook, 1931, Wayne County Historical Museum.

85. *John F. Krieger, Complainant v. Jenkins Television Corporation, Defendant,* March 11, 1933, No. 1011 in Equity Order of Sale, *New York Post,* July 3, 1935 (legal notices).

86. "De Forest Radio Bid Approved," *New York Times,* May 6, 1932, 29; "$500,000 Radio Bid Is Ordered Accepted," *Washington Post,* March 7, 1933, 3; Kenneth Bilby, *The General: David Sarnoff and the Rise of the Communications Industry* (New York: Harper and Row, 1986), 120–23.

87. "Television Assets Sold for $200,000," *Washington Post,* July 15, 1933, 10; "Television Assets Sold at Auction," *Washington Evening Star,* July 14, 1933, Jenkins Scrapbook, 1931, Wayne County Historical Museum.

88. Daniel Stashower, *The Boy Genius and the Mogul: The Untold Story of Television* (New York: Broadway Books, 2002), 166.

89. "Television Towers Go for Scrap," Accession File. The exact date W3XK shut down is unknown. Clearly, following the sale, it was no longer a priority to RCA. Grace Jenkins notes, in her diaries, that it was shut down for thirty days beginning June 7, 1932. Grace Jenkins Diaries, June 7, 1932, Virginia Roach Family Papers.

90. "De Forest Radio Loses $106,800 on Poor Buy," unidentified newspaper clipping, Box 40, File 1, Clark Collection.

91. Alan Douglas, *Radio Manufacturers of the 1920s,* vol. 1 (Chandler, Ariz.: Sonoran Publishing, 1988), 173–75; "Former Kalamazoo Man Now Radio Leader," 1.

92. "Television Station Burns in New Jersey," *Washington Post,* January 23, 1932, 3; "Radio Fire Destroys Jenkins Equipment," Accession File.

93. David Weinstein, *The Forgotten Network: DuMont and the Birth of American Television* (Philadelphia: Temple University Press, 2004), 10.

94. Grace Jenkins Diaries, February 1–2, 1932, Virginia Roach Family Papers.

95. Grace Jenkins Diaries, January 1 and January 11, 1932, Virginia Roach Family Papers. Jenkins also met with "Mr. Eastman" for the same purpose, on November 2, 1932.

96. Jenkins, *Boyhood of an Inventor,* xvii–xix. The copyright is listed as 1931, which is likely when it was written. It was early 1932 when Jenkins was delivering copies and accepting orders.

97. "Dr. Jenkins Dies: Noted Inventor," *Washington Evening Star,* June 6, 1934; "C. Francis Jenkins, Ill Here," *Washington Evening Star,* March 30, 1934, Jenkins Scrapbook, 1931, Wayne County Historical Museum.

98. Telegram from Grace Jenkins to the Jenkins Brothers Insurance Company, March 21, 1934, Jenkins Scrapbook, 1931, Wayne County Historical Museum.

99. Telegram from Grace Jenkins to Atwood L. Jenkins, June 6, 1934, Jenkins Scrapbook, 1931, Wayne County Historical Museum.

100. "Dr. Jenkins Dies: Noted Inventor," *Washington Evening Star,* June 6, 1934, Jenkins Scrapbook, 1931, Wayne County Historical Museum. Angina pectoris produces intense chest pain created by insufficient oxygen reaching the heart muscles.

101. "Dr. Jenkins to Get Aerial Tribute," *Washington Daily Star,* June 9, 1934, Accession File.

102. "Widow to Receive Jenkins' Estate," Jenkins Scrapbook, 1931, Wayne County Historical Museum.

103. "Jenkins Bubbles with Enthusiasm," Washington, D.C., June 30, 1929, Philo T. Farnsworth Scrapbooks, in possession of the author. See also "Statement of the Federal-American National Bank and Trust Company," December 31, 1931, Jenkins Scrapbook, 1931, Wayne County Historical Museum.

104. Grace Jenkins diaries, May 14 and July 22, 1932, Virginia Roach Family Papers. Grace's 1932 diary indicated a normal cash flow of bank and expense activity.

105. "Mrs. Grace Jenkins to Be Buried in Rock Hill Today," *Washington Post,* June 19, 1943, 12.

106. "Grace L. Jenkins Estate Valued Near $260,000," *Washington Post,* July 13, 1943, B1.

107. George Herbert Clark, "Notes on Jenkins Television: Washington" unidentified clipping (1947): 17, Virginia Roach Family Papers.

Chapter 12. American Visionary

1. "C. Francis Jenkins Dead: Television Expert," *New York Times,* June 7, 1934, 23.

2. "Jenkins Dead Here: Creator of Television," *Washington Herald,* June 7, 1934, Jenkins Scrapbook, 1931, Wayne County Historical Museum.

3. "C. Francis Jenkins," *Scientific American* (August 1934): 91.

4. "Dr. Jenkins Dies: Noted Inventor," *Washington Evening Star,* June 6, 1934, Jenkins Scrapbook, 1931, Wayne County Historical Museum.

5. "Jenkins Dead Here; Creator of Television," *Washington Herald,* June 7, 1934, Jenkins Scrapbook, 1931, Wayne County Historical Museum.

6. "Heart Attack Closes Career of Dr. Jenkins," *Washington Post,* June 7, 1934, 11.

7. "Admiral Bullard and Dr. Lee de Forest Are the Deans—Forty Is the Average Age of Engineering and Manufacturers," *New York Times,* July 24, 1927, 15.

8. "Radio Post," *Christian Science Monitor,* July 7, 1926, 6.

9. "Charles Francis Jenkins, Inventor, Former Richmond Resident, Dies at Capital," *Richmond Palladium and Sun-Telegram,* June 6, 1934, 1, 7.

10. "Notable Achievements of C. F. Jenkins," *American Friend,* June 14, 1934, 214.

11. "Jenkins Bubbles with Enthusiasm," *Christian Science Monitor,* June 30, 1929, Farnsworth Scrapbooks, 1929 newspaper clippings, Philo T. and Elma G. Farnsworth Papers, Ms0648, Special Collections Department, J. Willard Marriott Library, University of Utah.

12. Charles Francis Jenkins, "The Law of Free Movement," in *The Boyhood of an Inventor* (Washington, D.C.: National Capital Press, 1931), 217–27.

13. Charles Francis Jenkins, "The Value of a Hobby," undated address to the Men's Club, Calvary Church, Washington, D.C., in *Boyhood of an Inventor,* 251–59.

14. "Modernism vs. Fundamentalism in Religion," in *Contemporary Forum: American Speeches on Twentieth-Century Issues,* ed. Ernest J. Wrage and Barnet Baskerville (Seattle: University of Washington Press, 1962), 87–130.

15. Jenkins, "Evolution of Civilization," in *Boyhood on an Inventor,* 229–35.

16. Diaries of Grace Jenkins, April 3 and October 20, 1932, Virginia Roach Family Papers.

17. Jenkins, "Evolution of Civilization," 232.

18. Elizabeth Maxfield-Miller, Jenkins Genealogy: Jenkins Family, February 1984, pp. 4–5, Philip Jenkins Family Papers, Richmond, Indiana.

19. C. Francis Jenkins to his father, January 20, 1920. According to Lewis Janney, Amelia Earnhart was a guest on one of Jenkins' flights. Lewis Janney Family Files.

20. A. B. Graham, U.S. Department of Agriculture, to C. Francis Jenkins, June 23, 1931, Virginia Roach Family Papers.

21. Tribute certificate to Jenkins from the Board of Trustees of the American University, June 6, 1934, Virginia Roach Family Papers.

22. Jenkins, "The Value of a Hobby," 254–55.

23. Ibid., 256–57.

24. This is a reference to the Parable of the Ten Virgins (see Matthew 25), as well as a reflection of Jenkins' religiosity.

25. Chas. F. Jenkins to his cousin, July 20, 1891, Ann McKee Coffin Family Papers. The cousin is identified only as "cousin J," a common practice in Charles' and Grace's personal writing.

26. C. Francis Jenkins, "Washington, The City of Enchantment," WCPA-AM radio broadcast, September 26, 1924, qtd. in *Boyhood of an Inventor,* 237–43.

27. Emmet Dougherty, "Men Who Have Made Radio," *Radio News* (Canada), clipping, 16–17, Jenkins Scrapbook, 1927, Wayne County Historical Museum.

28. C. H. Claudy, "Motion Pictures by Radio," *Scientific American* 127 (November 1922): 320.

29. Daniel Stashower, *The Boy Genius and the Mogul: The Untold Story of Television* (New York: Broadway Books, 2002), 58–61.

30. "Radio Vision: Show First Time in History by Capital Inventor," *Washington Sunday Star,* June 14, 1925, 1:4.

31. Hugo Gernsback, "What Is Coming in Television," *Radio News* 10 (October 1928): 20.

32. Kenneth Bilby, *The General: David Sarnoff and the Rise of the Communications Industry* (New York: Harper and Row, 1986), 120.

33. "C. F. Francis Dead: Television Expert," *New York Times,* June 7, 1934, 23.

34. Albert Abramson, *Electronic Motion Pictures: A History of the Television Camera* (1955; reprint, New York: Arno Press, 1974), 9–10.

35. Ibid.

36. Charles Francis Jenkins, *Radio Pictures* (Washington, D.C.: Radio Pictures Corporation, 1925), 6.

37. "Jenkins Dead Here; Creator of Television," *Washington Herald,* June 7, 1934, Jenkins Scrapbook, 1931, Wayne County Historical Museum.

38. "Heart Attack Closes Career of Dr. Jenkins," *Washington Post,* June 7, 1934, 11.

39. The number of patents filed by Jenkins varies in different sources. Kraeuter lists 277. David W. Kraeuter, *Radio and Television Pioneers: A Patent Bibliography* (Metuchen, N.J.: Scarecrow Press, 1992). Shiers reports around 116 television patents filed to Jenkins' companies. George Shiers and May Shiers, *Early Television: A Bibliographic Guide to 1940* (New York Garland, 1997). Various online sources, including the Franklin Institute, report "more than four hundred." See appendix A, in which 283 patents are attributed to Jenkins.

40. Albert Abramson, "Pioneers of Television—Charles Francis Jenkins," *SMPTE Journal* (February 1986): 243.

41. Frank A. Butler, "Inventions of C. F. Jenkins Set the Stage for Television, *World Records and Roses,* June 16, 1982, 6, Box 2, C. Francis Jenkins Collection, Wayne County Historical Museum. Nothing further was found about the publication or its author.

42. Author's correspondence with Kumen Blake, Applications and Architecture Engineers Analog and Interface Produce Division of Microchip Technologies Inc., August 16 through August 31, 2011. Also author's correspondence with David Szasz, Narrow Band Television Association, January 22 through 26, 2009.

43. "Radio Post," *Christian Science Monitor,* July 7, 1926, 6.

44. Author's correspondence with Kumen Blake, Applications and Architecture Engineers, Analog and Interface Produce Division of Microchip Technologies Inc., August 16 through August 31, 2011.

Epilogue

1. Jenkins Television Organization Chart, Box 40, File 4, Clark Papers.

2. Neil Baldwin, *Edison: Inventing the Century* (New York: Hyperion, 1995), 272–73.

3. "Radio Post," *Christian Science Monitor,* July 7, 1926, 6.

Appendix A. U.S. Patents Issued to Charles Francis Jenkins

1. This listing places the filing date first, because these dates are prominent to the inventor. They mark the first pinnacle in the development of theory, ideas, and technology.

2. Patent-issue date is listed as April 4, 1912, in David W. Kraeuter, *Radio and Television Pioneers: A Patent Bibliography* (Metuchen, N.J.: Scarecrow Press, 1992), 187.

3. Patent-filing date is listed as September 11, 1922, in Shiers and Shiers' patent list. George Shiers and May Shiers, *Early Television: A Bibliographic Guide to 1940* (New York: Garland, 1997), 581, 587, 594, 609, and 615.

4. Patent-filing date is listed as April 12, 1923, in Abramson's patent list. Albert Abramson, "U.S. Television Patents of C. F. Jenkins," in "Pioneers of Television: Charles Francis Jenkins," *SMPTE Journal* (February 1986): 228–29.

5. Patent-filing date is listed as April 24, 1924, in Abramson's patent list (ibid.).
6. Patent-filing date is listed as March 1, 1925, in Abramson's patent list (ibid.).
7. Patent-issue date is listed as September 14, 1927, in Abramson's patent list (ibid.).
8. Patent-filing date is listed as October 30, 1928, in Abramson's patent list (ibid.).
9. Patent-filing date is listed as March 11, 1930, in Abramson's patent list (ibid.).
10. Patent number is listed as 18,425 in Kraeuter, *Radio and Television Pioneers,* 197.
11. Patent-filing date is listed as March 4, 1932, in Abramson's patent list, "U.S. Television Patents of C. F. Jenkins."
12. Patent-filing date is listed as June 24, 1930, in Abramson's patent list (ibid.).

A Selected Research Bibliography

Books

Abramson, Albert. *Electronic Motion Pictures: A History of the Television Camera.* 1955; reprint, New York: Arno Press, 1974.
———. *The History of Television, 1880 to 1941.* Jefferson, N.C.: McFarland, 1987.
———. *The History of Television, 1942 to 2000.* Jefferson, N.C.: McFarland, 2003.
———. *Zworykin: Pioneer of Television.* Urbana: University of Illinois Press, 1995.
Allen, Frederick Lewis. *Only Yesterday.* 1931; reprint, New York: Harper and Row, 1965.
Archer, Gleason L. *Big Business and Radio.* 1939; reprint, New York: Arno Press, 1971.
———. *History of Radio to 1926.* 1938; reprint, New York: Arno Press, 1971.
Baker, T. Thorne. *Wireless Pictures and Television.* New York: Van Norstrand, 1927.
Baldwin, Neil. *Edison: Inventing the Century.* New York: Hyperion, 1995.
Bardeche, Maurice, and Robert Brasillach. *The History of Motion Pictures.* Trans. Iris Barry. New York: W. W. Norton, 1938.
Barnouw, Erik. *The Golden Web: A History of Broadcasting in the United States, 1933–1953.* New York: Oxford University Press, 1968.
———. *Tube of Plenty: The Evolution of American Television.* 3d ed. New York: Oxford University Press, 1990.
Barry, John M. *The Great Influenza: The Epic Story of the Greatest Plague in History.* New York: Viking Penguin, 2004.
Baskerville, Barnet. *The People's Voice: The Orator in American Society.* Lexington: University Press of Kentucky, 1979.
Berger, Michael L. *The Automobile in American History and Culture: A Reference Guide.* Westport, Conn.: Greenwood, 2001.
Bilby, Kenneth. *The General: David Sarnoff and the Rise of the Communications Industry.* New York: Harper and Row, 1986.

Bowser, Eileen. *The Transformation of Cinema, 1907–1915.* New York: Charles Scribner, 1990.
Brayer, Elizabeth. *George Eastman: A Biography.* Baltimore: Johns Hopkins University Press, 1996.
Burns, Russell. *John Logie Baird: Television Pioneer.* London: Institution of Electrical Engineers, 2000.
Caldwell, Genoa, ed. *The Man Who Photographed the World: Burton Holmes: Travelogues, 1892–1938.* New York: Harry N. Adams, 1977.
Cherny, Robert W. *A Righteous Cause: The Life of William Jennings Bryan.* Boston: Little Brown, 1985.
Clark, R. W. *Edison: The Man Who Made the Future.* New York: Putnam, 1977.
Coe, Brian. *The History of Movie Photography.* Westfield, N.J.: Eastview Editions, 1981.
Costigan, Daniel M. *Electronic Delivery of Documents and Graphics.* New York: Van Nostrand Reinhold, 1978.
Crane, Frank. *Four Minutes: Essays.* Vol. 5. New York: William H. Wise, 1919.
Crouch, Tom D. *Wings: A History of Aviation from Kites to the Space Age.* New York: W. W. Norton, 2004.
Croy, Homer. *How Motion Pictures Are Made.* 1918; reprint, New York: Arno Press, 1978.
Douglas, Alan. *Radio Manufacturers of the 1920s.* 3 Vols. Chandler, Ariz.: Sonoran Publishing, 1988, 1989, 1995.
Drinkwater, John. *The Life and Adventures of Carl Laemmle.* 1932; reprint, New York: Arno Press, 1978.
Dunlap, Orrin E., Jr. *Radio's 100 Men of Science: Biographical Narratives of Pathfinders in Electronics and Television.* 1944; reprint, New York: Books for Libraries, 1970.
Eckhardt, Joseph P. *The King of the Movies: Film Pioneer, Siegmund Lubin.* Madison, N.J.: Fairleigh Dickinson University Press, 1997.
Edgerton, Gary R. *The Columbia History of American Television.* New York: Columbia University Press, 2007.
Emery, Edwin. *The Press and America: An Interpretative History of Journalism.* 2d ed. Englewood Cliffs, N.J.: Prentice-Hall, 1962.
Farnsworth, Elma G. *Distant Vision: Romance and Discovery on an Invisible Frontier.* Salt Lake City: Pemberly Kent Publishers, 1989.
Fullerton, John, and Astrid Soderbert Widding, eds. *Moving Images from Edison to the Webcam.* London: John Libbey, 2000.
Georgano, G. N. *Cars: Early and Vintage, 1866–1930.* London: Granger-Universal, 1985.
Gillilan, Strickland. *Sunshine and Awkwardness.* Chicago: Forbes, 1920.
Godfrey, Donald G. *Philo T. Farnsworth: The Father of Television.* Salt Lake City: University of Utah Press, 2001.
Godfrey, Donald G., and Frederic A. Leigh, eds. *Historical Dictionary of American Radio.* Westport, Conn.: Greenwood Press, 1998.
Gray, Charlotte. *Reluctant Genius: Alexander Graham Bell and the Passion for Invention.* New York: Arcade, 2006.

Herron, George D. *The Message of Jesus to Men of Wealth*. New York: Fleming H. Revell, 1891.
Hopwood, Henry V. *Living Pictures: Their History, Photo-Production, and Practical Working*. 1899; reprint, New York: Arno Press, 1970.
Jones, Clarence R. *Facsimile*. New York: Rinehart, 1949.
Kimes, Beverly Rae, et al. *Standard Catalog of American Cars, 1805–1942*. 3d ed. Iola, Wisc.: Krause Publications, 1996.
Klein, Gilbert. *Reliable Sources: 100 Years of the National Press Club*. Nashville: Turner Publishing, 2009.
Koszarski, Richard. *An Evening's Entertainment: The Age of the Silent Feature Pictures, 1915–1928*. Berkeley: University of California Press, 1990.
Kraeuter, David W. *Radio and Television Pioneers: A Patent Bibliography*. Metuchen, N.J.: Scarecrow Press, 1992.
Lawrence, D. H. *The Hopi Snake Dance*. Flagstaff, Ariz.: Peccary Press, 1980.
Lichty, Lawrence W., and Malachi C. Topping, eds. *American Broadcasting: A Source Book on the History of Radio and Television*. New York: Hastings House Publishers, 1975.
Magoun, Alexander B. *Television: The Life Story of Technology*. Westport, Conn.: Greenwood, 2007.
Marey, Etienne-Jules. *Movement*. London: William Heinemann, 1895.
McArthur, Tom, and Peter Waddell. *The Secret Life of John Logie Baird*. London: Century Hutchinson, 1986.
McConnell, Curt. *The Record-Setting Trips by Auto from Coast to Coast, 1909–1916*. Stanford, Calif.: Stanford University Press, 2001.
Musser, Charles. *The Emergence of Cinema: The American Screen to 1907*. Berkeley: University of California Press, 1994.
Nixon, Raymond. *Henry W. Grady: Spokesman of the New South*. New York: Knopf, 1943.
Parrington, Vernon L. *Main Currents in American Thought*. 3 vols. New York: Harcourt Brace, 1930.
Pinchot, Gifford. *The Fight for Conservation*. New York: Doubleday, Page, and Co., 1910.
Powell, John Wesley. *The Exploration of the Colorado River and Its Canyons*. 1875; reprint, New York: Penguin Classics, 2003.
Ramsaye, Terry. *A Million and One Nights: A History of the Motion Picture*. 3d ed. New York: Simon and Schuster, 1964.
Robinson, David. *From Peepshow to Palace: The Birth of American Film*. New York: Columbia University Press, 1995.
Shackleton, Robert. *Russell Conwell: Acres of Diamonds, His Life and Achievements*. New York: Harper and Bros., 1915.
Shiers, George, and May Shiers. *Early Television: A Bibliographic Guide to 1940*. New York: Garland, 1997.
Sova, Henry. *Communications Serials: An International Guide to Periodicals in Communication and the Performing Arts*. Virginia Beach: Socacom, 1992.

Stashower, Daniel. *The Boy Genius and the Mogul: The Untold Story of Television.* New York: Broadway Books, 2002.
Sterling, Christopher H., and John Michael Kittross. *Stay Tuned: A History of American Broadcasting.* 3d ed. Mahwah, N.J.: Lawrence Erlbaum, 2002.
Toll, David W. *The Complete Nevada Traveler: A Guide to the State.* Reno: University of Nevada Press, 1976.
Twain, Mark. *The Gilded Age, a Tale of Today.* New York: Harper and Bros., 1901.
Udelson, Joseph H. *The Great Television Race: A History of the American Television Industry, 1925–1941.* Tuscaloosa: University of Alabama Press, 1982.
Weinstein, David. *The Forgotten Network: DuMont and the Birth of American Television.* Philadelphia: Temple University Press, 2004.
Whyte, Adam Gowans. *Forty Years of Electrical Progress.* London: Ernest Benn, 1930.
Williams, Ralph, and John P. Wolkonowicz. *A. Atwater Kent: The Man, the Manufacturer, and His Radios.* Chandler, Ariz.: Sonoran Publishing, 2002.
Wrage, Ernest J., and Barnet Baskerville. *American Forum: Speeches on Historic Issues, 1788–1900.* Seattle: University of Washington Press, 1960.
———. *Contemporary Forum: American Speeches on Twentieth-Century Issues.* Seattle: University of Washington Press, 1962.

Periodicals and Newspapers

Systematically checked for references to Jenkins (1876–1934):
Boston Sun
Christian Science Monitor
New York Herald
New York Times
Philadelphia Evening Bulletin
Richmond Palladium-Item, Telegraph
Washington Herald
Washington Post

Selected Journal and Periodical Articles

"Aeronautics: Break." *Time,* July 25, 1927, 4.
"Aeronautics Notes." *Time,* August 1, 1927, 5.
"Alumni Dots and Dashes: C. Francis Jenkins." *The Coherer* 1.1 (December 1902): 8.
Alvarez, Alex. "The Origins of the Film Exchange." *Film History: An International Journal* 17.4 (November 2006): 431–65.
Arceneaux, Noah. "Radio Facsimile Newspapers of the 1930s and 40s: Electronic Publication in the Pre-Digital Age." *Journal of Broadcasting and Electronic Media* 55.3 (2011): 344–59.
Armat, Thomas. "My Part in the Development of the Motion Picture Projector." *Journal of the Society of Motion Picture Engineers* 24 (March 1935): 241–56.

Austin, C. "The Romance of the Radio Telephone." *Radio Broadcast* 1.1 (May 1922): 16.
Bayer, H. M. "Equipment for Television Experimenters." *Radio News* 10 (December 1928): 582.
Bell, Donald Joseph. "Motion Picture Film Perforation." *Transactions of the Society of Motion Picture Engineers* 2 (October 1916): 4–6.
Benjamin, Louise M. "Working It Out Together: Radio Policy from Hoover to the Radio Act of 1927." *Journal of Broadcasting and Electronic Media* 42.2 (Spring 1998): 221–36.
Bidwell, G. L. "Television Arrives." *QST* (July 1925): 9.
Bliss, Louis D. "Concerning the School." *The Coherer* 1.1 (December 1902): 1.
Brady, John. "Inventor of Radio Movies Also Invented Motion Pictures." *Boston Post*, January 28, 1923, A8.
———. "Radio Movies in Your Home," *Boston Post*, January 28, 1923, A3.
———. "RadioVision—New Marvel May Be Born in Hub," *Boston Post*, December 30, 1923, A7.
Bretz, Arthur G. "Chronophotography to Furnish Material for the Laboratory and Clinic." *Medical Record*, October 28, 1905, 706.
"C. Francis Jenkins Tells of First Projector." *Moving Picture World*, July 15, 1916, 418.
Caswall, Muriel. "What It Means to Be the Wife of a Great Inventor." *Boston Post*, January 13, 1924, Household Section, 1.
Chambers, Gordon A. "A Short History of Standardization in SMPTE." *SMPTE Journal* 85 (July 1976): 6.
"Chronophotography." *Photographic Times* 25 (July 6, 1894): 2.
Clark, George H. "C. Francis Jenkins—Television Inventor." *Radio-Craft* 19 (January 1948): 32.
Claudy, C. H. "Motion Pictures by Radio." *Scientific American* 127 (November 1922): 320.
———. "Two Hundred Thousand Photographs per Minute." *Scientific American* 126 (April 1921): 288, 297.
Cohn, Richard Steven. "Who Put the Magic in Movies?" *SMPTE Moving Image Journal* 115:2–3 (February/March 2006): 88.
Cooper, Esther. "Francis Jenkins as a Boy on the Farm." *American Friend*, June 14, 1934, 214.
Coppersmith, Jonathan. "The Failure of Fax: When Vision Is Not Enough." *Business and Economic History* 33.3 (Fall 1994): 272–75.
Croy, Homer. "Infant Prodigy of Our Industries: The Birth and Growth of the Motion Picture." *Harper's Monthly* 135.805 (August 1917): 349–58.
Davis, Watson. "The New Radio Movies." *Popular Radio* 4.6 (December 1923): 437, 443.
Dewhirst, Thornton P. "Practical Picture Transmission." *QST* (December 1925): 12.
Dunlap, Orrin E., Jr. "Seeing Around the World by Radio: The Development of a New Combination Photo-electric Cell and Vacuum Tube Has Created an 'Eye' for Wireless." *Scientific American* 134 (March 1928): 162.

DuPuy, William Atherton. "A Professional Inventor." *Scientific American* 106 (March 1912): 292–93.

———. "What Inventors Are Doing." *Scientific American* 106 (March 1912): 293.

Fox, Robert. "The John Scott Medal." *Proceedings of the American Philosophical Society* 112.6 (December 9, 1968): 416–30.

Gernsback, H. "Radio Vision." *Radio News* 4 (December 1923): 681, 823–24.

———. "What Is Coming in Television." *Radio News* 10 (October 1928): 299.

Gilfillan, S. C. "The Future Home Theater." *The Independent* 73 (October 3, 1912): 886–91.

Godfrey, Donald G. "Senator Dill and the 1927 Radio Act." *Journal of Broadcasting* 23.4 (Fall 1979): 77–89.

Godfrey, Donald G., and Alf Pratte. "Elma J. Gardner-Farnsworth: The Pioneering of Television." *Journalism History* 20.1 (Summer 1994): 75–79.

Godfrey, Donald G., and Val E. Limburg. "The Rogue Elephant of Radio Legislation: William E. Borah." *Journalism Quarterly* 67.1 (Spring 1990): 214–24.

Goodrich, Author. "Short Stories of Interesting Exhibits." *The World's Work: A History of Our Time* 2 (August 1901): 1054–96.

Gregory, Carl Louis. "Motion Picture Cameras." *Transactions of the Society of Motion Picture Engineers* 3 (April 6–7, 1917): 6.

Herndon, Charles Allen. "Motion Pictures by Ether Waves." *Popular Radio* 8 (August 1925): 112.

Hoffer, Thomas W. "Nathan Stubblefield and His Wireless Telephone." *Journal of Broadcasting* 16.3 (Summer 1971): 317–29.

Hubbard, Henry D. "Standardization." *Transactions of the Society of Motion Picture Engineers* 1 (1916): 5–9.

"Improvement in Radio Pictures." *Motion Picture News* 18 (October 1922): 2.

"Inventor of Radio Photographs Got Big Idea While Piloting Plane." *Philadelphia Evening Bulletin*, March 5, 1923, 2.

"Jenkins' Awards." *QST* (October 1925): 21.

Joy, John M. "Film Mutilation." *Journal of the Society of Motion Picture Engineers* 26 (November 1926): 5.

Kelkres, Gene G. "A Forgotten First: The Armat-Jenkins Partnership and the Atlanta Projection." *Quarterly Review of Film Studies* 9 (Winter 1984): 45–58.

Light, Jennifer S. "Facsimile: A Forgotten 'New Medium' from the Twentieth Century." *New Media and Society* 8 (June 2006): 355–78.

Lord, H. M. "The National Budget System and the Financial Situation Facing the United States." *National Municipal Review* 12.2 (1923): 61–66.

Massie, Keith, and Stephen D. Perry. "Hugo Gernsback and Radio Magazines: An Influential Intersection in Broadcast History." *Journal of Radio Studies* 9.2 (2002): 264–81.

Matthews, Glenn E. "Historic Aspects of the SMPTE." *Journal of the Society of Motion Picture and Television Engineers* 75 (September 1966): 856.

Mayes, Thorn. "History of the American Marconi Company." *Old Timer's Bulletin* 13.1 (June 1972): 11–18.

McGuire, P. D. "Impressions of Spring Meetings of Society of Motion Picture Engineers." *Motion Pictures Today,* May 22, 1926, 26.
"Motion Pictures by Radio, Latest Achievement of Science." *Rochester (N.Y.) Post Express,* October 12, 1922, 1–2.
"Mr. Edison Outdone." *Baltimore Sun,* October 3, 1895, 2.
"The Observation Automobile Coach." *Horseless Age,* September 18, 1901, 727.
"Ocean to Ocean Assignment." *Camera Craft: A Photographic Monthly of the Photographers Association of California* 18.8 (1911): 397.
Porter, L. C. "C. Francis Jenkins—Television Adventurer." *Journal of the Society of Motion Picture Engineers* 23 (September 1934): 126–30.
———. "President's Address." *Transactions of the Society of Motion Picture Engineers* 16 (May 1923): 21.
"Radio Pen Draws Pictures from the Air." *Popular Mechanics* 45.5 (May 1926): 705–6.
"Radio Post." *Christian Science Monitor,* July 7, 1926, 6.
Roach, Charles. "Visual Instruction in Community Center Work." *Educational Film Magazine* 2 (January 1920): 8.
"Seven-League Camera." *Time,* June 2, 1924, Science Page.
Shiers, George. "Television Fifty Years Ago." *Journal of Broadcasting* 19.4 (Fall 1975): 387–400.
"Single Service Paper Milk Bottles." *Scientific American* 97 (June 1907): 446.
Sivowitch, Elliot N. "A Technological Survey of Broadcasting's Pre-History, 1876–1929." *Journal of Broadcasting* 15.1 (Winter 1970–71): 1–20.
Stashower, Daniel. "A Dreamer Who Made Us Fall in Love with the Future." *Smithsonian* 21.5 (October 1979): 44–55.
"Thomson Radio Club." *Lynn (Mass.) Works News,* January 18, 1924, 20.
"Transmission of Motion Pictures by Radio Likely." *Ottawa Citizen,* October 3, 1923, 4.
"Twenty-Fifth Anniversary of the Society of Motion Picture Engineers." *SMPE Journal* 37 (July 1941): 3–5.
Verralls, J. M. "America and Television Progress." *Television* 2 (December 1929): 483.
"Washington Picture Men at Dinner." *Moving Picture World* 18.5 (October 1913): 477–79.
Weinberger, Julius, Theodore A. Smith, and George Rodwin. "The Selection of Standards for Commercial Radio Television." *Proceedings of the Institution of Radio Engineers* 17.9 (September 1929): 1584.
White, Sally L. "John Brisben Walker, the Man, and Mt. Morrison." *Historically Jeffco* 16.26 (2005): 4–8.
Wilkerson, Dan C. "Visible Radio Communication." *QST* (May 1925): 15.
Wilson, Otto. "Transmission of Photographs by Radio." *Wireless Age* 10 (1923): 67–68.
Winters, S. R. "Amateurs Take Up Radio Vision." *Radio Age* 5.4 (April 1926): 17–18.
———. "A Motor Operated by Radio or Static from the Air." *Radio News* 10 (May 1928): 1240.
———. "The Passing of 'NOF' as a Broadcast Station." *Radio News* 5 (March 1923): 1623, 1742–43.

———. "The Transmission of Photographs by Radio," *Radio News* 4 (April 1923): 1772–73.
"World Wide Radio Service Aids Weather Prediction." *Science News-Letter* 10.294 (November 27, 1926): 133–34.
Wright, Milton. "Successful Inventors—VIII: They Seek More than Money, Says One of Them." *Scientific American* 137 (August 1927): 140–41.

Government Documents

Attorney General's Opinion. Department of Justice, 35 Ops. Attorney General 126, July 8, 1926.
Federal Motion Picture Commission, Hearings before the Committee on Education, House of Representatives, 64th Congress, 1st Session, HR 456, 1916.
Federal Radio Commission, Annual Reports, Nos. 1–3, 1927–29.
Motion Picture Patents Company v. Independent Moving Picture Company of America, U.S. District Court for Southern District of New York, Equity No. 5-167, 1912.
The Radio Act of 1927. Public Law 632, 69th Congress, February 23, 1927.
Squier, George. *Multiplex Telephone and Telegraphy by Means of Electrical Waves* (1919).
U.S. Department of Commerce, Radio Division. *Radio Service Bulletin,* October 21, 1929, 151.
United States v. Zenith Radio Corporation et al. 12F.2d 614, U.S. District Court for Northern District of Illinois, April 16, 1926.

Unpublished Documents

Cooper, William H. "The Construction of the Jenkins Television Laboratory at Wheaton, MD." Beta Chapter of the Tau Beta Pi Association thesis, University of Maryland, April 7, 1931.
Eustis, John. "Armat and Jenkins, Part I: The Dispute, 1895–1896." In Box 2, Folder 3, Thomas Armat Papers. University Library Special Collections Division, Georgetown University, Washington, D.C.
———. "Thomas Armat, Charles Francis Jenkins, and Motion Pictures in Washington, 1894–1910." In the Thomas Armat Papers. Box 2 Folder 3. University Library Special Collections Division, Georgetown University, Washington, D.C.
Fifer, William Hartge. "The Inventions of Dr. C. Francis Jenkins." Beta Chapter of the Tau Beta Pi Association thesis, University of Maryland, January 10, 1930.
Hollenback, David Arthur. "Contributions of Charles Francis Jenkins to the Early Development of Television in the United States." Ph.D. dissertation, University of Michigan, 1983.

Oral History Interviews and Family Correspondence

Ann McKee Coffin (2003–11)
Dorothy Dove (2011)

Philip Jenkins and Barbara Jenkins (2002–11)
Lewis Raney (2003)
Virginia Roach (2002–3)
Robin Sproul and David Sproul (2002–11)

Family Papers

Ann McKee Coffin Family Papers, Carp Lake, Mich.
Lewis Janney Family Papers, Phoenix, Ariz.
Grace Jenkins Personal Diaries. The diaries are scattered among various family members.
Phil Jenkins Family Papers, Richmond, Ind.
Virginia Roach Family Papers. This is the most extensive family collection.

Archives and Manuscript Collections

Benjamin Franklin Institute, Philadelphia. The Rare Book Room houses Jenkins' applications for the Franklin Medals, information on charges from Armat, and the Sciences and Arts Committee deliberations.
California State Railroad Museum, Old Sacramento. This collection has historical railroad information that helps in understanding Jenkins' travels West.
District of Columbia Public Library. Washingtonian Division, D.C. Community Archives. Local history from 1769–2000.
Earlham College, Lilly Library, Friends Archive, Richmond, Ind. Jenkins attended Earlham College.
George Eastman House, Rochester, N.Y. This collection has Armat and Edison-related materials.
Georgetown University, Washington, D.C., University Library Special Collections Division. The Thomas Armat Papers.
The Henry Ford and Greenfield Village Benson Ford Research Center, Dearborn, Mich. This collection pertains to historic automobiles.
Herbert Hoover Presidential Library, West Branch, Iowa. This collection has a small file of correspondence with Jenkins.
Library of Congress, Motion Picture Conservation Center, Culpepper, Va. Copies of all the now-rare books Jenkins published.
Library of Congress, Washington D.C. The Alexander Graham Bell Family Papers. These papers contain correspondence between Bell and Jenkins.
National Archives, Life Saving Service, National Personal Record Center, St. Louis.
National Archives, Washington, D.C. RG 173, Records of the Federal Communications Commission, Radio Division, General Correspondence, 1929–31. This collection contains records of Jenkins' FCC licensing.
National Museum of American History, Archive Center, George H. Clark Radioana Collection.

Smithsonian Institutional Archives. Permanent Administration Files, 1902–35. This collection contains the Smithsonian Photographic Exhibit.
University of Maryland, College Park. Maryland Room Archives and Records. Maryland history.
Wayne County Genealogical Society, Richmond, Ind. This collection has extensive local and Jenkins-family history.
Wayne County Historical Museum, Richmond, Ind. This is the most extensive collection of Jenkins materials.
Yale University Library Manuscripts and Archives, New Haven, Conn. This collection has the Mars-listening-experiment files.

Selected Publications by C. Francis Jenkins

Presented in chronological order; full bibliographic listings utilized in this research are provided where possible.

BOOKS

Picture Ribbons. Washington, D.C.: Press of H.L. McQueen, 1897.
Animated Pictures. 1898; reprint, New York: Arno Press, 1970.
Radio Pictures. Washington, D.C.: Radio Pictures Corporation, 1922.
Radio Pictures. Washington, D.C.: Radio Pictures Corporation, 1925.
Vision by Radio, Radio Photographs, Radio Photograms. Washington, D.C.: National Capital Press, 1925.
Radiomovies, Radiovision, Television. Washington: D.C.: National Capital Press, 1929.
The Boyhood of an Inventor. Washington, D.C.: National Capital Press, 1931.

BROCHURES, TRACTS, PAMPHLETS

The Jenkins Automobile Company. Brochure. Wayne County Historical Museum.
The Jenkins Automobiles Steam-Propelled Machines. Brochure. Wayne County Historical Museum.
The Jenkins Observation Automobile. Brochure. Wayne County Historical Museum.
U.S. Crude-Old Burner for Book Stoves and Ranges. Tract. Wayne County Historical Museum.
Radio Pictures: Pictures by Radio. Pamphlet. Herbert Hoover Presidential Library.
Televison: The Eye of Radio. A brief introduction to the era of visual broadcasting. National Museum of American History.
Expression in Photography. Brochure. Wayne County Historical Museum. Reprinted in *Photographic Times* 31.12 (December 1899): 545–52.
The Phantoscope. Jenkins Phantoscope Co. Catalog. Washington, D.C.: 1905. Wayne County Historical Museum.
Handbook For Motion Picture Stereo-opticon Operators (with O. B. Depue). Washington, D.C.: Knega Company, 1908.

Motion Pictures in Teaching. Washington, D.C: Graphoscope Company, 1916.
Diathermy Compact Pads. Pamphlet. Washington, D.C.: November 3, 1933. Wayne County Historical Museum.

PERIODICAL ARTICLES

"Hobbies that Led to Inventions," n.d. Wayne County Historical Museum.
"Transmitting Pictures by Electricity." *Electrical Engineer* 18 (July 25, 1894): 62–63.
"Photochromographic Camera." *Photographic Times* 25 (July 1894): 2–3.
"Glow Lamp Receiver." *Cosmos* 29 (1894): 161.
"Controlling a Motor from a Distance." *Electrical World* 25 (1895): 320.
"Night Sittings." *Photographic Times* 28 (January 1896): 5–6.
"Blacklight Photographs." *Photographic Times* 28 (March 1896): 152.
"Measuring the Velocity of the 'Peacemaker's' Projectile." *Photographic Times* 28 (April 1896): 177.
"The Phantoscope." *Photographic Times* 28 (May 1896): 222–25.
"A Shutterless Camera." *Photographic Times* 28 (August 1896): 375.
"The Development of Chronophotography." *Photographic Times* 28 (October 1896): 449–54.
"A New Use for Stereoscopic Effects." *Photographic Times* 28 (December 1896): 571.
"Films." *Photographic Times* 29 (March 1897): 127–28.
"The Picture Ribbons Used in Chronophotography." *Photographic Times* 29 (June 1897): 259–60.
"Improved Kinetoscopic Camera and Printing Apparatus." *Scientific American* 76 (1897): 281.
"The Gates Double Microscope." *Photographic Times* 30 (January 1898): 4–7.
"The Phantoscope." *Journal of the Franklin Institute* 145 (January 1898): 78–79.
"Patentable Priority in Chronophotographic Apparatus." *Photographic Times* 30 (April 1898): 152.
"The Perforations." *Photographic Times* 30 (May 1898): 113–14.
"A Peculiarity of Chronophotography." *Photographic Times* 30 (June 1898): 263.
"Animated Pictures." *Photographic Times* 30 (July 1898): 289–97.
"Stereoscopic Pictures." *Photographic Times* 31 (August 1899): 310.
"Expression in Photography." *Photographic Times* 31 (December 1899): 545–49.
"How to Secure Expression in Photography." *Cosmopolitan* 27 (January 1901): 131–36.
"Photography as an Aid in Teaching the Deaf." *Photographic Times* 33.3 (March 1901): 121–22.
"A Simple Current Interrupter." *Electrical World and Engineer,* March 7, 1903, n.p. Wayne County Historical Museum.
"Multiplying on the Fingers." *Munsey's Magazine* (May 1903): 293–96.
"The Struggle for the Bottle." *Scientific American* 103 (November 1910): 423.
"Little Things That Make Inventions Big." *Scientific American* 106 (May 1912): 452.
"Rewarding the Inventor." *Scientific American* 109 (September 1913): 227.

"Motion Pictures by Wireless." *Moving Picture News* 8 (October 4, 1913): 17–18.
"Signs on the Bid Road." *Collier's,* January 10, 1914, 19.
"The Romance of Motion Pictures." *Scientific American Supplement* 79.2055 (May 22, 1915): 323.
"C. Francis Jenkins Tells of the First Projector." *Moving Picture World,* July 15, 1916, 418–19.
"A General Discussion of Standard and Safety Standard Films." *Educational Film* (January 1920): 13–14.
"The Transmission of Photographs by Radio." *Engineering News* 5 (October 1924): 96, 110–11.
"Award Announcement for Radio Suggestions." *QST* (May 1925): 18.
"Seeing across the Oceans by Radio." *Illustrated Mechanics* 10.4 (July 1925): 3, 12.
"Moving Pictures by Radio." *Minnesota Techno-Log* 6.1 (October 1925): 10–11, 24.
"Radio Vision." *Proceedings of the United States Naval Institute* 51 (November 1925): 2150–55.
"Broadcasting Weather Maps by Radio." *Monthly Weather Review* 54.10 (October 1926): 419–20.
"Radio Vision." *Proceedings of the Institute of Radio Engineers* 15.11 (1927): 958–68.
"Radio Finds Its Eyes" (with Georgette Carneal). *Saturday Evening Post,* July 27, 1929, 12–14, 129–30, 133–34.
"Radio Movies and the Theatre." *Motion Picture Projectionist* (January 1928): 13, 31.
"The Chronoteine Camera." *Journal of the Society of Automotive Engineers* 22 (February 1928): 200–202.
"Synchronism." *QST* 12 (September 1928): 38.
"The Jenkins Radiovisor." *Radio* (December 1928): 18.
"Television Broadcast Transmission." *Science* 70 (August 1929): 16.
"What Are the Facts about Television?" (with A. H. Lynch). *Radio News* 11 (August 1929): 124–27.
"The Drum Scanner in Radiomovies Receivers." *Proceedings of the Institute of Radio Engineers* 17 (September 1929): 1576–83.
"Televison as a Field for Standardization." *Commercial Standards Monthly* 8 (December 1931): 166.

SMPE

These Jenkins publications are all from the *Transactions of the Society of Motion Picture Engineers* and the *Journal of Motion Picture Engineers,* today known as the *Journal of the Society of Motion Picture and Television Engineers.*
"Chairman's Address." 2 (October 1916): 3.
"Condensers, Their Contour, Size, Location and Support." 2 (October 1916): 10.
"President's Address." 3 (April 6, 1917): 3–5.
"President's Address." 4 (July 16, 1917): 5.
"President's Address." 5 (October 8, 1917): 5–6.

"The Motion Picture Booth." 5 (October 8–9, 1917): 13–15.
"Condensers." 6 (April 8–9, 1918): 26–28.
"President's Address." 6 (April 6–7, 1918): 5–6.
"President's Address." 7 (October 18, 1918): 5–6.
"Society History." 7 (November 18, 1918): 6–9.
"Stereoscopic Motion Pictures." 9 (October 1919): 37.
"Continuous Motion Picture Machines." 10 (May 1920): 97–102.
"History of the Motion Picture." 11 (October 1920): 36–49.
"Continuous Motion Projector for Taking of Pictures at High Speed." 12 (May 1921): 126–31.
"100,000 Pictures per Minute." 13 (October 1921): 69–73.
"Prismatic Rings." 14 (May 1922): 65–73.
"Radio Photographs, Radio Movies, and Radio Vision." 16 (May 1923): 78–89.
"The Discrola." 16 (May 1923): 234–38.
"A Motion Picture Camera Making 3,200 Pictures per Second." 17 (October 1923): 77–80.
"Recent Progress in the Transmission of Picture by Radio." 17 (October 1923): 15–17, 84–85.
"Radio Movies." 21 (May 1925): 7–12.
"Jenkins Chronoteine Camera for High Speed Motion Studies." 25 (September 1926): 25–30.
"Radio Movies and the Theater." 29 (July 1927): 45–52.
"Pantomime Pictures by Radio for Home Entertainment." 12 (April 1928): 110–16.
"Transmission of Movies by Radio." 12 (September 1928): 915–20.
"The Development of Television and Radiomovies to Date." 14 (March 1930): 344–49.
"The Engineer and His Tools." 15 (August 1930): 260–62.
"A Silhouette Studio." 15 (September 1930): 381–84.
"Television Systems." 15 (October 1930): 445–50.

Index

2XBU, 140, 246
3XK. *See* W3XK

airplane, 3, 67, 82–85, 127; aerial television, 158–59, 169; airmail, 111; camera mount 83; catapult (landing and take off), 84–85; recreational flying, 171–72; warfare, 83
Akeley, Carle E., 113
amateur photography: visual literacy, 62–64
AM broadcasting, 114, 140, 139
American Radio Relay League (ARRL), 115–18, 131–33, 183; contests for, 117; QST, 116–18; selling to, 113, 128, 135–36, 142–47, 157, 176
American Telephone and Telegraph (AT&T), 102–4, 113, 122, 132
American University, 172
Amstutz, Noah, 96
Annabelle-the-Dancing Girl. *See* Buchan, Annabelle Witfort Moore
Armat, Hunter, J. 31
Armat, Thomas J., 16–17, 42, 46–51, 75; adversary, 24–28, 132; agreement with, 29, 34; chronology, 38–40; Cotton States Exhibit, 30–32; demonstrations, 31, 34, 39; Edison, 34–35, 49, 53; inventions of, 24, 28, 33; Jenkins-Armat Phantoscope, 21, 26, 29, 33–35, 49, 52; Oscar, 49; partners with, 21–33, 25–32, 34–39, 176; patent holder, 33–34, 47–48, 52–53; profits, 48; protests of, 24, 31, 34–43, 46–49, 52–53; rebuke, 49

Armat Moving Picture Company, 23, 31
Associated Press, 129
Atwater, A. Kent, 130
Atwater, W. O., 36
audion tube, 96, 151, 163, 254n15. *See also* De Forest, Lee; De Forest Radio Corporation
automotive, 78–82; accessories, 88–89; electric car, 80–81; freight truck, 79; Premier Automobiles, 91; resistance to 82; Steam Trap, 78–80; tour bus, 79–80. *See also* Jenkins Automobile Corporation

Baird, John L., 103, 122–23, 127, 139
Baldwin, George, 78
Barkely, W. J., 157
Battle, Mildred, 155
Berlin, Edouard, 97
Bell, Alexander Graham, 23, 36, 51–53, 109, 135; funding, 25–26, 38, 42. *See also* Jenkins demonstrations
Bell, Donald J., 171
Bell and Howell Corporation, 171
Bell, Mabel Gardiner Hubbard, 54–55
Benson, Louis, 156
Berliner, Emile, 36
Bidwell, G. L., 133
Black Maria, 16. *See also* Edison, Thomas A.
Bliss, Louis D., 28, 39; introduces Armat, 28–29
Bliss School of Electricity, 17, 166, 203n10; Armat enrolls, 24, 28, 39; Jenkins enrolls, 27–28, 38

Boyce, J. D., 24, 38
Brady, John, 102
Bretz, Author G., 55
Breuninger, Lewis, 11–13, 199n51
Brockett, Paul, 71
Brown, Nat I., 70
Bryan, William Jennings, 106, 130
Buchan, Annabelle Witfort Moore, 1, 45
Burgess, George M., 129
Bush, Mrs. Philo L., 25, 38

Cahill, Jeremiah Edward, 60, 216n51
Casler, Herman, 58–60
C.C. Kerr and Company, 150
censorship, 69–70
Chicago, Ill., 9, 147, 198n37; exhibits in, 58, 118, 241n93; motion picture industry in, 65–66, 71–72, 163–65, 218n83, 246n59, 255n25; World's Fair, 16, 24, 39
Chinese, 105, 109–10; Embassy, 109, 236n17
Chiquita, 80–81
Chroney, Joseph, 155
Chronophotography, 55, 205n32, 213n10
Chronoteine camera, 67
Clark, Florence A., 162
Clark, George H., 133
Clark, John R., 155
clubs, 63, 142; Chevy Chase Country Club, 26, 38; Capital City Camera Club, 38, 63; National Press Club, 109, 236n13; Thomson Radio Club, 112; Radio Engineers Club, 146
Colomo, Arelia, 156
Colomo, Rosemarie, 156
Columbia Milk Bottle Corporation. *See* Jenkins Paper Milk Bottle Company
Columbia Phonograph Company, 19, 31–33, 58
contests. *See* programming
Conwell, Russell Rev., 3
Cook, George, 155
Coolidge, Calvin, Pres., 74, 161
Costello, John F., 156
Cotton States Exhibition, 30–31, 39, 44, 52
Cowling, Herford Tynes, 75
Croy, Homer, 41, 49

Davenport, Dorothy, 155–56
Davis, H. P., 143
Davis, Stephen B., 129
Davis, Watson, 111–12

De Forest, Lee, 23–24, 75, 96–97, 143, 251n48
De Forest Radio Corporation, 149–52, 158, 163–65; bankruptcy, 164; lawsuits, 159; Passic, N.Y., 157–58; patents, 150; RCA competition, 164; stock, 163–64
Demeny, George, 15
Denby, Edwin, 107
Department of Commerce, 114, 117, 129, 132; licensing, 135–35, 138–39, 173
Depue, Oscar B., 65
Dickson, W. F. L., 16
Dieudonne, C. C., 60
Dillinger, J. H., 114
Discrola, 126, 178; Discrola Inc., 106, 127–28, 153, 159, 176, 258n67; patents, 244n38
Dodge, J. Smith, 140, 249n29
Donovan, William, J., 139
drum scanner, 99, 123, 137–38, 141–42, 148, 154–55, 175, 187. *See also* Jenkins, C. Francis
DuMont, Allen B., 157, 161, 165, 256n45

Earlham College, 6, 9, 159
Eastman, George, 15, 75, 113
Eastman Kodak, 16, 136
Edgerton, D. C., 107
Edison, Thomas A., 16, 23–25, 36, 75, 182; company, 29, 30–32, 39, 58, 60–61, 64–66, 102; Kinetoscope, 16, 21, 24, 27, 31, 58, 60–61; Vitascope, 35, 40, 48, 52, 61, 65. *See also* Armat, Thomas J.
entrepreneur, 48–49, 72, 77, 175

Farnsworth, Philo T., 23–24, 75, 103, 112, 173, 175; electronic scanning, 99, 122–23, 139, 252n60, 252n64; Farnsworth Television, 162, 181
Federal Radio Commission, 119, 139, 157–59, 173, 251n50; Jenkins testifies, 139–40. *See also* Jenkins demonstrations
Fessenden, Reginald A., 23, 97
Field, Carter, 109
fire: Atlanta, Ga., 31–32, 39, 207n60; projector safety, 51–61–62, 73
Foley, Paul, 129
Fowle, E. F., 114
Frank, Glen, 74
Franklin Institute, 34–37, 39–43, 47–52, 105; Jenkins awards, 22, 36–37, 52, 113, 177, 209n83

Freeman, James P., 17–18, 26–27, 38, 42, 45–46
Friedman, William F., 114–15

Gammon, Frank R., 24–25, 34–35, 46, 58, 203n8
Garside, James W., 149–52, 157–65
General Electric, 105, 113, 122, 132, 224
Gernsback, Hugo, 111–12, 127, 174, 224
Gilded Age, 4, 69, 81
Gillett, E. K., 70
Gillilan, Stricfkland, 77
Goldsmith, Alfred N., 145, 174
Graphoscope Corporation, 40, 62, 176
Great Depression, 4, 128, 173, 181–82; effects of, 122, 150, 161–65, 169, 176–77
Greene, Wallace, 27
Gregory, Carl, R., 72, 221n22
gyroscope, 83, 106

Hardesty, H. P., 136
Harding, Warren G., 145, 108
Hardy, Arthur, C., 4, 102
Hayden, John J., 25, 45
Hays, William H., 107
Hearst, William Randolph, 98
Henderson, 109–10, 145
Herndon, L. C., 132
Hertz, Heinrich, 95
high-speed camera, 51–52, 67–68, 82, 104, 109, 112–13, 152, 178
Hill, Guy, 143–44, 251n50
Hoadley, George A., 105
Holmes, Burton, 65, 173, 218n83
Hooper, Stanford C., 107–8, 118, 143, 251n50
Hoover, Herbert: as Commerce Secretary, 101, 105–8, 113, 144, 251n51; as President, 157, 162; radio conferences, 130, 139, 245n50
Hopi Snake Dance, 51, 55–57, 63, 215n30
Hopkins, John D., 58
Hubbard, Henry, D., 46, 71–72, 111, 144
Hunter, Vera, 155, 161–62

Independent Motion Picture Company, 144, 174
independent producers, 64–65, 70, 173
Ives, Herbert E., 75
investors, 20, 60, 84, 87, 149–51, 164–65, 177, 181

Janney, Lewis L., 144–45, 215n25
Japanese Embassy, 105, 109–10
Jenkins, Alfred William (brother), 7
Jenkins, Alice (sister), 7, 61
Jenkins, Alvin (brother), 7–9, 197n33
Jenkins, Amasa (father), 6–8, 166, 171–72
Jenkins, Atwood (brother), 7–8, 39, 166, 171; demonstrates for Francis, 19, 43–46
Jenkins Automobile Burner Company, 88
Jenkins Automobile Corporation, 78–81. See also automotive
Jenkins, C. Francis, 4, 16–17, 53, 169; bankruptcy, 59, 164, 176; Board of Directors, Capital City Camera Club, 166; education of, 6, 9, 170–71; education technology, 53–61, 64, 75, 100–101, 169, 174–76, 182; family values, 6, 10–13, 20–21, 71, 171–72; family tree, 7; farm growing up, 5–9, 170; financing, 150; fraud charges, 41–47, 176; future vision, 101–2, 130, 111, 177–78, health issues, 61, 77, 85, 91, 156, 163–66, 169, 177, 182; honorary doctorate, 159,170; Life Saving Service, 10–12, 21, 26, 31–33, 38, 53, 72, 104, 172; Mars listening, 114–15; mechanical television, 3, 9, 42, 96–99, 122–23, 137–38, 169, 171–78, 182; patents, joint, 33, 36–37, 41, 45, 52; patents, list of, 185–91; patents referenced, 193–94; patents, sale of, 87, 176; Quaker foundations, 6; Park Savings Banks, 159–60; romantic, 170–72; science and religion, 170–71; solo flight, 83–84; sports menace 130; technical contributions in film, 65–68, 72; workaholic, 61, 77, 163–66, 169, 183. See also Armat protest; Bell, Alexander Graham; drum scanner; Franklin Institute; Jenkins demonstrations; lawsuits; motion pictures; optical scanning; prismatic rings; patents, legal actions; Radiovision; religion; sports
Jenkins, Charles Milton (cousin), 18, 33; jewelry store demo, 1–2, 18–19, 43–44
Jenkins, Grace Hannah Love-Jenkins (wife), 11–13, 59, 165–66; death, 166; family, 20, 171; family demonstrations 17–21; flying, 83, 159; and Francis' health, 165–66; patents, 59, 185; travel, 61, 83, 93–94, 159, 171–72; wedding, 13, 19; wedding gift, 13, 39
Jenkins, Olive (sister), 7
Jenkins, Robert (grandfather), 6–7

Jenkins demonstrations: for Bell, 17–40, 54–55; family, 1–2, 17–21, 28, 38–39; Federal Radio Commission, 143–46; first demonstrations, 124–27, 129–30, 143–48; Franklin Institute, 35–36; friends, 20, 26, 38; Gernsback, 111–12, 127; Industrial Exhibition, 155–56; National Aeronautical Association, 82; National Press Club, 109; newspapers, 108–9, 119; Richmond, Ind., 31–34, 39; sabotaged, 58; SMPTE for, 107–9; U.S. Navy, 19, 107, 118–19,126, 129–31; U.S. Post Office, 61, 101, 109–11, 118, 126; windmill 90, 129–31, 135, 182, 194. *See also* Cotton States Exhibit; Jenkins Laboratories; programs; motion pictures; Pure Food Exhibit; television; W3XK

Jenkins Kerosene Gas-Burner Company, 88, 176

Jenkins Laboratories, 4, 67, 71, 123–28, 143–45; aerial experiments, 158; Christmas tree holder, 89; closing of, 166–69; demonstrations from, 118, 129–30; joins with De Forest, 149–53; pocket calculator, 89; talking signs, 89; staff of, 162; wireless transmissions, 63, 110, 114, 118, 174–76. *See also* 3XK; W3XK

Jenkins Paper Milk Bottle Company, 4, 77, 85–86, 91, 174; Columbia Milk Bottle Corporation, 176, 197

Jenkins Phantoscope Company, 57–60, 78, 176. *See also* Phantoscope

Jenkins' Radio Pictures Corporation. *See* Radio Pictures Corporation

Jenkins Television Corporation, 4, 23; assets, 149–51, 160, 164–65, 254n3; Edison, 40; lawsuits, 159–60; management, 157, 163, RCA take over, 164–77; receivers, 145–58; trade war, 162–63; stock market crash, 161–64. *See also* Garside, James W.; Jenkins Laboratories; Sarnoff, David; 3XK; W3XK; W3XJ

Jones, Robert, 67–68
Joy, John, M., 107

Kinetographic camera 16, 21, 28, 39, 42, 52
Kinetoscope, 24, 27, 31, 58–60, 126
Korn, Arthur, 96–100
Kramme, John, 60
Kaemmle, Carl, 66, 174

Lathan, Woodville, 16
lawsuits, 23–24, 40, 47, 132, 159–60, 165

Lee Edwin, 27
licensing stations, 136, 139–40; royalties on 149–50
Liquid Paper Package Company, 87–88
Lord, Arthur D., 159–60
Lord, H. M., 111
Lubin, Sigmund, 171, 218n91
Lubin Manufacturing, 66
Lumiere brothers (Auguste and Louis), 16

MacCracken, William P., 145
Marconi, Guglielmo, 23, 96, 100, 230n16
Marey, Etienne, 15
Mars, listening. *See* Jenkins, C. Francis
Martin, C. F. 118
Marveloscope, 59–61
Marvin, Charles, F., 105, 234n73
McKinley, William, 81
mechanical television, 3–5, 139. *See also* Jenkins, C. Francis: mechanical television; Nipkow, Paul
Miles, Herbert J., 66, 71, 173
military, 20, 70–72, 104–8, 116, 119, 127–29, 135, 158, 173; control over radio, 97; Jenkins' airplane, 83; Jenkins' launching planes 84–84; Mars, 113–14; Signal Corps 82, 114, 117, 144. *See also* Jenkins, C. Francis: Mars, listening; U.S. Navy
milk bottles, 1, 77, 85–88, 174; paper bottles 86–87, 176. *See also* Jenkins Paper Milk Bottle
Millay, Kathleen, 142
Moffett, W. A., 106
Morrison, Annie, 13, 20–21, 27, 38
Morrison, Mary E. Love, 20, 27
Morse, Samuel, 96, 100, 104, 116, 131
motion pictures, 53–55, 111–12; in education, 60–61, 64, 74; for home, 57–58. *See also* Jenkins, C. Francis; Jenkins demonstrations; programs; Radiovision; Society of Motion Picture Engineers; television
Motion Picture Board of Trade, 69
Motion Picture Company, 64–65, 174
Motion Picture Patent Company, 64, 70
Mutoscope, 58–60, 66, 215n38. *See also* Casler, Herman

National Association of the Motion Picture Industry, 69–70
National Board of Censorship, 70
National Bureau of Standards, 71, 111, 114, 126–29, 144

INDEX · 283

National Forest Service, 53–54
National Press Club, 109, 236n13
NEA Service, 129–30
nickelodeon, 1, 15–17, 29, 32, 50–53, 57–68, 78, 127–28, 176, 182
Nipkow, Paul, 96, 104, 122–25, 175
North American Newspaper Alliance, 108, 236n10

optical scanning, 123–24, 182; optical-electronic, 4, 164, 173

Packard Motor Cars, 136
Pan American Exposition, 79–80
paper bottle. *See* milk bottles
patents: legal actions, 4; infringement, 23, 66; interference, 23, 33, 132; litigation, 35, 64. *See also* Jenkins, C. Francis: patents
Patrick, M. M., 106
Paul, Robert W., 16
peep show, 16–17, 52–53, 176
Pershing, John J., 106, 234n83
Phantoscope, 17–22, 26–29, 32–37, 51–53, 59, 177, 205n32, 206n38; Armat controversy, 38–49, 132, 169; Phantoscope Company, 57–58, 60, 78, 176; teaching with, 53–58
Photographic Expression, 51, 62–64
Pickard, Samuel, 143–44
Pinchot, Gofford, 53–54, 108, 214n14
prismatic rings, 99, 104–5, 108, 132, 137–38; prisms, 123–28
Powell, John W., 53
producers, 65–66
programs, programming: contests, 155–56; first, 153–54; networking, 159; production of, 131, 156; schedules of, 133, 136, 146, 155, 157, 233n65; silhouettes, 158; sound and visual, 159; W3XK inaugural, 147–49. *See also* W3XK
Prohibition, 171
Pure Food Exposition, 20

Quaker. *See* religion

Radio Act, 138–39
Radio Corporation of America (RCA), 24, 114, 143, 150, 152; corporate, 122, 132–33, 145–46, 181–82, 230n16, 238n50; Jenkins Television sold to, 136, 162–64, 173–74, 176–77
Radio Engineers Club, 147
Radio Manufactures Association, 147

radio pen transmitter, 118
radio photography, 103, 113
Radio Pictures Corporation, 106, 113–14, 117, 127–28, 153, 176, 238n50
Radio Trust, 163. *See also* Radio Corporation of America (RCA)
Radio World's Fair, 155
Radiovision, 46, 67–68, 74, 95–99, 101–7, 110–16, 124, 127, 152, 160, 176; defense, 104; defined, 121; demonstrations of, 107, 110–23, 118; RadioMovies, 152, 160; receivers, 133, 142, 176; receiver kits, 132, 136, 146, 149, 157, 161; scanning, 101; weather 118. *See also* Jenkins demonstrations; television
Raff and Gammon, 24, 31
Raff, Norman C., 25, 33–35, 46, 58, 203n8
railroad, 8–9, 94, 101, 170; Armat inventions, 24, 28, 38; gravity railway device, 89; *Great Train Robbery*, 61; railways, 24, 94; travel, 9–10, 55–57, 94, 163, 198n37
religion, 71, 171; faith, 13, 160, 171; Methodist, 11, 13, 171; Quaker, 1, 5–6, 10–11, 61, 170–73. *See also* Jenkins, C. Francis
Roaring Twenties, 4–5, 12, 31, 61, 91, 95, 100, 122, 138, 171, 181, 191
Robinson, Ira E., 143, 251n50
Robinson, S. S., 107
Roosevelt, Franklin, D., 83, 224n34
Rounds, Edward. W., 83
royalties: to Armat, 37, 212n136; to Jenkins, 87; 149–50

Sarnoff, David, 24, 122, 143, 162, 164
Scranton Novelty Company, 89
screens. *See* television
Selfridge, Thomas, 82
Shadick, G. J., 117
Shortwave and Television Laboratories, 163
Smithsonian, 114, 173; photography exhibit, 37, 40–49
Society of Automotive Engineers, 67
Society of Motion Picture and Television Engineers (SMPTE): Armat conflict, 40, 46, 50; Jenkins addresses, 72, 98, 111, 126, 141, 177, 267–77; Jenkins' leadership, 70–71, 74–75; Society of Motion Picture Engineers (SMPE), 69–71, 174; standards, 71–73, 147
sports, 67, 100, 130; baseball, 102, 111, 141; football, 102, 130; golf, 67–68, 219n107; polo, 102–3; menace to, 130

284 · INDEX

Squires, George O., 143
standards: film, 46, 69–70, 107; radio and television, 111, 114, 126–28, 146–48, 155; SMPTE, 71–73. *See also* National Bureau of Standards
Stefford, Ellory 155,
Stuart, Hugh, 25, 38
Stubblefield, Nathan B., 96
Sullivan, Frank R., 114

Tabb, H. A., 24–25, 34
Taylor, A. Hoyt., 107, 117
Taylor, D. W., 129
tel-autographs, 97
television, 101; actors, 145; ARRL, 131–3; defined, 121; demonstrations of, 103, 124–31, 143–48, 156; demonstrations for home, 57–58; Depression, 162; electronic, 123; experimental, 104, 117, 122–28, 132–40, 145, 157–58, 182; motion in, 124, 129–31, 152, 187; optical electronic, 98, 125, 140, 164, 171; receivers, 135, 140, 147–49, 163; screen size, 98, 123, 140–41, 157, 183; stations, 135–40. *See also* mechanical television; programs, programming; W3XK
Terrill, W. D., 129
theater, 1, 16, 21, 37, 66, 58, 169; fire, 31, 61; projection 178; televison for 123, 141, 157, 164, 183. *See also* television
Thomas, Edward Clifton, 90
Thomas, Luke, 9
Thomson, Elihu, 105
Todd, David, 114–15
Tolman, R. P., 46–48
Tompkins, J. T., 197
travel, 1, 3, 53, 65, 170–72, 198n37; cross country, 81–94; exhibitors, 65, 233n70; films, 58, 218n83; Hopi, 51, 55–57; ocean to ocean, 91, 228; western, 9–10, 17, 23, 42

Uncle Tom's Cabin, 66
U.S. Navy, 72, 96–97, 116, 133, 139; amateur radio 133; Indian Head Naval Proving Ground, 19; landing catapult, 84; plane, 83–85; NOR radio, 104–9, 121, 233n65; Research Lab 105, 117; wireless, 1–4, 97, 110

U.S. Post Office, 110; flying, 85; radio mail, 107, 109–11. *See also* Jenkins demonstrations
U.S.S. Kittery, 118–19
U.S.S. Trenton, 119

Vitascope, 61, 65; Armat's, 35, 40, 48; Edison's, 35, 48, 52, 86

W2XBS, 181
W2XCD, 157, 165, 181
W3XK (3XK), 145–47, 149; 3XK, 133–40, 145–46, 148–49,153; Wheaton, Md.: program schedule, 153–59; 162–64, 181–83
W8XAY, 181
WAAW, 111
Washington, D.C., 172–73; Armat Theater in, 37; industrial exhibition 155–56; Jenkins Phantoscope Company, 58; laboratory, 125–27; SMPE, 74. *See also* Jenkins Laboratory; W3XK
WEAF, 104
weather, 104, 127; Bureau, 97, 105; wireless maps, 105, 118–19; 150
Western Television Corporation, 103, 113, 132, 143, 163–65, 181
WGBS, 157
WGI, 109–10
Wireless Association of America, 111
wireless pictures: motion, 124, 131; photographs, 95–105, 174, 182. *See also* Chinese Embassy; Japanese Embassy; Radiovision; weather
Wissman, Bertram, 156
WNAC, 112
WOAW, 111
World War I, 4, 61, 72, 83, 90, 100, 116–18, 122
WRC, 114, 130
Wright brothers, 82

Ziegemeier, Henry R., 107
Zworykin, Vladmir, 75; electronic scanning, 112, 122–23, 139, 173–75, 252n60; RCA 145, 181–82; Westinghouse, 103. *See also* Farnsworth, Philo T.

DONALD G. GODFREY is a broadcast educator, professional broadcaster, and historian. His many works include *Philo T. Farnsworth: The Father of Television* and the *Historical Dictionary of American Radio*. Godfrey is a past president of the national Broadcast Education Association (BEA), a former editor of the *Journal of Broadcasting and Electronic Media*, and served as president of the National Council of Communication Associations (CCA).

THE HISTORY OF COMMUNICATION

Selling Free Enterprise: The Business Assault
   on Labor and Liberalism, 1945–60   *Elizabeth A. Fones-Wolf*
Last Rights: Revisiting *Four Theories*
   of the Press   Edited by *John C. Nerone*
"We Called Each Other Comrade": Charles H. Kerr
   & Company, Radical Publishers   *Allen Ruff*
WCFL, Chicago's Voice of Labor, 1926–78   *Nathan Godfried*
Taking the Risk Out of Democracy: Corporate Propaganda versus
   Freedom and Liberty   *Alex Carey; edited by Andrew Lohrey*
Media, Market, and Democracy in China: Between the Party Line
   and the Bottom Line   *Yuezhi Zhao*
Print Culture in a Diverse America   Edited by *James P. Danky*
   and Wayne A. Wiegand
The Newspaper Indian: Native American Identity
   in the Press, 1820–90   *John M. Coward*
E. W. Scripps and the Business of Newspapers   *Gerald J. Baldasty*
Picturing the Past: Media, History, and Photography
   Edited by Bonnie Brennen and Hanno Hardt
Rich Media, Poor Democracy: Communication Politics
   in Dubious Times   *Robert W. McChesney*
Silencing the Opposition: Antinuclear Movements and the Media
   in the Cold War   *Andrew Rojecki*
Citizen Critics: Literary Public Spheres   *Rosa A. Eberly*
Communities of Journalism: A History of American Newspapers
   and Their Readers   *David Paul Nord*
From Yahweh to Yahoo!: The Religious Roots
   of the Secular Press   *Doug Underwood*
The Struggle for Control of Global Communication:
   The Formative Century   *Jill Hills*
Fanatics and Fire-eaters: Newspapers and the Coming
   of the Civil War   *Lorman A. Ratner and Dwight L. Teeter Jr.*
Media Power in Central America   *Rick Rockwell and Noreene Janus*
The Consumer Trap: Big Business Marketing
   in American Life   *Michael Dawson*
How Free Can the Press Be?   *Randall P. Bezanson*
Cultural Politics and the Mass Media:
   Alaska Native Voices   *Patrick J. Daley and Beverly A. James*

Journalism in the Movies   *Matthew C. Ehrlich*
Democracy, Inc.: The Press and Law in the Corporate Rationalization
   of the Public Sphere   *David S. Allen*
Investigated Reporting: Muckrakers, Regulators, and the Struggle
   over Television Documentary   *Chad Raphael*
Women Making News: Gender and the Women's Periodical Press
   in Britain   *Michelle Tusan*
Advertising on Trial: Consumer Activism
   and Corporate Public Relations in the 1930s   *Inger Stole*
Speech Rights in America: The First Amendment, Democracy,
   and the Media   *Laura Stein*
Freedom from Advertising: E. W. Scripps's
   Chicago Experiment   *Duane C. S. Stoltzfus*
Waves of Opposition: The Struggle
   for Democratic Radio, 1933–58   *Elizabeth Fones-Wolf*
Prologue to a Farce: Democracy and Communication
   in America   *Mark Lloyd*
Outside the Box: Corporate Media, Globalization,
   and the UPS Strike   *Deepa Kumar*
The Scripps Newspapers Go to War, 1914–1918   *Dale Zacher*
Telecommunications and Empire   *Jill Hills*
Everything Was Better in America: Print Culture
   in the Great Depression   *David Welky*
Normative Theories of the Media   *Clifford G. Christians,
   Theodore L. Glasser, Denis McQuail, Kaarle Nordenstreng,
   Robert A. White*
Radio's Hidden Voice: The Origins of Public Broadcasting
   in the United States   *Hugh Richard Slotten*
Muting Israeli Democracy: How Media and Cultural Policy
   Undermine Free Expression   *Amit M. Schejter*
Key Concepts in Critical Cultural Studies
   Edited by Linda Steiner and Clifford Christians
Refiguring Mass Communication: A History   *Peter Simonson*
Radio Utopia: Postwar Audio Documentary
   in the Public Interest   *Matthew C. Ehrlich*
Chronicling Trauma: Journalists and Writers
   on Violence and Loss   *Doug Underwood*
Saving the World: A Brief History of Communication
   for Development and Social Change   *Emile G. McAnany*

The Rise and Fall of Early American Magazine Culture   *Jared Gardner*
Equal Time: Television and the Civil Rights Movement
   Aniko Bodroghkozy
Advertising at War: Business, Consumers, and Government
   in the 1940s   *Inger L. Stole*
Media Capital: Architecture and Communications
   in New York City   *Aurora Wallace*
Chasing Newsroom Diversity: From Jim Crow
   to Affirmative Action   *Gwyneth Mellinger*
C. Francis Jenkins, Pioneer of Film and Television   *Donald G. Godfrey*

The University of Illinois Press
is a founding member of the
Association of American University Presses.

Composed in 10.5/13 Adobe Minion Pro
by Lisa Connery
at the University of Illinois Press
Manufactured by Thomson-Shore, Inc.

University of Illinois Press
1325 South Oak Street
Champaign, IL 61820-6903
www.press.uillinois.edu